High Temperature Glass Melt Property Database for Process Modeling

High Temperature Glass Melt Property Database for Process Modeling

Edited by
Thomas P. Seward III
Terese Vascott

Published by
The American Ceramic Society
735 Ceramic Place, Suite 100
Westerville, Ohio 43081

For information on ordering titles published by The American Ceramic Society, or to request a publications catalog, please call 614-794-5890.

Order Code: G092

ISBN 1-57498-225-7

Contents

Preface viii

Acknowledgements x

Chapter 1. Introduction 1

T.P. Seward III and T. Vascott

Chapter 2. Commercial and Statistically Designed Glass Compositions 27

T. Vascott and T.P. Seward III

Chapter 3. Gas Solubility and Water Diffusion in Glasses and Melts of Commercial Compositions 47

J.E. Shelby

Gas Solubility 47

Water Diffusion 65

Chapter 4. Solubility and Diffusion of SO_3 in Soda-Lime-Silicate Glass Melts at High Temperatures 75

O.A. Prokhorenko

Chapter 5. Radiative Thermal Conductivity of Melts 95

O.A. Prokhorenko

Chapter 6. The Measurement of the Density and Surface Tension of Glass Melts Using the Sessile Drop Method 119

A.G. Clare, A. Kucuk, D.R. Wing, and L.E. Jones

Chapter 7. Viscosity of Commercial Glasses 131

A. Fluegel and A.K. Varshneya; J.E. Shelby; P. Hrma, C.A. See, O.P. Lam, and K.B.C. Minister

High-Temperature Viscosity of Commercial Glasses 133

P. Hrma, C.A. See, O.P. Lam, and K.B.C. Minister

Parallel Plate Viscometry 151

A. Fluegel and A.K. Varshneya

Transformation Range Viscosity Measurements 163

J.E. Shelby

Chapter 8. Electrical Resistivity 173

A.K. Varshneya, T. Vascott, R. Karuppanan, and J.M. Jones

Chapter 9. Statistical Analysis of Viscosity, Electrical Resistivity, and Further Glass Melt Properties 187

A. Fluegel, D.A. Earl, A.K. Varshneya, and D. Öksoy

Chapter 10. Summary 257
T.P. Seward III
Appendix I. Glass Properties 261
J.E. Shelby
Appendix II. Glass Compositions in Mol %. 275
Appendix III. Units, Conversions, and Molecular Weights 285
Author Index 289
Keyword Index 290

Preface

Glass is a material of great scientific and technical interest. The glass industry is currently presented with many opportunities to expand its impact on our way of life and to improve its own sustainability. It is also facing many constraints: Economic pressures to decrease cost, financial pressures to improve asset (capital) utilization, customer demands for improved quality, pressures from competitive materials, energy costs and conservation efforts, environmental concerns and regulations, increased global competition and more.

To surmount these difficulties, the industry is concurrently following two technological pathways. One seeks to improve process efficiencies and yields, with simultaneously lessening energy consumption, decreasing environmental impact and improving product quality. Improved return on capital investment and increased earnings per share of investment are also goals. The second pathway seeks "high-value-added" products and new, innovative and hopefully highly profitable uses for glass. Both pathways are important and likely to persist.

For the convenience of readers wishing to import some or all of the data into their proprietary computer programs, we have posted the data from chapters 2, 5, 6, 7, 8, and 9 on the American Ceramic Society Web Site at http://www.ceramics.org/glassmelt.

This book directly supports the first of these pathways and indirectly the second. It provides new glass melt property data for several sets of glass compositions potentially useful to the commercial glass industry. A perhaps-unique aspect of the reported research is that measurements of a variety of meaningful properties have been made on the same, relatively large set of compositions. The overall goal was to develop a high-temperature melt properties database with sufficient reliability to allow mathematical modeling of glass melting and forming processes for improved product quality, improved efficiency and lessened environmental impact. It was initiated by the United States glass industry through the NSF Industry/University Center for Glass Research (CGR) at Alfred University and supported in part by the United States Department of Energy Industrial Technologies Program.

Six different types/families of glass were studied because of their important commercial value: Container, float, fiberglass (E- and wool-types), low-expansion borosilicate, and color TV panel glasses. CGR member companies supplied production-quality glass from all six families upon which was measured, as a function of temperature in the molten state, density, surface tension, viscosity, electrical resistivity, infrared transmittance, gas solubility and gas diffusivity.

Because knowledge of the compositional dependencies of melt viscosity and electrical resistivity are extremely important for glass furnace design and operation, these properties were studied more extensively. About 140 composition variations, covering all six types/families of glass, were statistically designed and melted on a laboratory scale. Viscosity and electrical resistivity were then measured as a function of temperature.

The measurements were completed in February 2003 and are reported here along with the experimental techniques and methodologies used. Statistical analysis of the compositional dependencies of the melt properties, particularly viscosity and electrical resistivity, is also presented.

While it is probably not essential for modelers of glass melt behavior to understand in depth the science involved, simple application of data to mathematical models without understanding of the limitations of its applicability can produce misleading results. Therefore, the introductory chapter of this book briefly reviews the underlying principles and directs the reader to more extensive references. The introductory chapter also includes a brief discussion and historical perspective of the glass industry and some of its modeling needs.

The authors and editors hope that the work presented will be valuable for glass scientists and engineers not only in their endeavors to model glass melting and forming processes, but also in the creation of new and useful glass materials and products.

T.P. Seward III

T. Vascott

Alfred, New York

February, 2005

Acknowledgements

This book and the information it contains resulted from a multi-investigator, multi-disciplined effort conducted over multiple years, thus many people and organizations deserve recognition. The order in which the editors chose to do so follows the order that the work is reported in the book, and not necessarily the order of the importance of the contributions.

We begin with the drafters of original DOE proposals, Drs. L. David Pye, William C. LaCourse and Innocent Joseph. Without their vision, enthusiasm and recognition of the needs of the glass industry and the capabilities of the faculty, students and facilities at Alfred University, the work would never have begun.

We follow with the United States Department of Energy and its Office of Industrial Technologies (now the Industrial Technologies Program) who funded a major portion of the reported work through grant #DE-FG07-96EE41262. In particular we are grateful for the assistance and support of Elliott Levine, Theodore Johnson, Ramesh Jain, Edward Gallagher and Matae McCray. Together they inspired us and helped maintain our motivation and enthusiasm. Elliott's tireless encouragement and support during the concluding stages of the project proved indispensable.

The key Industry Contacts consisted of the Industrial Liaison Board of the NSF Industry/University Center for Glass Research at Alfred University and the technical representatives of the member companies. The Center membership at the start of the project in 1997 consisted of AFG Industries Inc., Air Liquide America Corp, Asahi Glass Co., Ltd., Cardinal FG, CertainTeed Corp., Corning Incorporated, D. George Harris & Associates (U.S. Silica), Ford Motor Co., Johns Manville, Libbey, Inc., Nippon Electric Glass Co., Ltd., Osram Sylvania, Inc., Owens Corning, Owens-Brockway, Inc., PPG Industries, Inc., Philips Corporation, Pilkington plc, Praxair, Inc., Schott Corporation, U.S. Borax Inc., Vitro, and Westinghouse Savannah River Co. Individuals deserving of special thanks include Dr. Andrea Kielpinski Sadler (Westinghouse), chair of the CGR Modeling Research Subcommittee, Dr. Frank Woolley (Corning), William Horsfall (Corning), Dr. Manoj Choudhary (Owens Corning), Dr. James Shell (Techneglas) and W. Alan Poulos (Owens Brockway).

Further industrial and government agency support was received through the US Department of Energy G-PLUS (Glass Program Laboratory User Services) program, which, through the membership of Schott Glass Technologies (Duryea, PA) in the Glass Manufacturing Industry Council (GMIC), provided funding for most of the viscosity measurements conduced at the Pacific Northwest National Laboratories (PNNL).

The people who designed the experiments and made the measurements deserve most of the credit. These were primarily Alfred University faculty members, students and post-doctoral researchers. They are listed here according to the functions they performed:

Project oversight
Principle Investigators Dr. William LaCourse (1996-97) and Dr. Thomas Seward (1997-2003).

Statistical design
Principal Investigator: Dr. Dolun Oksoy (AU).
Glass melting – Post-doctoral Researcher: Dr. Alexander Fluegel; Graduate assistants: Yi Zheng, Rebecca Neill; Undergraduate Assistant: Ian Spinelli; Corning Engineering Laboratory Services (CELS), subcontractor.

Chemical Analysis
Corning Engineering Laboratory Services (CELS) and Integrex Testing Services, Inc., subcontractors.

Radiative conductivity
Principal Investigator: Dr. Oleg Prokhorenko; Senior Scientist: Dr. Marina Chistokolova; Technical Advisor: Dr. Oleg Mazurin (all, Laboratory of Glass Properties, St. Petersburg, Russia).

Surface tension and density
Principal Investigator: Dr. Alexis Clare (AU); Co-investigator: Dr. Linda Jones (AU); Graduate Assistants: Ahmet Kucuk and Douglas Wing.

Gas solubility and diffusion
Principal Investigator: Dr. James E. Shelby (AU); Research Scientist: Dr. Melissa Mesko; Graduate Assistants: Brian Kenyon, Kirk Newton, Douglas Rapp and Christopher Tournour; Undergraduate Assistants: Penny Schader, Rachel Alley. At the Laboratory for Glass Properties, St. Petersburg, Russia, Principal Investigator: Dr. Oleg Prokhorenko; Senior Scientist: Dr. Marina Chistokolova.

Viscosity – Rotating spindle
Principal Investigator: Dr. Alexis Clare (AU); Graduate assistants: Saritha Rajamma and Juergen Walker; Pacific Northwest National Laboratory (PNNL) Principal Investigator: Dr. Pavel Hrma; Student Assistants C.A. See, O.P. Lam, and K.B.H. Minister.

Viscosity – Parallel plate
Principal Investigator: Dr. Arun Varshneya (AU); Post-doctoral Researcher: Dr. Alexander Fluegel; Undergraduate Research Assistants: Cory Bishop and Jessica Torrey.

Viscosity – Beam bending
Principal Investigator: Dr. James Shelby (AU); Research Scientists: Dr. Melissa Mesko and Dr. Holly Shulman; Post Doctoral Researcher: Dr. Daniel Edson.

Electrical resistivity
Principal Investigators: Dr. Arun Varshneya (AU); Research Scientists: Terese Vascott and Douglas Korwin; Post Doctoral Researcher: Dr. Ramesh Karuppannan; Undergraduate Research Assistants: Heather Neil and Jeffrey Jones.

Glass physical properties
Principal Investigator: Dr. James Shelby (AU); Research Scientists: Dr. Melissa Mesko and Dr. Holly Shulman; Undergraduate Research Assistants: Sara Scheffler, Kelly Murphy and Cleo Shelby.

Statistical analysis
Principal Investigators: Dr. Dolun Oksoy and Dr. David Earl; Research Scientist: Dr. Alexander Fluegel; Graduate Research Assistant: Rebecca Neill.

Sample management and data control
CGR Assistant Director for Technology Terese Vascott.

Special thanks to Dolun Oksoy for his involvement and enthusiasm from even before this project began. Without his statistical designs the project would never have gotten off the ground. We wish him well as he returns to his native Turkey as an Alfred University emeritus faculty member. Also special thanks to research scientist Alexander Fluegel whose enthusiasm and belief in the value of what we were doing maintained him through the stages of melting glass batches, measuring properties and analyzing data. His commitment to the statistical data analysis continued long after the project had officially ended. Indeed, it was Alexander who "carried the ball" in the stages of preparing Chapter 9.

While many of the people mentioned above were also chapter authors, we wished to explicitly recognize some of their other roles here at the start. To anyone we may have omitted, we apologize for our memories and our record keeping.

It is also fitting to provide here some information about the NSF Industry-University Center for Glass Research (CGR) and the New York State College of Ceramics at Alfred University to help put in perspective why the work was performed at these institutions.

Throughout its history, the New York State College of Ceramics at Alfred University has been oriented to the needs of the world beyond academia. Ever since Governor Theodore Roosevelt signed it into being in 1900 as a statutory college of New York State, the school has devised ways to respond to technological and economic needs of industry. In the words of then Alfred University President B.C. Davis, it did so "without detriment to the classical, academic curriculum". The first school director, Charles Fergus Binns, promoted the understanding of glass science. In 1932 the College established the first collegiate department of glass technology in the United States under Dr. Samuel R. Scholes Sr. Between 1949 and 1966, Dean John F. McMahon strengthened the linkages between industry and the College. That glass science has emerged as a separate field of study is in no short measure due to conferences held at or organized by members of the College, beginning in 1958.

The NSF Industry-University Center for Glass Research (CGR) at Alfred University has continued this tradition since 1985 under the successive directorships of Dr. L. David Pye (founding director), Dr. William C. LaCourse and Dr. Thomas P. Seward III. The CGR is a multi-university cooperative research center that has continuously operated under guidelines established by the National Science Foundation. Its members represent all segments of the glass industry and its mission is "to advance the field of glass science and engineering through research, education, and technology exchange driven by the cooperative efforts of academe, industry and government." In addition to the lead site at Alfred University, the CGR now maintains sites for research on glass surfaces, interfaces and coatings at the Pennsylvania State University and research on refractories used in glassmaking at the University of Missouri-Rolla.

T.P. Seward III
T. Vascott
Alfred, New York
February 2005

Chapter 1

INTRODUCTION

Thomas P. Seward III and Terese Vascott
New York State College of Ceramics at Alfred University
Alfred, NY 14802

ABSTRACT

The goal of the study reported here was to develop a high-temperature melt properties database having sufficient comprehensiveness and reliability to allow mathematical modeling of glass melting and forming processes for improved product quality, improved efficiency and lessened environmental impact. The study was initiated by the United States glass industry through the National Science Foundation Industry/University Center for Glass Research at Alfred University (CGR) [1, 2], with funding from the United States Department of Energy (DOE) [3]. This introductory chapter reviews key events involving the DOE Office of Industrial Technologies (now the DOE Industrial Technologies Program, DOE-ITP) and the CGR that led to the generation of this melt properties database. The chapter includes a brief discussion of the glass industry, some of its modeling efforts and the key melt properties for which reliable data are needed for modeling of glass melting processes. It presents the methodology used in choosing the glass compositions studied and in selecting the properties measured. It briefly discusses current state-of-the-art understanding of the property dependencies on temperature and composition, and refers the reader elsewhere for more in-depth discussions. It also presents the statistical methods used in designing the experiments and analyzing the results.

BACKGROUND

The glass industry is an important component of the U.S. economy, employing more than 150,000 people in skilled jobs, and generating more than 21 million tons of glass products each year with an estimated value of $22 billion [4].

Glass manufacturing is capital-intensive. In 1996, the cost of a new float glass facility was estimated to be about $150 million. The corresponding figures for fiberglass, container glass and color TV tube bulbs were $80, $100 and $300 million, respectively. Furnace rebuilding costs run into millions of dollars [5]. It is imperative that any proposed changes in furnace design have near 100% assurance of success. Lost production resulting from a poor furnace design can ultimately cost far more than the construction. Radical changes in design are almost always considered "high risk." On the positive side, design changes that improve production yields or result in even a 5% decrease in fuel consumption per ton of glass melted would have important economic and environmental benefits for the industry [5]. Consequently modeling of the glass-melting process to predict performance has become a necessity for any new furnace design.

Glass manufacturing is also very energy intensive. In July 1996, the U.S. Department of Energy (DOE) Office of Industrial Technologies (now the DOE Industrial Technologies

Program) and the NSF Industry/University Center for Glass Research (CGR) conducted a workshop at Alfred University on "Modeling in the Glass Industry" [6]. Two of the primary needs identified for virtually all modeling of the glass melting process were 1) reliable data on high-temperature melt properties and 2) improved sensors for in-line measurement of process variables. Also deemed important was an "improved understanding" of the fundamental principles of combustion and the fluid flow models themselves.

Table I lists the key properties for glass furnace modeling identified at the Workshop. They are important for quantifying glass flow, energy transfer, electrical heating and gas exchange in bubbles. Mathematical models are sensitive to these properties to varying degrees and the accuracy to which these properties are known varies considerably. Table II, also generated at the workshop, places the properties of Table I on a matrix of model sensitivity versus measurement accuracy. Clearly, some of the most sensitive properties are considered to be the least accurately measured. The concentrations of dissolved gases in the melt, their solubilities (Henry's law constants) and their diffusivities top the list; equilibrium constants for gas-producing reactions within the melt are a close second.

Prior to the Workshop, faculty members at Alfred University and the industrial representatives to the CGR agreed to address the need for improved data on glass melt properties and proposals were submitted to the DOE for grant support to help develop such a melt properties database [7, 8]. Because reliable gas solubility and diffusivity data is such a pressing need, the Center agreed to fund faculty researchers for additional research in this area independent of the DOE proposal. In fact, the CGR has long recognized this need and has a history of funding projects in this area. The DOE grant (DE-FG07-96EE41262) was awarded in the fall of 1996 and three CGR-funded projects related to gas solubility were started at the same time. In effect, the investigations discussed in this publication represent a DOE - Alfred University - CGR cost sharing effort. Figure 1 illustrates how the research activities evolved and are interrelated. The measurements were completed in February 2003.

A BRIEF REVIEW OF GLASS FURNACE MODELING

A traditional glass-melting furnace can be viewed in its simplest form as composed of three discrete spaces, or volume regions: 1) the combustion region, which contains the combustion flames and the gases produced by combustion as well as those evolved from the melting batch and molten glass surface, 2) the molten glass, which contains gas bubbles, dissolved gases and sometimes un-dissolved batch particles, and 3) a batch layer of varying thickness that occupies part of the boundary region separating the combustion region from the molten glass. These three regions are enclosed in a complicated refractory container that is not entirely inert and whose behavior cannot be ignored. Each region is in constant motion, not always steady state.

Historically, engineers have mathematically modeled each region in isolation for various purposes and with varying degrees of success. The greatest attention was given to the molten glass region, perhaps because it seemed more mathematically tractable, but certainly because the prediction of mixing action and motion of bubbles and un-dissolved batch particles was key to predicting the quality of glass produced. Efforts to couple the models for the various regions were hampered by a lack of necessary property data, in particular good spectral heat transfer data. Reference [9] (Viskanta) reviews the history of mathematical modeling of glass melting furnaces prior to about 1994.

TABLE I. KEY PROPERTIES REQUIRED FOR FURNACE MODELING *

GLASS FLOW	Viscosity	ν
	Density	ρ
	Volume expansion coefficient	β
	Surface tension	γ(melt-atmosphere)
ENERGY TRANSFER	Effective thermal conductivity	$k_{eff} \sim (16n^2\sigma T^3)/3\ \alpha_r$**
	Absorption coefficient	α_λ, α_r
	Heat capacity	C_p
	Surface tension	γ(melt-atmosphere)
GAS EXCHANGE WITH BUBBLES		
	Concentration of species "i" in glass	c_i
	Solubility	$S_i = c_{sat,i}/Pi$
	Diffusivity of species "i"	D_i
	Equilibrium constants for gas producing reactions	K_i
ELECTRICAL	Resistivity	ρ_{elec}
BATCH MATERIALS	Enthalpy (to melt)	ΔH_{batch}
	Thermal conductivity	k_{batch}
	Emissivity	ε_{batch}
REFRACTORIES	Thermal conductivity	$k_{refractory}$
	Emissivity	$\varepsilon_{refractory}$

* Adapted in part from a presentation by William W. Johnson, Corning Inc., at "Modeling in the Glass Industry" Workshop. Surface tension and refractory-related properties were added to the table by CGR members following the Workshop.

** n = refractive index, • = Stephan's constant, T = absolute temperature (K)

Over the last decade or so, with the aid of more powerful computer technology and more advanced CFD (computational fluid dynamics) techniques, attention has been given to modeling the combustion and batch-layer spaces and coupling these regional models into a single simulation model applicable to the entire furnace. References [10] and [11] offer two examples. A more recent example of this effort is the DOE supported activity started in 1998 at Argonne National Laboratory, Purdue University, and Mississippi State University with Techneglas, Libbey, Osram-Sylvania, Pilkington-LOF, Owens Corning and Visteon. The status of that project in 2003 is reported in reference [12] (Golchert). Also, several commercial CFD codes for glass furnace modeling have been developed and are available.

With production in some plants at 300 to 600 ton/day or more, manufacturing problems must be diagnosed and resolved as quickly as possible [5]. Consequently, recent activity has been directed to utilizing rapid modeling and process simulation for process control, e.g. references [13, 14], as well as for process design.

TABLE II. MODEL SENSITIVITY VS. PROPERTY MEASUREMENT ACCURACY

Model Sensitivity to property	Accuracy of Present Day Measurements +/- 10 %	+/- 25%	+/- 50 - 100%
Very Sensitive		Gas reaction equilibrium constants	Gas concentrations in glass Gas solubilities Gas diffusivities
Sensitive	Viscosity Specific Heat Density Electrical resistivity Enthalpy to melt batch Batch thermal conductivity	Thermal expansion Radiative absorption coefficient Effective thermal conductivity Emissivity of batch Thermal conductivity of refractory Emissivity of refractory	Radiative absorption coefficient (when low)
Insensitive	None	None	None

Glass melting process modeling and simulation is extremely complicated and many physical and thermodynamic properties of the various gaseous, liquid and solid species must be known before the models can be applied and validated. For the combustion region, the information needed includes knowledge of the gaseous species present (including nitrogen, oxygen, fuel and combustion products), heats of combustion, concentrations of radiative gas species present, such as CO_2, H_2O, and carbonaceous soot, their spatial distributions, and the emissivities of the refractory container surfaces. For the batch space, one should know the size distribution of the batch particles, their compositions, densities and specific heats, reaction paths, reaction temperatures, heats of reactions, reaction products, and more.

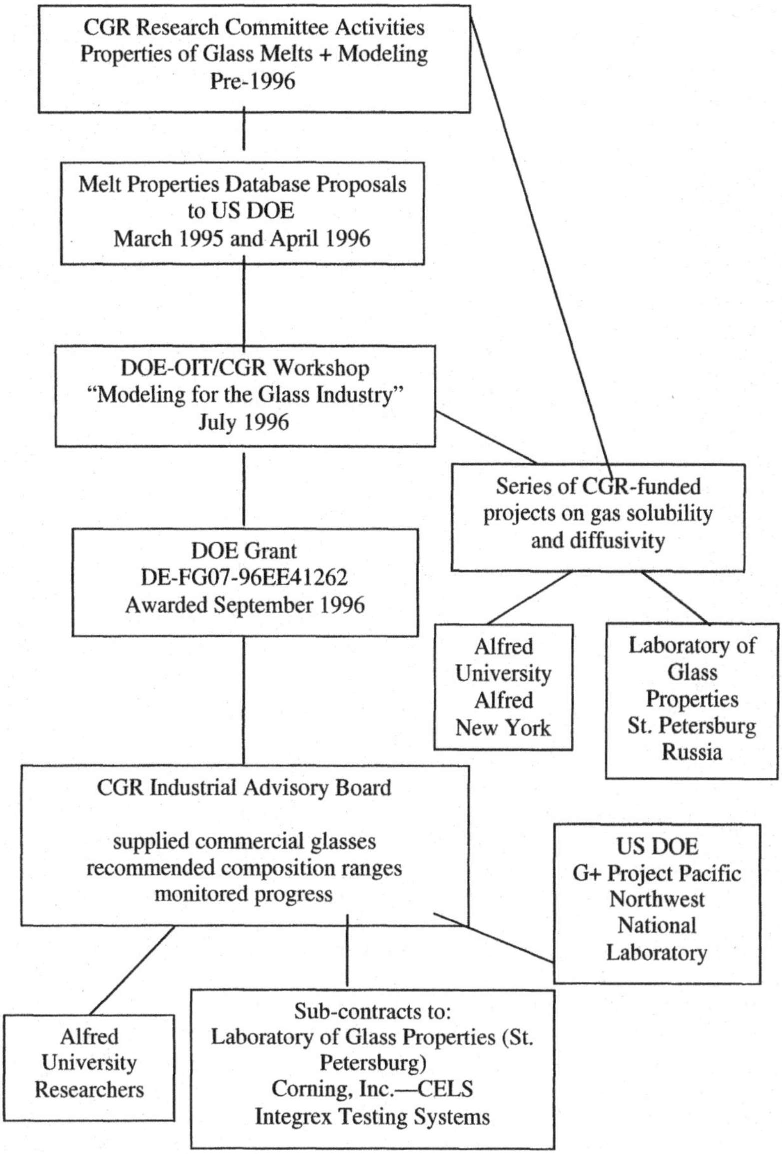

Figure 1. Schematic of US DOE—Alfred University—CGR research activities

For the melt, to model heat transfer and mass flow (including convective motion), properties such as viscosity, density, heat capacity, electrical resistivity and radiative thermal conductivity (the radiative component of heat transfer) need to be known. To model fining, gas solubilities, diffusivities, equilibrium constants, surface tension and more must be known. The study reported in this book focuses on the melt properties. The importance of most of these properties will be reviewed in this introduction, and again in the chapters dealing with the corresponding measurements.

OBJECTIVES AND SCOPE OF STUDY

In keeping with the background discussed above, the main objective of the proposed work was to develop a comprehensive and reliable data base for commercial glass forming melts that would allow full use of numerical simulation models by a broad cross-section of the glass industry for the purpose of achieving energy savings, improving product quality, increasing productivity, and meeting present and future environmental regulations. To insure that the data would be useful for this application, an industrial advisory group from the CGR was closely involved in every aspect of this project.

Glass Compositions Studied - Six different types/families of glass were studied because of their important commercial value: container, float, fiberglass (E- and wool-types), low-expansion borosilicate, and color TV panel glasses. CGR member companies provided bulk quantities (25 to 50 pounds) of production-quality commercial glass from all six families. These glasses were chemically analyzed and provided long-term base glass "standards" for the project (Table III). For each glass family, a focus group of industry technical representatives specified the key composition variables and ranges they wished to see investigated. This information became the basis for the experimental design, and for each family of glasses a set of 24 compositions was statistically designed as described below. These glasses, 144 total, were then melted on a laboratory scale and chemically analyzed as described in Chapter 2.

Melt Properties Studied - The melt properties selected for investigation are a subset of those listed in Table I. They are shown in Table IV along with the chosen measurement techniques and the proposed temperature (or property) ranges for study. [Properties in Table I related to glass batch and furnace refractories were not part of the present study.]

Table V reclassifies the properties of Table IV into three classes according to their dependencies upon different composition variables: I) Those dependent primarily on the bulk glass composition (usually the major components) and temperature; II) those additionally dependent upon minor chemical components of the glass, such as coloring ions, and upon their redox state; and III) those with more complex composition dependencies. Viscosity, density, heat capacity (not measured in the present study) and electric conductivity generally fall in class I. The effect of ferrous/ferric ion equilibrium on radiative thermal conductivity is a good example for class II. As an example for class III, gas solubility depends upon the identity of the gas specie being studied, the melt composition, temperature and, for gases that dissolve chemically in the melt (reactive species), the partial pressures of other reactive gases (such as H_2O and O_2) present in and surrounding the melt. Surface tension shows similar dependencies.

Density, surface tension, viscosity, electrical resistivity and optical transmittance (visible and near infrared, to determine high temperature radiative conductivity) were measured as a function of temperature in the molten state for all six base glasses. With CGR cost sharing, gas solubility and diffusivity were studied in each of these glasses as well.

Only viscosity and electrical resistivity were measured for all 144 experimental glass compositions. This was a limitation imposed by time, funding, and the great difficulty of most of the other property measurements.

TABLE III. BASE GLASS COMPOSITIONS AS ANALYZED (WEIGHT PER CENT)

Base Glass ID	SiO_2	Al_2O_3	B_2O_3	MgO	CaO	SrO	BaO	Li_2O	Na_2O	K_2O
container	74.38	1.28		0.20	10.51	0.01	0.10		13.30	0.29
float	72.92	0.13		3.84	8.74				13.74	0.03
TV panel	61.53	2.06			0.05	9.19	9.23	0.010	7.64	7.55
E-glass	56.60	14.80	7.04	4.33	18.20				0.63	
LE borosilicate	[80.98]	2.19	12.60	0.008	0.023			0.010	4.09	0.055
wool	[62.66]	3.31	4.90	3.46	8.27				16.00	0.83

Base Glass ID	TiO_2	CeO_2	ZrO_2	PbO	ZnO	Fe_2O_3	Sb_2O_3	F	FeO	SO_3
container	0.008					0.036			0.008	0.21
float	0.017					0.109			0.042	
TV panel	0.430	0.280	1.390	0.000	0.51	0.037	0.30		0.004	
E-glass	0.580					0.350		0.220	0.100	
LE borosilicate		0.005				0.026			0.016	
wool						0.31			0.085	0.17

TABLE IV. PROPERTIES AND MEASUREMENT TECHNIQUES

PROPERTY	TECHNIQUE	MAX. TEMPERATURE or PROPERTY RANGE
Viscosity	Rotating Spindle Parallel Plate Beam Bending	10^0 to 10^3 Pa·s 10^4 to 10^{10} Pa·s 10^{10} to 10^{13} Pa·s
Density (Thermal Expansion)	Sessile/Pendant Drop with Image Analysis	1450°C
Surface Tension	Sessile/Pendant Drop with Image Analysis	1450°C
Gas Solubility + Diffusivity	Saturation and Outgasing with Specialized Techniques	1400°C
Electrical Resistivity	Two Point Probe	1450°C
Radiative Conductivity	Optical Absorption	1450°C

TABLE V. CLASSIFICATION OF PROPERTY DEPENDENCIES

PROPERTIES	DEPENDENCIES
Viscosity Density Heat capacity Electrical Resistivity	Glass composition (except trace components) Temperature
Radiative heat conductivity	Glass composition (incl. trace components) Redox state (partial O_2 pressure) Temperature
Gas solubility Surface tension	Glass composition (including dissolved gases) Redox state Atmosphere composition Temperature

STATISTICAL DESIGN OF THE EXPERIMENTS

We based the choices of experimental glass compositions on two requirements: First, to provide ranges of certain oxide components deemed particularly important to the industrial representatives based on their manufacturing needs, and second, to provide composition variations that would allow statistical modeling of property behavior as a function of composition.

From the perspective of property modeling, for a glass melt with i oxide components, a property value P can be written as:

$$P(T) = \Sigma\, f_i(T) \cdot x_i \tag{1}$$

where x_i is the amount (mole fraction, weight fraction, etc.) of component i, T is temperature and $f_i(T)$ represents the contribution of component i to property P (often called partial molar or partial specific values). However, $f_i(T)$ is not necessarily a linear function for large variations of an oxide, nor does such a representation explicitly allow for multiple component interactions. As a first approximation, $f_i(T)$ is represented by a linear coefficient, $P_i(T)$. This approach provides a first order basis for designing and analyzing composition variation experiments and has been used to successfully describe viscosity-composition relationships in soda-lime-silica glasses by Lakitos [15] and Öksoy [16]. Examples of this approach applied to other glass melt properties include electrical conductivity (Sasek [17]), surface tension (Varshneya [18] and Kucuk et al. [19]) and heat capacity ([20] Priven and Mazurin).

Such models are often most useful to describe property variation as a function of composition variation about a particular base composition, such as might be encountered in maintaining property control in a manufacturing operation. Reference [16] shows the use of "spider diagrams" derived from such statistical modeling to demonstrate property sensitivity as a function of composition and temperature.

As originally proposed, the experimental program reported in this book would involve selecting a centroid composition for each system (glass family) to be studied. Variations of sufficient magnitude would be made around this centroid to effect property changes large enough to be measured conveniently and analyzed statistically. Once the measured incremental changes of a given property around a centroid composition were known, a composition-property database would be constructed.

Since the chosen commercial glass compositions have many (five or more) components, full factorial designs would have required the melting and measuring of too many glasses. Instead we planned to use 2^{k-p} or 3^{k-p} fractional factorial designs, where k stands for the number of components in the glass, p represents the number of components whose dependence would not be considered in the design and 2 and 3 refer to the number of levels assigned to each variable. As experimental design proceeded, practical considerations further limited us to 2^{k-p} designs (two levels high and low values only). Selection of composition ranges and experimental designs is discussed in Chapter 2. Chapter 9 presents and discusses the statistical analysis of the experimentally determined property-composition-temperature relationships for viscosity, electrical resistivity and several other properties.

MELT PROPERTIES STUDIED

The materials science and geo-science literature contains extensive discussions of the properties of silicate glasses and melts and their behaviors as functions of composition and temperature. While it is probably not necessary for engineers who model glass melt behavior to understand extensively the science involved, simple application of data to mathematical models without some understanding of the limitations of its applicability can sometimes produce misleading results. Therefore, the next portion of this chapter briefly reviews current understanding of the property dependencies, suggests some application limitations and directs the reader to more extensive references.

Gas Solubility and Diffusion

Gases such as CO_2, SO_2 and N_2 are generated by chemical reactions within the batch layer (carbonate, sulfate and nitrate decomposition) and captured in the melt, either as bubbles or as dissolved species. Ar, N_2, CO_2 and H_2O present in the combustion atmosphere above the melt also become dissolved into the melt to varying degrees. If the melt is saturated or supersaturated with some or all of these gases, bubbles (called seeds or blisters) will be present in the melt. The process of removing these bubbles is called fining (or refining), and requires additional time beyond that needed for reacting and melting the batch materials. Glass-melting furnaces must be designed to allow sufficient time for fining to occur, and designed to minimize this time. Calculations of the required times involve knowledge of the gas solubilities and diffusivities. Consideration of these properties is further complicated because some gases dissolve physically into the melts (retain their own molecular identities) while other gases react partially or fully with the glass constituents, becoming new molecular species.

For example, the gases He, Ne, Ar and N_2 dissolve physically in glass melts, that is, the gas molecules occupy interstices in the glass structure. The larger the atom or molecule, the less its solubility. Reactive gases like H_2, O_2, H_2O, CO_2 and SO_2 behave differently. They dissolve both physically and chemically. They can dissolve as molecules, they can form bonded species within the melt or they can react with the melt to change the oxidation state of other ions. For example, H_2O dissolves primarily by reacting with Si-O-Si bonds in the network to generate pairs of Si-OH bonded species. Oxygen can react with multivalent metal ions, entering the glass structure by oxidizing these elements, or leaving the structure when these elements are thermally reduced, as in the fining reactions involving antimony and arsenic oxides described below. In commercial melts, CO_2 and SO_2 may be present as species chemically bound to the network, for example at sites of non-bridging oxygens, being released only slowly, making measurements of their equilibrium concentrations extremely difficult. A recent book by James Shelby discusses these complications more fully [21].

During fining, O_2, SO_2, and Cl_2 can be generated within a glass melt by chemical fining agents deliberately introduced to the batch to promote gas generation at certain melt temperatures, for the purpose of liberating gases that will cause small bubbles to grow and thereby rise to the surface more rapidly. More detailed discussions of fining may be found in references [22] (Cable) and [23] (Beerkens).

Solubility, S, is defined in this book as the concentration of a gas specie dissolved at equilibrium in the glass (physically and chemically) per gram of glass per atmosphere of applied pressure (#/g·atm) (See Table II). Therefore, the concentration of gas species in the melt in chemical and thermal equilibrium is proportional to the applied pressure (or partial pressure for atmospheres containing multiple species). As defined, S is the Henry's law proportionality constant. It is, of course, a function of melt composition and temperature. The proportionality holds at the atmospheric pressures encountered in commercial glass melting, and pressures many orders of magnitude higher. However, it holds rigorously only for gases where the dissolved species are the same in both the gas phase and the solution. For reactive species such as CO_2 and SO_2, any deviations from the linear dependence on pressure should be experimentally verified.

The literature contains other definitions of solubility, for example, Ostwald solubility, which is the ratio of the concentration of the gas specie in the glass or melt to the concentration of that specie in the surrounding atmosphere. These two solubility definitions are related through the ideal gas law and the density of the glass.

Gas solubility measurements are extremely time consuming, and according to Shelby [21] published data can differ by almost one order of magnitude. In Shelby's studies (Chapter 3), the measurements of He, Ne, and Ar solubilities are straightforward: saturate the melt at the desired temperature and pressure, then drive off the dissolved molecules at a higher temperature under vacuum, measuring the quantity of gas released. Unavoidably however, as the atomic or molecular size increases, the solubility and the accuracy of its measurement decrease. Consequently, Shelby and co-workers were unable to measure nitrogen solubility in the glasses because the concentration levels obtainable at atmospheric saturation pressures were below the detection limit of their system. (The detection limit is affected in part, by the amount of N_2 residual in the system from the laboratory air.)

Among the reactive species, water solubility in commercial glasses is sufficiently high that its measurement is also rather straightforward, provided suitable care is taken. On the other hand, determinations of CO_2 and SO_2 solubility are considerably more difficult.

Shelby has recently reviewed the literature related to CO_2 in glass melts [24] and concluded that for silicate melts, CO_2 solubility as carbonate increases with increasing NBO (non-bridging oxygen) concentration and is influenced by the nature of the modifier cation species which cause the NBOs to form. On the other hand, the solubility of molecular CO_2 may decrease with increasing NBO, the net solubility trend being a combination of the two. Shelby's team developed and used in this study a fluxing technique to more efficiently outgas CO_2 from the melt samples. Shelby has pointed out that the extremely low solubilities of carbon dioxide (from 0.2×10^{16} to 2.0×10^{16} molecules/gram-atmosphere) in the relatively high silica commercial glass compositions studied made these measurements extremely difficult and their accuracies less than what might be desired. (By comparison, CO_2 solubilities in the more highly modified melts commonly reported in the literature range from approximately 10^{18} to 10^{22}, on the same scale.) To put these concentrations in perspective, 10^{16} molecules (per gram of glass) released into the atmosphere at 1450°C and one atmosphere pressure would occupy about 2.4×10^3 cm^3 of volume. That is still a lot of potential bubble volume!

The situation for sulfur is even more complicated. Sulfur can exist in glass melts as elemental sulfur, sulfides (S^{2-}), sulfur dioxide (SO_2), sulfite (SO_3^{2-}), and sulfate (SO_4^{2-}). Factors affecting the solubility of sulfur in glass melts include composition of the melt, partial pressure of the vapors of sulfur compounds present in the atmosphere, partial pressure of oxygen, partial pressure of water (H_2O), and temperature. The CGR funded Oleg Prokhorenko and his co-workers, in St Petersburg, Russia, from mid-1997 to mid-2000 to study solubility and diffusivity in this complicated system. The results of these studies are reported in Chapter 4. Very low rates of exchange of dissolved SO_2 gas with the atmosphere challenged the normal techniques for equilibration. Prokhorenko concluded that transport of the sulfur species between melt and atmosphere is controlled not only by the diffusion process, but by reactions at the interface as well. In fact, sulfate layers tend to form at the melt-vapor interface. Prokhorenko's team investigated several techniques to overcome this difficulty and developed them to the point whereby reliable saturation and desorption are achieved. They also developed special diffusion couples for measuring the diffusion coefficients. Mathematical models of the desorption processes at free surfaces and into rising bubbles were developed as well.

Prokhorenko reports dissolved levels of SO_3 in soda-lime-silica melts of about 0.1 wt%, at 1/100 atmosphere partial pressure, which translates to about 10^{20} molecules/gram. The pressure dependence of this solubility was not determined.

Gas Diffusivity

Determinations of gas diffusivity in a melt are difficult because they generally involve measuring concentration changes within a sample, or a surrounding chamber, as a function of time (as the gas is absorbed or given off by the melt) to at least an order of magnitude greater precision than that of the concentration measurement itself. Such determinations often involve weight gain or weight loss measurements, or some method of quantitative chemical analysis. For N_2, CO_2 and the inert gas species of this study and the sample sizes and saturation pressures accessible to the investigators, dissolved concentrations approached the detection limit. Hence measurement of the small changes needed to determine diffusivities was impossible.

However, for gases that react with the melt to cause change in some signature property, such as infrared spectral absorption bands (OH bonds) for dissolved water, or spectral absorption of some melt component due to its oxidation or reduction, these property changes can be used to monitor gas diffusivity. Changes in infrared absorption were employed by James Shelby to measure the diffusivity of water in five of the studied commercial glasses, as reported in Chapter 3.

For SO_2, the solubility in commercial soda-lime-silica glass melts is relatively high allowing direct chemical analysis. As reported by Oleg Prokhorenko in Chapter 4, electron microprobe analysis of special diffusion couples allowed determination of diffusion coefficients in this system.

Radiative Conductivity of Melts

Oxide glasses and their melts tend to transparency at visible and near infrared spectral wavelengths, hence they transmit much of the high temperature thermal energy spectrum. Such materials are called diaphanous. Heat radiation is continuously absorbed and re-radiated throughout the body, be it glass or melt. For any given volume region, the heat content is increased by radiation absorption and decreased by radiation emission, the two effects balancing when steady state heat flow is achieved. It is often useful discuss this process in terms of an effective radiative thermal conductivity, K_{eff} as listed in Table I and repeated here.

$$K_{eff} = (16n^2\sigma T^3)/3\alpha_r \qquad (2)$$

where n is the refractive index of the melt, σ is the Stephan-Boltzmann constant ($5.67\times10^{-8}W/m^2K^4$), T is absolute temperature and α_r is an "average" absorption coefficient of the melt. By average is meant a weighted average over the wavelength spectrum of the radiation source (for example, black or gray body). α_r can vary greatly among commercial glasses, especially those containing coloring ions. In the Rosseland approximation [25], $1/\alpha_r$ is replaced in Equation 2 by an effective mean free path (see the discussion below and in Chapter 5).

However, it must be remembered that the effective conductivity described by Equation 2 is not a property of the material alone, but a property of the material (through refractive index and absorptivity) and the radiation spectrum (as expressed through Stephan's law). This is discussed more fully in the section on photon conductivity below.

Of course, transfer of thermal energy into and within a glass melt can occur via three mechanisms: conduction, convection and radiation. However, for large volumes of transparent glass at commercial glass melting temperatures, normal conduction occurs effectively only near boundaries with the surroundings, the bulk of the heat transfer occurring primarily by radiation throughout the bulk of the melt. Heat transfer by convection can occur near the boundaries and, under certain conditions, elsewhere within the melt. So, while we present and discuss data on radiative heat transfer, normal conduction and convection cannot be ignored in modeling and simulation. Gardon et al. discuss the relative effectiveness of conduction versus radiation and its dependence on thermal path length in references [26, 27] and Pierrot discusses and models the relative roles of radiation and convection in reference [28].

In traditional fuel-fired glass melting tanks, the molten glass is heated primarily by thermal radiation from the hot radiative gas species and carbonaceous soot present in the combustion chamber and from the hot refractory walls. The heating of the melt occurs when the radiant energy is absorbed by the melt and propagated through it. Absorption and re-radiation in glasses (or glass melts) are wavelength dependent properties. So, for modeling, these properties must be known as functions of both wavelength and temperature.

The relative contributions of thermal vibrations and radiation to heat transfer depend on temperature. It is well known that in the range of about 500°C to 1000°C, both thermal acoustic vibrations (phonons) and radiated light energy (photons) contribute significantly to thermal conductivity. At temperatures below about 500°C, silicate melts are essentially opaque to the most intense portions of the long wavelength black body radiation spectra characteristic of those temperatures. Hence thermal vibrations (phonons) dominate the thermal conductivity at these temperatures. Above about 1000°C, the commercial glass melting regime, glass is considerably transparent to the most intense wavelengths and radiative (photon) conductivity is the dominant component of heat transfer. The following paragraphs describe why this behavior is so.

Phonon conductivity – The propagation of thermal energy through a material, whether it be solid, liquid or gas, can be viewed as occurring by collisions between molecules (as in a gas) or between energy waves and the structure of the material through which they propagate (for condensed phases such as solids or liquids). For condensed phases, important factors for heat propagation are the density of the energy present, the velocity of the waves (or quantized particles), and the average distance between collisions, called mean-free-path (MFP). Phonon conductivity can be expressed as

$$K = 1/3 \cdot C_v \cdot v \cdot \Lambda, \qquad (3)$$

where C_v is heat capacity at constant volume (representing the energy density within the material), v is the average velocity of sound waves, and Λ is the phonon MFP.

For disordered materials such as glasses and melts, the MFP for phonon conduction is found to be a few molecular diameters. Kittel [29, 30] calculates the phonon MFP of vitreous silica (fused quartz) at 100°C to be about 7.8 Angstroms, the same order of magnitude as the dimensions of a single SiO_2 tetrahedron (7 Å). He also calculates MFPs for a crown glass and a flint glass as 4.2 and 4.8 Angstroms, respectively, indicating a decrease in MFP with increasing complexity of the glass. Kingery [31, 32] calculates the MFP for fused silica, Corning code 0080 (soda-lime-silica) and Corning code 7740 (Pyrex brand borosilicate) glasses as 5.2, 4.3 and 3.3 Å, respectively, again indicating that the phonon MFP decreases as the complexity of the vitreous structure increases.

Because the MFP for phonons is approximately the same for all oxide glasses and appears independent of phonon wavelength and phonon density, the temperature dependence of thermal conductivity is primarily determined by the heat capacity. Thus, following the behavior of the heat capacity, the phonon component of the thermal conductivity of glasses increases rapidly from low (cryogenic) temperatures, ultimately leveling off (between 500 and 1,000°C) at a value corresponding to the heat capacity given by the Dulong and Petit law (3R/gram-atom). It has been suggested that the phonon MFP will decrease further at temperatures above the glass transition because of increased disorder in the melt. Hence, the measured phonon conductivity below the glass transition should be considered an upper limit to the phonon contribution. For most glasses, this is about 1 W/m^2. As will be shown in Chapter 5, this value is much less than the values determined for radiative heat transfer, and can safely be eliminated from most (but not all) calculations of heat transfer in glass melts.

Photon conductivity - The mean free path (MFP) for photonic (radiative) heat transfer is very much wavelength-dependent. For radiation of any given wavelength, the MFP is essentially the inverse of the absorption coefficient (cm^{-1}). Oxide glasses that contain few or no transition metal impurities and no absorbing particles are almost completely transparent throughout the visible spectrum. For these wavelengths, the photon MFP is much greater than the thickness of the glass in most situations, the radiation passing through the glass to be absorbed by the container walls. In the mid- and far-infrared regions (wavelengths greater than about 4.5 microns), all oxide glasses are essentially opaque, the MFP being on the order of a few molecular diameters. Hence thermal energy at those long wavelengths is absorbed at the surface of the glass (or melt) and conducted through the interior via phonon conduction.

Between these limits, the absorption and MFP depend greatly on melt composition, wavelength and temperature. Hence it is necessary to know (measure) the wavelength dependence of the absorption as a function of temperature for each melt composition. Of course, for calculation of heat transfer one must also know the wavelength distribution of the heat source, as determined by such factors as burner flame temperatures and luminosity and refractory crown temperatures and emissivities, remembering that the higher the temperature, the more the peak intensity of black (or gray) body radiation moves to shorter (visible) wavelengths. Discussion of these effects can be found in references [18] (Varshneya) and [32] (Kingery, et al.).

The most effective and dependable way to determine the photonic component is by measuring the absorption spectra of the glass melt in the wavelength region between 0.6 and 3.8 μm at the anticipated melt temperatures. Having done this, the key factors controlling radiative heat transfer (an effective, wavelength-weighted mean-free-path length and the effective radiative conductivity, K_{eff},) can be calculated using the Rosseland approximation [25], valid when the MFP for radiation is less than the dimensions of the glass. Equation 2 and the Rosseland approximation are not valid for optically thin glass layers ($t \ll$ MFP), such as near boundaries. For these regions modifications more advanced mathematical approaches have evolved, e.g., [33] [34]. Oleg Prokhorenko discusses radiative thermal conductivity more fully and presents experimental results in Chapter 5.

Thermal Diffusivity – The concepts of thermal conductivity described above are most applicable to steady state heat-flow situations; that is, where the temperature gradients within the medium are fixed. When temperatures (and their gradients) vary with time, account must be taken of the varying amount of heat stored within the medium through its heat capacity. In such situations the concept of thermal diffusivity is most useful. In direct analogy with the diffusion equations used to analyze and model mass transport, heat flow in one dimension is described by

$$dT/dt = \kappa\, d^2T/dx^2 \qquad (4)$$

where κ is called the thermal diffusivity. Thermal conductivity and thermal diffusivity are related as

$$\kappa = K/\rho C_p \qquad (5)$$

where ρ is density and C_p is specific heat at constant pressure.

It was originally proposed that this study measure thermal diffusivity (phonon and photon contributions together) using a temperature wave technique, specifically a modified Angstrom technique, as done by Van Zee and Babcock [35] and Jain and Pye [36]. However considerations of the difficulties of performing the measurements at those high temperatures and interpreting them led us to abandon the method. Lack of high temperature heat capacity data needed to relate thermal diffusivity and thermal conductivity would also have been problematic. In any case, there is evidence that for moderate deviations from steady state, heat transfer within the melt can be adequately described in terms of radiative thermal conductivity [37]. (Of course, convection must also be accounted for when present.)

Note: If heat sources and heat sinks are present within the melt, Equation 4 must be modified accordingly. Solutions to steady state, periodic and transient heat flow problems are found in many texts, including the classic one by Carslaw and Jaeger [38].

Surface Tension, Density and Thermal Expansion

Knowledge of surface tension as a function of melt composition, temperature and furnace atmosphere is important for understanding bubble growth and shrinkage, such as encountered during the fining process and at other bubble-generating locations, for example refractory walls and joints, and in foaming action at the melt surface. Surface tension is also a key factor in determining the "equilibrium" thickness of a layer of molten glass floating on a tin bath. Surface tension is important in understanding bubble behavior because the gas pressure within a bubble consists quantitatively of two terms, the first, the pressure resulting from the hydrostatic "head" (height of the glass above the bubble) acted on by gravity, and the second, the pressure resulting from the surface tension, given by $2\gamma/r$, where γ is the surface tension (dynes/cm or joules/m^2; 10^3 dyn/cm = 1 mN/m = 1 J/m^2) and r is the bubble radius.

Convective flow in glass melters, with its associated mass transport, thermal energy transport and mixing, is driven by melt density gradients arising from temperature gradients. Knowledge of a melt's density as a function of temperature is essential to modeling such convection. Extrapolation of glass thermal expansion into the melting temperature range is generally insufficiently accurate because it tends to underestimate the expansion of the melt.

In the present study, density and surface tension were treated together, since the same apparatus, and often the same set of measurements, were used to determine both. Sessile and pendant drop techniques were used. The sessile drop technique for measuring density and surface tension of liquids (a stationary drop on a horizontal, non-wetting substrate) was developed in the 19th century by Bashforth and Adams [39]. More recently, the technique has been significantly improved by use of computerized image analysis and curve fitting [40, 41]. Density is calculated from the measured volume of the drops and the known masses. Thermal expansion, if needed, can be calculated from the temperature dependence of the density, provided the data are sufficient.

Surface tension is a property of the interface between a liquid and the atmosphere in contact with it. Hence, the surface tension of a melt can be expected to vary, depending on the potentially reactive constituents present in the atmosphere, such as oxygen, water vapor, CO_2 and SO_2. Because the atmosphere above the melt and that inside a bubble are not necessarily the same, the surface tension values governing a rising bubble and those controlling fluctuations (ripples and waves) and refractory wetting at the melt's upper surface may not be the same. This complicates the modeling of how a bubble breaks when it reaches the surface, and how foams form and dissipate at the surface. These latter processes are further complicated when the melt surface composition differs from the bulk composition because of volatization, as in certain borosilicate and fluorine-containing glasses.

The literature suggest that surface tensions for most binary alkali silicate glasses range between 225 and 350 mN/m, but can be higher for alkaline earth silicates [42]. Kucuk and coworkers provide a method for surface tension estimation based on statistical analysis of much literature data. A rule of thumb is that most commercial silicate glasses have a surface tension around 300 mN/m and have a negative dependence on temperature (decrease with temperature increase) of less than 10% from 1200°C to 1500°C [43] (Bansal & Doremus).

However, as mentioned above, surface tension can be dependent on reactive components in the atmosphere above the melt; typically 71% N_2, 14% H_2O, 12% CO_2 and 3% O_2 (by volume) in a air-gas fired furnace, 64% H_2O, 31% CO_2, 4% N_2, and 1% O_2 in a oxy-gas fired furnace, and 79% N_2, 21% O_2 and ~1% Ar in an electrically heated furnace. In prior publications by Alexis Clare and co-workers, the effects of relative humidity and other atmospheres (reducing and oxidizing) were investigated [44, 45], as were the effects of iron redox (Fe2+/Fe3+) [45]. Surface tension measurements previously reported elsewhere [46] for four of the six commercial base glasses of this project show that surface tension is affected by the oxygen and water content of the atmosphere. For soda-lime-silica glass at 1400°C, γ in wet vs. dry air decreases by about 5 % for as little as 4% water concentration. Chapter 6 by Alexis Clare, et al. further discusses the surface tension and density measurements related to the present study.

Heat Capacity

Heat capacity is a measure of the quantity of thermal energy absorbed or released by a volume of material when its temperature changes. Thus, the amount of energy required to heat a given volume of molten glass to a higher temperature depends on its heat capacity. Heat capacity is generally expressed as the amount of heat required to raise the temperature of a body one K (or one °C) per unit of mass or per mole of atoms, e.g., J/(kg·K) or J/(mol·K).

Oxide glasses possess no electronic conductivity nor do they consist of molecules with rotational degrees of freedom, so the heat capacity is principally comprised of thermal vibration and structural rearrangement components. The thermal vibration contribution is well described at temperatures below the glass transition by the Debye theory [47]. It increases in magnitude from low temperatures toward the Dulong and Petit limit of 3R/gram-atom as the glass transition is approached, where R is the gas constant, 8.314 J/(mol·K)*. For crystalline ceramic oxides, this limit is reached at about 1000°C, almost independent of their composition and crystal structure. Based on the nature of the chemical bonds, we would expect similar behavior for oxide glasses, if the glass transition did not intervene. The heat capacity of oxide glasses generally reaches 80% to 95% of the Duong and Petit value at the glass transition. As the glass transition is passed, the heat capacity of most commercial glasses increases by an additional 15 to 30%, over a narrow temperature interval, depending on the glass composition (and the network structure). This increase is related to changes in the molecular-level structural configurations accessible to the melt, as characterized by what is known thermodynamically as configurational entropy.

* **NOTE:** According to basic kinetic theory, each atom possesses 1/2 k_BT of kinetic energy and 1/2 k_BT of potential energy per degree of freedom (where k_B is the Boltzmann constant, 1.381 x 10-23 J/K). For three degrees of vibrational freedom, the thermal energy is thus 3 k_BT/atom, or 3N k_BT/mol (where N is Avagadro's number). Therefore C_P = 3NkB/gram-atom = 3R/gram-atom = 24.94 J/gram-atom = 5.96 cal/gram-atom.

So, above the glass transition, one would expect the vibrational component of the heat capacity to increase until the Dulong-Petit limit is reached and that a configurational entropy contribution should be added to it. The configurational component is not easily predictable. Above the glass transition, the rate of loosening of the network might be expected to diminish with temperature, thus gradually decreasing the heat capacity. However, beyond the changes associated with the general loosening of the network (increase in free volume), other configurational changes can occur. For example, borosilicate glasses, for which boron coordination can change from three-fold to four-fold, often have higher heat capacities (on a joule per gram basis) than soda-lime-silicates. Alumina additions tend to decrease the heat capacity by tightening up the network. Taking all this into account, above the glass transition the Cp can be constant, or can have either slightly positive or slightly negative dependence on temperature. This being said, the range of heat capacities of melts is not great, generally between 1.2 to 1.6 J/(g·K), for temperatures between 1000°C and 1500°C. The theoretical foundations for heat capacity of silicate melts are discussed in References [30] (Kittel); [32] (Kingery, Chapter 12); [18] (Varshneya, Chapter 11). A recent review and discussion of published heat capacities of glass melts and methods for their calculation has been offered by Priven and Mazurin [20].

Experimental - Heat capacity is generally measured using calorimetric techniques such as drop calorimetry and DSC (differential scanning calorimetry), however the reliability of the data for glass melts above the temperature range of 800°C to 1000°C is not good, as reported in the literature. We have had similar experience at Alfred. The difficulties for DSC usage include the gradual wetting of the sample container by the softening sample which changes the thermal contact conditions during measurement. (Similar difficulty is not encountered with high-temperature crystalline materials.) We concluded that the accuracy of our measurements would be no better than for estimates based on theory. Hence no new measurements are presented here. Instead, we offer a method of calculation and some guidelines based on the literature.

The heat capacity of an oxide glass can be estimated as follows: First, the Dulong and Petit limit for the heat capacity of the glass can be calculated knowing the molecular content of the glass. To that can be added an estimated increment for the configurational entropy, which for many commercial glasses lies between 15 and 30 percent.

As a demonstration, in Table VI, for each base glass we determine the mole fraction of each of the component oxides and the number of atoms per oxide. From these we calculate the gram formula weight, the number of atoms per gram formula weight of glass, and the vibrational heat capacity of those atoms ($3k_B$/atom). Knowing the gram formula weight and the density of the melt, we calculate C_p per gram of glass and C_p per cm^3 of glass. Finally, for several of the glasses, we multiply those figures by factors taken from the literature [48] and listed in the next paragraph to approximate the contributions of the configurational entropy.

TABLE VI. HEAT CAPACITY CALCULATED FROM THEORETICAL VIBRATIONAL C_p AND EXPERIMENTAL ΔC_p AT THE GLASS TRANSITION

Base Glass	gfw (g)	atoms/gfw (mol)	atoms/g (mol/g)	C_p (vib.) (J/g·K)	ΔC_p (%)	C_p (total) (J/g·K)	density* (g/cm^3)	C_p (total) (J/cm^3·K)
Container	60.25	2.901	0.0481	1.201	21	1.45	2.32	3.36
Float	58.94	2.854	0.0484	1.209	21	1.46	2.28	3.33
Wool	60.82	2.978	0.0490	1.224	20	1.47	2.28	3.35
TV	70.38	2.920	0.0415	1.036	20	1.24	2.50	3.11
E-glass	62.29	3.027	0.0486	1.214	24	1.51	2.40	3.62
LE Borosilicate	61.80	3.249	0.0526	1.313	15	1.51	2.13	3.22

*Estimated density at 1275 °C

Data for Pyrex™ (Corning code 7740, similar to our low expansion borosilicate glass), an E-glass, and a soda-lime silica glass reported in the literature [48] (J/(gram-atom·K) are:

	Tg (°C)	C_p (glass)	C_p (liquid)	% Increase
Pyrex	871	23.68	27.28	15
E-Glass	996	24.87	30.87	24
Soda-lime-silica	850	24.45	29.63	21

So, based on Table VI, we offer the following guide: for soda-lime-silica container and float glasses, use C_p = 1.40 to 1.5 J/(g·K) and for low-expansion (Pyrex-type) borosilicates and E-glass, use 1.5 J/(g·K). If we assume the ΔC_p for wool and TV glass to be about 20% (similar to soda-lime-silica), the calculated C_p will be as show in Italics in Table VI.

As pointed out by Privin and Mazurin [20], care must be taken when comparing composition dependencies. Because heat capacities can be reported on a weight basis or on a mole basis, (J/(kg·K) vs. J/(mol·K), opposite trends are sometimes reported. Further, dense glasses contain fewer atoms per gram than do less-dense glasses. Substituting a heavy atom for a lighter one (such as by substituting among the alkaline earths or by substituting lead for an alkaline earth) increases the density, but does not proportionately change the number of atoms or moles per unit volume.

Viscosity

Background: Glasses are generally melted in large furnaces at temperatures considerably above 1200°C, where the melt viscosity is on the order of 100 Poise (10 Pa·s); they are formed into useful objects (e.g., sheet, tubing, containers, CRT bulbs, etc.) at lower temperatures, corresponding to viscosities near 1,000 Pa·s (10,000 Poise). Glass is tempered at viscosities near 10^9 Pa·s (10^{10} Poise) and annealed at viscosities near 10^{12} Pa·s (10^{13} Poise). Figure 2 indicates more clearly the viscosity ranges used in a variety of glass manufacturing steps.

Understanding how the viscosity of a melt depends upon its composition and temperature is extremely important for designing and operating most glass manufacturing processes, whether at the higher temperatures required for melting and fining, or at the lower ones required for forming useful objects. Accurate data or reliable predictive models for viscosity as a function of temperature over this range are essential. Unfortunately, composition and temperature dependence of viscosity is quite complex and no fundamental model is yet capable of predicting the behavior in a sufficiently useful quantitative manner.

Viscous flow of a glass melt requires the extensive breaking and reestablishing of chemical bonds within the molecular structural network. The simplest reaction rate theory for describing thermally activated processes would predict an Arrhenius type behavior where log-viscosity varies inversely with absolute temperature [49] (Glasstone, et al.). No oxide glass former shows such Arrhenius-like behavior, although pure vitreous silica comes closest over most of its useful viscosity range. The viscosity of complex silicates (such as the commercial glasses of this study) can be approximated as Arrhenius-like over narrow viscosity ranges somewhat above the glass transition and again over narrow ranges in the operating range of glass melting furnaces (10 to 10^3 Pa·s), but the slopes (activation energies) in these two regions are considerably different, and these differences vary significantly with melt composition. For a soda-lime-silica glass, the apparent activation energy at about 10^{11} Pa·s can be about 100 kcal/mole while at 10 Pa·s it is about 50 kcal/mole.

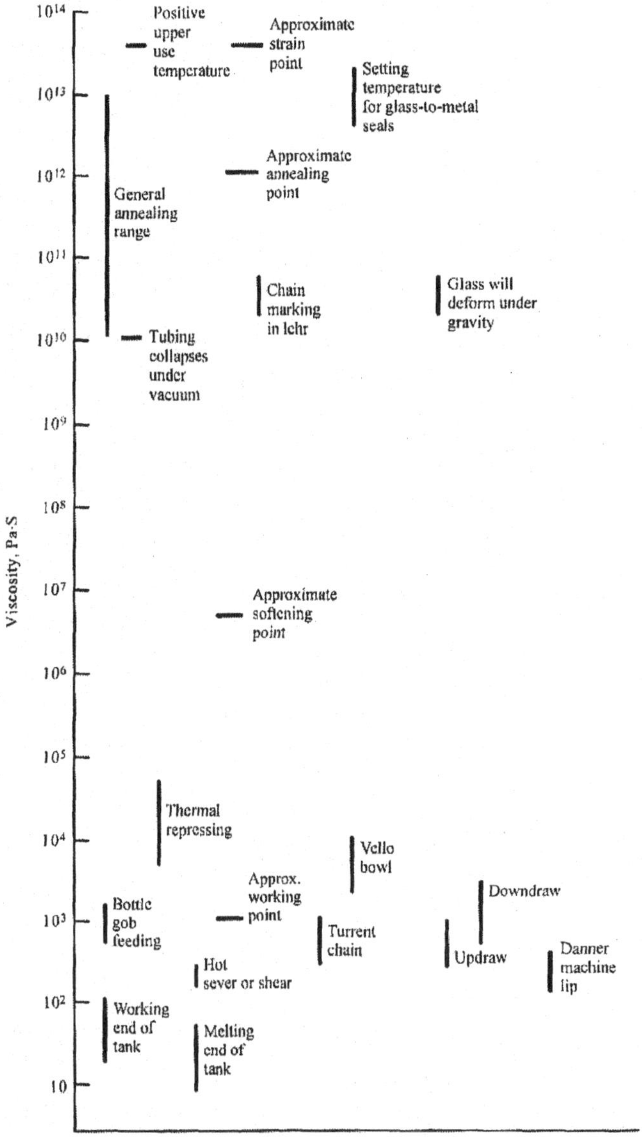

Figure 2 – Approximate viscosity values for forming and processing methods (Pa·s x 10 = P). (Courtesy of Corning Incorporated)

Other theoretical models, based on the "free volume" available for flow [50] (Turnbull and Cohen) or "excess entropy" [51] (Adam and Gibbs), use three adjustable parameters to fit the varying curvature. Each of these models can be described by an equation of the form

$$\eta = E \exp [F/(T - T_0)], \tag{6}$$

where η is viscosity, T is absolute temperature and E, F and T_0 are constants. This can be rewritten as

$$\log \eta = -A + B/(T - T_0) \tag{7}$$

where A and B are constants. In this form the equations are identical to the empirical relation commonly known as the Vogel-Fulcher-Tammann (VFT) equation, or simply the Fulcher equation [52]. Because of the well-accepted effectiveness of this equation, glass manufacturers often tabulate the VFT constants in their product literature. We will use Equation 7 in our data analysis, but it should be remembered that this procedure constitutes an empirical fitting of data and the constants are not based on absolute theoretical principles. Reviews of the theoretical basis for these behaviors can be found in Reference [18] (Varshneya), Chapter 9 and Reference [32] (Kingery, et al.), Chapter 14.

In short, for modeling glass melting and forming processes, reliable experimental data are essential. Because the viscosity range of interest to the CGR industrial members was so broad, 10^0 to 10^{12} Pa·s (10^1 to 10^{13} Poise), and because no single measurement technique is effective over this range, we employed three techniques: Rotating spindle (1 – 10^3 Pa·s), parallel plate (~ 10^6 – 10^9 Pa·s), and beam bending (~ 10^{10} - 10^{12} Pa·s). These techniques are briefly described in Ref. [18] (Varshneya), pp. 190-194. Pavel Hrma, Alexander Fluegel, Arun Varshneya and James Shelby report the measurement results in separate sections of Chapter 7 and Alexander Fluegel and co-authors discuss statistical modeling of the viscosity-composition relationships in Chapter 9.

Electrical Resistivity

Electricity can be used to melt glass in several different ways: electric arcs, plasma arcs, external resistance heating and internal resistance heating. In the latter case, the electric current is caused to flow through the glass by applying a voltage to the melt through electrodes (often referred to as Joule heating) or by inductively generating eddy currents within the melt via an alternating current electromagnetic field. Electrode-powered Joule heating generally uses alternating current at power frequencies of 50 or 60 Hz, or somewhat higher. Induction heating usually employs radio frequencies (KHz and MHz), although operation at commercial power frequencies has been reported [53].

Electrical boosting of glass melters is commonplace and many fiberglass, specialty and optical glasses are melted entirely by electricity. All-electric "cold top" melters have great advantage for containing volatile components such as alkalis, boron, heavy metals and halides, because the melt surface is typically covered by a floating batch blanket many centimeters thick that serves to condense the volatile components in its cooler upper regions, thus decreasing batch costs and unwanted environmental effects. Increased fossil fuel prices, more stringent environmental restrictions and improved designs for electrically-heated finers could extend this trend. The principles of electric melting and the history of its commercial applications prior to about 1984 can be found in References [54-57].

Although silicate glasses are excellent electrical insulators at room temperature and up to a few hundred degrees C (but below the glass transition), it is well known that they become relatively good electrical conductors at melt temperatures. This is because of dramatically increased mobility of singly- and to some extent doubly-charged network modifier ions at melt temperatures.

Heat is generated by the flow of electricity through the melt (ionic conduction) and the conversion of electrical energy to heat through the resistance of the melt, commonly known as Joule heating. The power dissipated as heat in the melt is given by

$$P = V \cdot I \tag{8}$$

where P is power in watts, V is electromotive force in volts and I is electric current in amperes. Using ohms law ($V = I \cdot R$), the power can be written as

$$P = I^2 R = E^2/R \tag{9}$$

where R is resistance in ohms.

Resistivity, r, is an "intensive" property of matter. It is expressed in units of ohm-cm. The resistance of a particular volume of material is "extensive" and depends on its shape and volume. For example, the lengthwise resistance of a cylinder of material is given by

$$R = r\,(L/A) \tag{10}$$

where L is the cylinder length and A is its cross-sectional area. Thus the resistivity is numerically equal to the resistance of a cube of the material, of any arbitrary edge dimension, when electrical contact is applied uniformly to opposing faces.

The effective resistance of a melter, or a section of a melter, is determined by the resistivity of the glass melt and the designed electrical path of the furnace (electrode locations, surface areas and geometries). Too high a design resistance requires excessive voltages, with associated physical dangers and opportunity for electrolytic activity at the furnace walls and joints. Too low a resistance requires excessive current, with increased capital expense in transformers, conductors, electrodes and control systems, and increased electrode wear. Therefore, the resistivity of the melt greatly affects electric melter design.

Melt resistivity is a function of composition and temperature. For alkali-containing glasses, the conductivity (inverse of resistivity, 1/r, in units of Siemens/m or $ohm^{-1}cm^{-1}$) is roughly proportional to the sodium ion concentration, when no other alkalis are present. Other alkalis have varying effects when substituted for sodium on a cation-for-cation basis. Potassium, rubidium and cesium are progressively poorer conductors. This is related to their increasing ionic size with increasing atomic number. Lithium is a more effective conductor, although the effect of its smaller size is largely offset by the strong electric field bonding it to the network. For alkali-free glasses, doubly charged ions such as calcium, magnesium and other second column members of the periodic table may conduct, but much less effectively than alkalis, except perhaps at very high temperatures.

The temperature dependence of conductivity varies greatly with glass composition, being less for soda-lime-silica glasses than for borosilicates and alkaline earth aluminosilicates. The inherent decrease of resistivity with temperature presents challenges to control system designers, because as temperature increases, resistivity drops. For a constant voltage source, therefore, the current and the ensuing power dissipated in the melt increase, heating the glass even more, and vice versa. This is an inherently unstable situation (unlike the tungsten filament in an incandescent lamp, for which the resistivity increases with temperature). The slope of the resistivity versus temperature curve, at the operating temperature, is sometimes referred to as the deregulation angle [55].

Knowledge of melt electrical resistivity, as a function of composition and temperature, is clearly essential for effective furnace modeling and design. Unfortunately there is no universal model for predicting the behavior, and one cannot directly extrapolate the temperature dependence of a glass's electrical conductivity into its melt region. At temperatures below the glass transition, the temperature dependence of conductivity can often be well-fit to an Arrhenius-type equation, such as that proposed by Rasch and Hinrichsen [58]. That is generally not the case at higher temperatures, as the data reported in Chapter 8 and its analysis in Chapter 9 will show. Conductivity of glass melts (above the glass transition temperature) is sometimes fit to an equation of the type:

$$\sigma = \sigma_0 \exp\,[-aT + bT^2 + \ldots] \qquad (11)$$

where σ is conductivity and a, b, etc. are fitted constants [55]. Excellent data fits are often obtained using a three-parameter equation of the same form as the VTF equation for viscosity. This is to be expected because the same increase in free volume with temperature that enhances fluidity (decreases viscosity) also enhances ionic transport (electrical conduction). Reviews of the general state of understanding of ionic conductivity in silicate glasses and melts can be found in references [32], [18], [43] and [59].

Glass melt resistivity is generally measured experimentally using alternating current frequencies of 1 kHz or greater to avoid electrode polarization effects. However, at the power frequencies used in most commercial applications, electrode polarization within the melter is possible, with the current through the melter being less than that predicted by the applied voltage, the melter geometry and the melt resistivity. Nevertheless, the power generated within a volume of glass is still determined by the current squared times the resistance, as determined by the melter geometry and the melt resistivity.

In Chapter 8, Arun Varshneya and Terese Vascott discuss measurements of electrical resistivity in the range of 950 to 1450°C for about 150 glass compositions of container, float, low-expansion borosilicate, TV panel, and wool and textile fiberglass-types, using a simple two-probe method, and report the ranges of resistivity and general composition and temperature dependent behavior observed.

Statistical Modeling

In Chapter 9, Alexander Fluegel, Arun Varshneya, David Earl and Dolun Öksoy present statistical modeling of the composition and temperature dependencies of viscosity, resistivity and several other properties. This modeling serves at least three purposes: it helps identify defective glass samples and data sets so that they may be removed from the database, it provides a method whereby furnace and process design engineers can estimate melt properties when their compositions do not coincide with those studied here, and it provides a basis for identifying potential gaps in the data and composition regions deserving further study. Further, the analysis should provide a basis for evaluating data published elsewhere and possibly for incorporating those data into an expanded database.

Glass Physical Properties

During the course of this project James Shelby and several of his co-workers undertook to measure important glass physical properties on all the available glasses from this study, using supplemental funding from the CGR and the PI's (principal investigator) discretionary research funds. These properties include density, refractive index, coefficient of thermal expansion, dilatometric softening point, glass transition temperature, dissolved water concentration and electrical conductivity at 200°C and 300°C. Although this work lay outside the scope of the original grant, and has been reported in part elsewhere[60], it is presented in an Appendix to this book for completeness and to show another way in which the database begun with the DOE grant will continue to expand.

Quality Assurance

The glass melting and data collection conducted at Alfred University for this study were carried out in accordance with a Quality Assurance Plan adapted from an earlier DOE-approved quality assurance program in place at the University (Institute of Glass Science and Engineering). The format of this Plan is patterned after a highly successful QA program developed at the New York State College of Ceramics at Alfred University during the execution of a contract with the USDOE West Valley Demonstration Project.

The objectives of the Plan were to ensure the performance of reliable, verifiable and traceable scientific research; to ensure that accurate, retrievable records would be available for future inspection; and to provide documentation showing objective evidence that the research had been carried out as planned. Implementation of the Plan required formal work assignments, rigorous sample tracking and control, instrument calibration, well-documented data collection and analysis procedures, and centralized record keeping. For sub-contract work performed outside the University, documentation of quality assurance was the responsibility of the sub-contractor.

REFERENCES

1. Seward, T.P., *The NSF Industry-University Center for Glass Research: An Overview.* Ceramic Engineering and Science Proceedings, 1999. **20**(1): p. 215-226.
2. Seward, T.P., *Aspects of the Glass Melt Properties Database Investigations at Alfred University.* Ceramic Engineering and Science Proceedings, 2001. **22**(1): p. 149-163.
3. Anonymous, DE-FG07-96EE41262 Modeling of Glass Making Procsses for Improved Efficiency. 1996, U.S. Department of Energy.
4. Anonymous, *Glass: A Clear Vision for a Bright Future.* 1996, U.S Department of Energy.
5. Sewell, P., *Industrial Perspective of Modeling in a Modern Float Glass Plant.* The GlassResearcher: Bulletin of Glass Science and Engineering, 1997. **6**(2): p. 1-3.
6. LaCourse, W.C., *DOE, CGR Collaborate on Modeling Workshop.* The GlassResearcher: Bulletin of Glass Science and Engineering, 1997. **6**(2): p. 1-2.
7. Pye, L.D., Development of a Composition-Property Data Base for the Computer Modeling of Glass Melting Processes - A proposal submitted to the United States Department of Energy. 1995, Alfred University: Alfred, NY.
8. LaCourse, W.C. and I. Joseph, Modeling of Glass Making Processes for Improved Efficiency - A proposal submitted to the Unites States Department of Energy. 1996, Alfred University: Alfred, NY.

9. Viskanta, R., *Review of Three-Dimensional Mathematical Modeling of Glass Melting.* Journal of Non-Crystalline Solids, 1994. **177**: p. 347-362.
10. Schnepper, C., et al., Coupled Combustion Space-Glass Bath Modeling of a Float Glass Melting Tank Using Full Oxy-Combustion. Ceramic Transactions, 1997. **82**: p. 219-224.
11. Hoke, B.C. and R.D. Marchiando. Using Computational Fluid Dynamics Models to Assess Melter Capacity Changes when Converting to Oxy-Fuel. in 18th International Congress on Glass. 1998. San Francisco, California.
12. Chang, S.-L., B. Golchert, and C.Q. Zhoy, *Computer Modeling of Glass Furnace Flow and Heat Transfer.* Ceramic Transactions, 2004. **141**: p. 399-411.
13. Op den Camp, O., et al. The Use of Glass Process Simulation for Control of Glass Quality in Glass Production. in XX International Congress on Glass. 2004. Kyoto, Japan: The Ceramic Society of Japan.
14. Schill, P. Progress in Mathematical Simulation and Advanced Control of Glass Aggregates. in XX International Congress on Glass. 2004. Kyoto, Japan: The Ceramic Society of Japan.
15. Lakatos, T., L.-G. Johansson, and B. Simminsköld, Viscosity-temperature relations in the glass system SiO_2-Al_2O_3-Na_2O-K_2O-CaO-MgO in the composition range of technical glasses. Glass Technol., 1972. **13**: p. 88-95.
16. Oksoy, D., L.D. Pye, and E.N. Boulos, *Statistical analysis of viscosity-composition data in glassmaking.* Glastech. Ber. Glass Sci. Technol., 1994. **67**(7): p. 189-195.
17. Sasek, L. and M. Knotek, Chem. Technol. Silik., 1985. **L13**: p. 91-106.
18. Varshneya, A.K., *Fundamentals of Inorganic Glasses.* 1994, San Diego, CA: Academic Press, Inc. 570.
19. Kucuk, A., A.G. Clare, and L.E. Jones, An estimation of the surface tension for silicate glass melts at 1450°C using statistical analysis. Glass Technol., 1999. **40**(5): p. 149-153.
20. Priven, A.I. and O.V. Mazurin, *Heat Capacity of Glass Melts*, in *Handbook on the Properties of Glass-Forming Melts Properties*, L.D. Pye, I. Joseph, and A. Montenaro, Editors. In press, CRC Press: Boca Raton, Florida. p. Chapter 8.
21. Shelby, J.E., *Handbook of Gas Diffusion in Solids and Melts.* 1996, Materials Park, OH: ASM International. 240.
22. Cable, M., *Principles of Glass Melting*, in *Glass: Science and Technology - Processing I*, D.R. Uhlmann and N.J. Kreidl, Editors. 1984, Academic Press, Inc.: Orlando, Florida. p. 1-44.
23. Beerkens, R.G.C., *Analysis of Advanced and Fast Fining Processes for Glass Melts.* Ceramic Transactions, 2004. **141**: p. 3-34.
24. Shelby, J.E., *Carbon Dioxide Solubility in Silicate Melts.* Ceramic Transactions, 1998. **82**: p. 51-56.
25. Rosseland, S., Theoretical astrophysics: atomic theory and the analysis of stellar atmospheres and envelopes. 1936, Oxford, England: Clarendon Press.
26. Gardon, R., *A Review of Radient Heat Transfer in Glass.* J. Am. Ceram. Soc., 1961. **44**(7): p. 305-312.
27. Chui, G.K. and R. Gardon, *Interaction of Radiation and Conduction in Glass.* J. Am. Ceram. Soc., 1969. **52**(10): p. 548-553.
28. Pierrot, L. Accuracy of the Rosseland Approximation in Toy Models of Glass Tanks. in XX International Congress on Glass. 2004. Kyoto, Japan: The Ceramic Society of Japan.
29. Kittel, C., *Interpretation of the Thermal Conductivity of Glasses.* Physical Review, 1949. **75**(6): p. 972-974.

30. Kittel, C., *Introduction to Solid State Physics*. 4th ed. 1971, New York: John Wiley & Sons, Inc. Fig. 27, p. 231.
31. Kingery, W.D., *Heat-Conductivity in Glass*. J. Am. Ceram. Soc., 1961. **44**(7): p. 302-304.
32. Kingery, W.D., H.K. Bowen, and D.R. Uhlmann, *Introduction to Ceramics*. Second ed. Wiley Series on the Science and Technology of Materials. 1976, New York, NY: John Wiley & Sons. 1032.
33. Howell, J.R. and M.E. Goldstein, *Effective slip coefficients for coupled conduction-radiation problems*. J. Heat Transfer, 1969. **91**: p. 165-166.
34. Prokhorenko, O.A., Personal Communications to Center for Glass Research. 1999-2001.
35. Van Zee, A.F. and C.L. Babcock, *A Method for the Measurement of Thermal Diffusivity of Molten Glass*. J. Am. Ceram. Soc., 1951. **34**(8): p. 244-250.
36. Jain, A. and L.D. Pye, Automated Thermal Diffusivity Measurements of Glasses and Their Melts, in Advances in the Fusion of Glass- Proceedings of the 1st International Conference on Advances in the Fusion of Glass, D.F. Bickford, et al., Editors. 1988, The American Ceramic Society: Westerville, Ohio. p. 40.1-40.13.
37. Charnok, H., Experimental and Theoretical Comparison of Radiation Conductivity Predicted by Steady-State Theory With That Effective Under Periodic Temperature Conditions. J. Am. Ceram. Soc., 1961. **44**(7): p. 313-317.
38. Carslaw, H.S. and J.C. Jaeger, *Conduction of Heat in Solids*. Second ed. 1959, London: Oxford University Press.
39. Bashforth, F. and J.C. Adams, *An Attempt to Test the Theories of Capillary Action*. 1883, Cambridge, England: Cambridge University Press.
40. Rotenborg, Y., L. Burokova, and A.W. Neumann, Determination of Surface Tension and Contact Angle from the Shapes of Axisymmetrical Fluid Interfaces. J. Coll. Inter. Sci., 1983. **93**(1): p. 169-183.
41. Kucuk, A., A.G. Clare, and L.E. Jones, *Density and Surface Tension of Glass Melts as a Function of Composition at 1400°C*. Ceramic Transactions, 1998. **82**: p. 287-298.
42. Shartsis, L. and S. Spinner, *Surface Tension of Molten Alkali Silicates*. J. Res. Natl. Bur. Std. (U.S.), 1951. **46**(5): p. 385-390.
43. Bansal, N.P. and R.H. Doremus, *Handbook of Glass Properties*. 1986, Orlando, Florida: Academic Press, Inc. 680.
44. Kucuk, A., A.G. Clare, and L.E. Jones, Differences between the Surface and Bulk of Glass Melts. I. Compositional Differences and Influences of Volatilization on Composition and Other Physical Properties. J. Non-Crystalline Solids, 2000. **261**: p. 28-38.
45. Kucuk, A., A.G. Clare, and L.E. Jones, Differences between the Surface and Bulk of Glass Melts. Part 2. Influence of Redox Ratio on the Surface Properties of Silicate Melts. Phys. Chem. Glasses, 2000. **41**(2): p. 75-80.
46. Kucuk, A., A.G. Clare, and L.E. Jones, *Influence of Various Atmospheres on Surface Properties of Silicate Melts*. Glastech. Ber. Glass Sci. Technol., 2000. **73**(5): p. 123-129.
47. Debye, P., *Zur Theorie der spezifischen Wärmen (On the Theory of Specific Heat)*. Annalen der Physik und Chemie, 1912. **39**(3): p. 789-839.
48. Richet, P., et al., *Configurational heat capacity and entropy of borosilicate melts*. J. Non-Crystalline Solids, 1977. **211**: p. 271-280.
49. Glasstone, S.K., K.J. Laidler, and H. Eyring, *The Theory of Rate Processes*. 1941, New York: McGraw-Hill Book Company.
50. Turnbull, D. and M.H. Cohen, *Free-Volume Model of the Amorphous Phase: Glass*

Transition. J. Chem. Phys., 1961. **34**(1): p. 120-125.

51. Adam, G. and J.H. Gibbs, On the Temperature Dependence of Cooperative Relaxation Properties in Glass-Forming Liquids. J. Chem. Phys., 1965. **43**(1): p. 139-146.
52. Fulcher, G.S., *Analysis of Recent Measurements of the Viscosity of Glass*. J. Am. Ceram. Soc., 1925. **8**(6): p. 339-355.
53. Carrillo, E., M.A. Barron, and J. Gonzalez, *Electromagnetic Induction Heating in Molten Glass at 60 Hz with No Susceptors*. Ceramic Transactions, 2004. **141**: p. 45-523.
54. Penberthy, L., *Electric Melting of Glass*, in *The Handbook of Glass Manufacture*, F.V. Tooley, Editor. 1975, Books For Industry (Ashlee Publishing): New York, NY. p. 387-400.
55. Stanek, J., *Electric Melting of Glass*. Glass Science and Technology, 1. 1977, Amsterdam: Elsevier Scientific Publishing Co.
56. Scarfe, F., *Electric melting, a review*. Glass Technol., 1980. **21**(1): p. 37-50.
57. Steitz, W.R., C.W. Hibscher, and G.R. Mattocks, *Electric Melting of Glass (Update)*, in *The Handbook of Glass Manufacture*, F.V. Tooley, Editor. 1984, Books for the Glass Industry Division, Ashlee Publishing Co., Inc.: New York, NY. p. 400.1 - 400.28.
58. Rasch, E. and F.W. Hinrichsen, Über eine Beziehung zwischen Elektrischer Leitfähigkeit und Temperatur (On the Relationship between Electrical Conductivity and Temperature). Zeitschrift für Elektrochemie, 1908. **14**(5): p. 41-48.
59. Mazurin, O.V. and O.A. Prokhorenko, *Electrical Conductivity of Glass Melts*, in *Handbook on the Properties of Glass-Forming Melts*, L.D. Pye, I. Joseph, and A. Montenaro, Editors. In press, CRC Press: Boca Raton, Florida. p. Chapter 9.
60. Shelby, J.E., et al., *Properties of Soda-Lime-Silica Glasses*. Ceramic Transactions, 2004. **141**: p. 371-378.

Chapter 2

COMMERCIAL AND STATISTICALLY DESIGNED GLASS COMPOSITIONS

Terese Vascott and Thomas P. Seward III
New York State College of Ceramics at Alfred University

COMMERCIAL GLASS COMPOSITIONS

The glasses that form the basis of this study represent six key composition areas: Container, E glass, float, low-expansion borosilicate, TV panel, and wool fiber glass. Bulk quantities (25 to 50 pounds) of commercial glass in each of the six composition areas were supplied by CGR member companies. These glasses were chemically analyzed and served as the 'base glasses' for the project. The analyzed compositions are shown below in Table I. As silica was not reported for low-expansion borosilicate and wool glasses, it is shown here as a calculated difference from 100%.

TABLE I. BASE GLASS COMPOSITIONS AS ANALYZED (WEIGHT PER CENT)

Base Glass ID	SiO_2	Al_2O_3	B_2O_3	MgO	CaO	SrO	BaO	Li_2O	Na_2O	K_2O
container	74.38	1.28		0.20	10.51	0.01	0.10		13.30	0.29
float	72.92	0.13		3.84	8.74				13.74	0.03
TV panel	61.53	2.06			0.05	9.19	9.23	0.010	7.64	7.55
E-glass	56.60	14.80	7.04	4.33	18.20				0.63	
LE borosilicate	[80.98]	2.19	12.60	0.008	0.023			0.010	4.09	0.055
wool	[62.66]	3.31	4.90	3.46	8.27				16.00	0.83

Base Glass ID	TiO_2	CeO_2	ZrO_2	PbO	ZnO	Fe_2O_3	Sb_2O_3	F	FeO	SO_3
container	0.008					0.036			0.008	0.21
float	0.017					0.109			0.042	
TV panel	0.430	0.280	1.390	0.000	0.51	0.037	0.30		0.004	
E-glass	0.580					0.350		0.220	0.100	
LE borosilicate		0.005				0.026			0.016	
wool						0.31			0.085	0.17

A focus group of industry technical representatives to the CGR provided information about the key composition variables and ranges the industrial members wished to see investigated. This information, shown below in Table II, provided the basis for experimental design of the composition variations to be further investigated.

TABLE II. DOE GLASS MELT PROPERTIES PROJECT: Composition Ranges (Wt. %)

Component	Container	E-glass	Float	Low-exp. Borosilicate	TV panel	Wool
SiO_2	65-75	52-60	70-74	70-81	58-64	56-66
B_2O_3		0-9		10-15		3-9
Al_2O_3	1-3	12-16	0.1-1.6	2.7	1.3-3.5	0-6
MgO	0-3	0.5-4.5	3-4		0-1.5	1-5
CaO	7-12	16-24	7-9	0-2	0-3.5	5-11
SrO					1-10	
BaO				0-2	2-13	
Li_2O	0-1				0-0.5	
Na_2O	11-15	0-2	12-15	4-8	6-9	13-17
K_2O	0-2	0-0.5	0.05-0.8	0-3	6-9	0-2
Fe_2O_3	0-0.4	0-0.8	0.1-1.5			0-0.6
Cr_2O_3	0-0.3					
TiO_2	0-0.5	0-1			0.1-0.5	
CeO_2					0-0.7	
ZrO_2					0-3	
PbO					0-3	
ZnO					0-1.5	
As_2O_3					0-0.3	
Sb_2O_3					0.2-0.6	
SO_3	0-0.3					0-0.2
F		0-0.6			0-0.7	0-0.6

STATISTICALLY DESIGNED GLASS COMPOSITIONS

Factorial designs for the glass compositions used in this project project were developed by Dolun Oksoy (Professor of Management Science and Information Systems, Alfred University) based on Plackett-Burman Design. They are coded as pbxxyy; where pb stands for Plackett-Burman, xx stands for number of factors, and yy stands for number of runs (number of different compositions). The six experimental composition sets were designed as follows: container glass (CO) - pb1124; E-glass (E) - pb1024; float (FL) - pb1224; low-expansion borosilicate (LO) - pb724; color TV panel glass (TV) - pb1724; and wool-type fiberglass (WO) - pb1024. These composition sets are identified as target compositions in Tables III-VIII. Because of the design structure, some target SiO_2 values fell outside the range specified in Table II in order to accommodate the desired range for other oxides.

TABLE III. Statistically Designed Container Glass Compositions (Weight Fraction)

Sample ID container	SiO_2 target	SiO_2 analyzed	Al_2O_3 target	Al_2O_3 analyzed	MgO target	MgO analyzed	CaO target	CaO analyzed
1	0.810	0.8111	0.01	0.0114	0	0.0006	0.07	0.0694
2	0.729	0.7388	0.01	0.0101	0	0.0010	0.12	0.1177
3	0.700	0.6902	0.01	0.0116	0	0.0010	0.12	0.1234
4	0.728	0.7340	0.01	0.0109	0.03	0.0295	0.07	0.0681
5	0.707	0.7097	0.01	0.0112	0.03	0.0308	0.07	0.0677
6	0.691	0.6903	0.01	0.0113	0.03	0.0324	0.12	0.1164
7	0.718	0.7206	0.03	0.0312	0	0.0006	0.07	0.0677
8	0.753	0.7505	0.03	0.0324	0	0.0006	0.07	0.0680
9	0.685	0.6821	0.03	0.0319	0	0.0011	0.12	0.1214
10	0.739	0.7397	0.03	0.0318	0.03	0.0314	0.07	0.0696
11	0.647	0.6488	0.03	0.0312	0.03	0.0315	0.12	0.1157
12	0.703	0.7059	0.03	0.0305	0.03	0.0284	0.12	0.1213
13	0.732	0.7450	0.01	0.0114	0	0.0001	0.07	0.0716
14	0.788	0.7953	0.01	0.0109	0	0.0006	0.07	0.0679
15	0.696	0.7002	0.01	0.0105	0	0.0010	0.12	0.1203
16	0.750	0.7513	0.01	0.0102	0.03	0.0280	0.07	0.0695
17	0.682	0.6873	0.01	0.0111	0.03	0.0345	0.12	0.1149
18	0.717	0.7105	0.01	0.0115	0.03	0.0335	0.12	0.1205
19	0.744	0.7375	0.03	0.0320	0	0.0006	0.07	0.0699
20	0.728	0.7298	0.03	0.0311	0	0.0010	0.12	0.1219
21	0.707	0.7110	0.03	0.0325	0	0.0008	0.12	0.1186
22	0.735	0.7312	0.03	0.0291	0.03	0.0279	0.07	0.0689
23	0.706	0.7123	0.03	0.0321	0.03	0.0279	0.07	0.0745
24	0.625	0.6294	0.03	0.0313	0.03	0.0315	0.12	0.1140

TABLE III CONT'D.

Sample ID container	Li_2O target	Li_2O analyzed	Na_2O target	Na_2O analyzed	K_2O target	K_2O analyzed	Fe_2O_3 target	Fe_2O_3 analyzed
1	0		0.11	0.1063	0	0.0025	0	0.0003
2	0		0.11	0.1005	0.02	0.0208	0	0.0003
3	0.01	0.0093	0.15	0.1531	0	0.0002	0.004	0.0046
4	0		0.15	0.1462	0	0.0002	0.004	0.0043
5	0.01	0.0091	0.15	0.1471	0.02	0.0197	0	0.0004
6	0.01	0.0089	0.11	0.1072	0.02	0.0209	0.004	0.0046
7	0		0.15	0.1466	0.02	0.0204	0.004	0.0046
8	0.01	0.0092	0.11	0.1085	0.02	0.0208	0.004	0.0046
9	0.01	0.0091	0.15	0.1472	0	0.0001	0	0.0003
10	0.01	0.0091	0.11	0.1068	0	0.0003	0	0.0005
11	0		0.15	0.1472	0.02	0.0212	0	0.0005
12	0		0.11	0.1034	0	0.0002	0.004	0.0039
13	0.01	0.0056	0.15	0.1340	0.02	0.0213	0	0.0002
14	0.01	0.0092	0.11	0.1051	0	0.0006	0.004	0.0043

TABLE III CONT'D.

Sample ID container	Li_2O target	Li_2O analyzed	Na_2O target	Na_2O analyzed	K_2O target	K_2O analyzed	Fe_2O_3 target	Fe_2O_3 analyzed
15	0		0.15	0.1446	0.02	0.0209	0.004	0.0038
16	0		0.11	0.1096	0.02	0.0206	0.004	0.0040
17	0		0.15	0.1525	0	0.0002	0	0.0003
18	0.01	0.0092	0.11	0.1091	0	0.0003	0	0.0006
19	0		0.15	0.1479	0	0.0002	0	0.0004
20	0		0.11	0.1015	0	0.0003	0.004	0.0038
21	0.01	0.0099	0.11	0.0998	0.02	0.0211	0	0.0004
22	0		0.11	0.1087	0.02	0.0207	0	0.0003
23	0.01	0.0107	0.15	0.1379	0	0.0001	0.004	0.0044
24	0.01	0.0096	0.15	0.1474	0.02	0.0202	0.004	0.0044

TABLE III CONT'D.

Sample ID container	Cr_2O_3 target	Cr_2O_3 analyzed	TiO_2 target	TiO_2 analyzed	SO_3 target	SO_3 analyzed
1	0	0	0	0.0001	0	0.0001
2	0.003	0.0026	0.005	0.0047	0.003	0.0024
3	0.003	0.0031	0	0.0001	0.003	0.0029
4	0.003	0.0029	0.005	0.0050	0	0.0002
5	0	0	0	0.0001	0.003	0.0029
6	0	0	0.005	0.0051	0	0.0002
7	0	0	0.005	0.0051	0.003	0.0028
8	0.003	0.0030	0	0.0001	0	0.0002
9	0	0	0.005	0.0052	0	0.0002
10	0.003	0.0030	0.005	0.0051	0.003	0.0011
11	0.003	0.0029	0	0.0001	0	0.0002
12	0	0	0	0.0001	0.003	0.0026
13	0.003	0.0028	0.005	0.0054	0	0.0001
14	0	0	0.005	0.0050	0.003	0.0014
15	0	0	0	0.0001	0	0.0003
16	0.003	0.0030	0	0.0001	0.003	0.0023
17	0	0	0.005	0.0050	0.003	0.0024
18	0.003	0.0028	0	0.0001	0	0.0002
19	0.003	0.0032	0	0.0001	0.003	0.0025
20	0.003	0.0028	0.005	0.0050	0	0.0004
21	0	0	0	0.0001	0.003	0.0029
22	0	0	0.005	0.0052	0	0.0003
23	0	0	0	0.0001	0	0.0001
24	0.003	0.0030	0.005	0.0050	0.003	0.0027

TABLE IV. STATISTICALLY DESIGNED E GLASS COMPOSITIONS (WEIGHT FRACTION)

Sample ID E glass	SiO_2 target	SiO_2 analyzed	B_2O_3 target	B_2O_3 analyzed	Al_2O_3 target	Al_2O_3 analyzed	MgO target	MgO analyzed
1	0.7150	0.6920	0	0	0.12	0.1220	0.005	0.0054
2	0.6540	0.6570	0	0	0.12	0.1500	0.045	0.0153
3	0.5570	0.5570	0	0	0.12	0.1120	0.045	0.0503
4	0.6310	0.6250	0	0	0.16	0.1500	0.005	0.0056
5	0.5700	0.5760	0	0	0.16	0.1530	0.005	0.0058
6	0.5360	0.5290	0	0	0.16	0.1470	0.045	0.0588
7	0.5860	0.5960	0.09	0.0868	0.12	0.1120	0.005	0.0053
8	0.5220	0.5300	0.09	0.0849	0.12	0.1120	0.005	0.0053
9	0.4790	0.4980	0.09	0.0863	0.12	0.1200	0.045	0.0393
10	0.4890	0.4800	0.09	0.0859	0.16	0.1530	0.005	0.0046
11	0.5100	0.5000	0.09	0.0898	0.16	0.1500	0.045	0.0468
12	0.5370	0.5380	0.09	0.0883	0.16	0.1490	0.045	0.0464
13	0.5940	0.5890	0	0	0.12	0.1120	0.005	0.0055
14	0.6210	0.6170	0	0	0.12	0.1110	0.005	0.0067
15	0.6420	0.6320	0	0	0.12	0.1220	0.045	0.0420
16	0.6520	0.6570	0	0	0.16	0.1480	0.005	0.0199
17	0.6090	0.6040	0	0	0.16	0.1480	0.045	0.0455
18	0.5450	0.5390	0	0	0.16	0.1460	0.045	0.0518
19	0.5950	0.6460	0.09	0.0883	0.12	0.0933	0.005	0.0044
20	0.5610	0.5530	0.09	0.0881	0.12	0.1140	0.045	0.0474
21	0.5000	0.5280	0.09	0.0855	0.12	0.1110	0.045	0.0418
22	0.5740	0.5710	0.09	0.0891	0.16	0.1530	0.005	0.0058
23	0.4770	0.4740	0.09	0.0889	0.16	0.1540	0.005	0.0049
24	0.4160	0.4130	0.09	0.0926	0.16	0.1600	0.045	0.0438

TABLE IV CONT'D.

Sample ID E glass	CaO target	CaO analyzed	Na_2O target	Na_2O analyzed	K_2O target	K_2O analyzed	Fe_2O_3 target	Fe_2O_3 analyzed
1	0.16	0.1570	0	0	0	0	0	0
2	0.16	0.1580	0	0	0.005	0.0052	0	0.0059
3	0.24	0.2470	0.02	0.0205	0	0	0.008	0.0080
4	0.16	0.1700	0.02	0.0205	0	0	0.008	0.0081
5	0.24	0.2540	0.02	0.0204	0.005	0.0056	0	0
6	0.24	0.2530	0	0	0.005	0.0053	0.008	0.0084
7	0.16	0.1680	0.02	0.0204	0.005	0.0054	0.008	0.0079
8	0.24	0.2460	0	0	0.005	0.0050	0.008	0.0084
9	0.24	0.2260	0.02	0.0205	0	0	0	0
10	0.24	0.2220	0	0	0	0	0	0
11	0.16	0.1680	0.02	0.0203	0.005	0.0053	0	0
12	0.16	0.1680	0	0	0	0	0.008	0.0080
13	0.24	0.2500	0.02	0.0201	0.005	0.0054	0	0
14	0.24	0.2510	0	0	0	0	0.008	0.0081
15	0.16	0.1600	0.02	0.0213	0.005	0.0051	0.008	0.0081
16	0.16	0.1490	0	0	0.005	0.0051	0.008	0.0050
17	0.16	0.1690	0.02	0.0209	0	0	0	0
18	0.24	0.2460	0	0	0	0	0	0
19	0.16	0.1350	0.02	0.0195	0	0	0	0
20	0.16	0.1730	0	0	0	0	0.008	0.0082
21	0.24	0.2270	0	0	0.005	0.0049	0	0
22	0.16	0.1710	0	0	0.005	0.0053	0	0
23	0.24	0.2200	0.02	0.0213	0	0	0.008	0.0077
24	0.24	0.2460	0.02	0.0213	0.005	0.0046	0.008	0.0083

TABLE IV CONT'D.

Sample ID E glass	TiO_2 target	TiO_2 analyzed	F target	F analyzed	Sample ID E glass	TiO_2 target	TiO_2 analyzed	F target	F analyzed
1	0	0	0	0	13	0.01	0.0099	0.006	0.0052
2	0.01	0.0100	0.006	0.0017	14	0	0	0.006	0.0054
3	0.01	0.0100	0	0	15	0	0	0	0
4	0.01	0.0101	0.006	0.0052	16	0.01	0.0100	0	0.0026
5	0	0	0	0	17	0	0	0.006	0.0049
6	0	0	0.006	0.0051	18	0.01	0.0098	0	0
7	0	0	0.006	0.0052	19	0.01	0.0080	0	0
8	0.01	0.0098	0	0	20	0.01	0.0105	0.006	0.0047
9	0	0	0.006	0.0039	21	0	0	0	0
10	0.01	0.0094	0.006	0.0031	22	0	0	0.006	0.0047
11	0.01	0.0101	0	0	23	0	0	0	0
12	0	0	0	0	24	0.01	0.0101	0.006	0.0059

TABLE V. STATISTICALLY DESIGNED FLOAT GLASS COMPOSITIONS (WEIGHT FRACTION)

Sample ID float	SiO_2 target	SiO_2 analyzed	Al_2O_3 target	Al_2O_3 analyzed	MgO target	MgO analyzed	CaO target	CaO analyzed
1	0.7467		0.0200		0.0300		0.0700	
2	0.7187	0.7219	0.0010	0.0022	0.0300	0.0261	0.0700	0.0709
3	0.7235		0.0010		0.0300		0.0700	
4	0.6940		0.0010		0.0300		0.0900	
5	0.6957		0.0010		0.0400		0.0900	
6	0.6745	0.6704	0.0200	0.0210	0.0400	0.0367	0.0900	0.0940
7	0.6942	0.6901	0.0200	0.0196	0.0400	0.0362	0.0900	0.0904
8	0.6770		0.0200		0.0400		0.0900	
9	0.7025	0.6980	0.0200	0.0198	0.0400	0.0367	0.0700	0.0730
10	0.6760		0.0200		0.0300		0.0900	
11	0.6860		0.0010		0.0400		0.0700	
12	0.7007		0.0200		0.0300		0.0900	
13	0.7132	0.7049	0.0010	0.0022	0.0400	0.0377	0.0900	0.0974
14	0.7052		0.0200		0.0400		0.0700	
15	0.6807	0.7006	0.0200	0.0184	0.0300	0.0244	0.0700	0.0647
16	0.7192		0.0010		0.0300		0.0900	
17	0.7265		0.0010		0.0400		0.0900	
18	0.7022		0.0200		0.0400		0.0700	
19	0.7190		0.0200		0.0300		0.0700	
20	0.7017	0.6901	0.0010	0.0023	0.0300	0.0278	0.0900	0.0983
21	0.7352	0.7321	0.0010	0.0022	0.0400	0.0378	0.0700	0.0725
22	0.7225	0.7149	0.0200	0.0212	0.0300	0.0280	0.0900	0.0913
23	0.7430		0.0010		0.0400		0.0700	
24	0.7725		0.0010		0.0300		0.0700	

TABLE V CONT'D.

Sample ID float	MnO_2 target	MnO_2 analyzed	Na_2O target	Na_2O analyzed	K_2O target	K_2O analyzed	Fe_2O_3 target	Fe_2O_3 analyzed
1	0.0025		0.1200		0.0050		0.0010	
2	0		0.1500	0.1423	0.0050	0.0055	0.0150	0.0146
3	0.0025		0.1500		0.0200		0.0010	
4	0		0.1500		0.0200		0.0010	
5	0		0.1500		0.0200		0.0010	
6	0		0.1500	0.1512	0.0200	0.0218	0.0010	0.0012
7	0		0.1200	0.1189	0.0200	0.0207	0.0150	0.0141
8	0.0025		0.1500		0.0050		0.0150	
9	0.0025	0.0027	0.1200	0.1175	0.0200	0.0209	0.0150	0.0142
10	0.0025		0.1500		0.0050		0.0150	
11	0.0025		0.1500		0.0200		0.0150	
12	0.0025		0.1200		0.0200		0.0010	
13	0		0.1200	0.1228	0.0050	0.0059	0.0150	0.0148
14	0.0025		0.1500		0.0050		0.0010	

TABLE V CONT'D.

Sample ID float	MnO_2 target	MnO_2 analyzed	Na_2O target	Na_2O analyzed	K_2O target	K_2O analyzed	Fe_2O_3 target	Fe_2O_3 analyzed
15	0		0.1500	0.1334	0.0200	0.0202	0.0150	0.0127
16	0.0025		0.1200		0.0200		0.0150	
17	0.0025		0.1200		0.0050		0.0010	
18	0		0.1500		0.0050		0.0010	
19	0		0.1200		0.0200		0.0150	
20	0.0025	0.0023	0.1500	0.1440	0.0050	0.0056	0.0150	0.0148
21	0.0025	0.0027	0.1200	0.1210	0.0200	0.0214	0.0010	0.0013
22	0		0.1200	0.1213	0.0050	0.0059	0.0010	0.0012
23	0		0.1200		0.0050		0.0150	
24	0		0.1200		0.0050		0.0010	

TABLE V CONT'D.

Sample ID float	Cr_2O_3 target	Cr_2O_3 analyzed	Co_3O_4 target	Co_3O_4 analyzed	TiO_2 target	TiO_2 analyzed	SO_3 target	SO_3 analyzed
1	0.0004		0		0.0005		0	
2	0		0		0.0100	0.0106	0	0.0003
3	0		0.00015		0.0005		0	
4	0.0004		0		0.0100		0	
5	0		0.00015		0.0005		0	
6	0.0040	0.0039	0		0.0005	0.0006	0	0.0003
7	0		0		0.0005	0.0006	0	0.0004
8	0		0		0.0005		0	
9	0		0		0.0100	0.0104	0	0.0003
10	0		0.00015		0.0100		0	
11	0.0004		0.00015		0.0100		0	
12	0.0004		0.00015		0.0100		0	
13	0.0040	0.0001	0.00150	0.0020	0.0100	0.0109	0	0.0004
14	0.0004		0.00015		0.0005		0	
15	0.0040	0.0038	0		0.0100	0.0094	0	0.0003
16	0		0.00015		0.0005		0	
17	0.0004		0		0.0100		0	
18	0		0.00015		0.0100		0	
19	0.0004		0.00015		0.0005		0	
20	0.0040	0.0040	0		0.0005	0.0006	0	0.0004
21	0		0		0.0100	0.0107	0	0.0003
22	0		0.00150	0.0020	0.0100	0.0104	0	0.0003
23	0.0004		0.00015		0.0005		0	
24	0		0		0.0005		0	

TABLE V CONT'D.

Sample ID float	Se target	Se analyzed	Sample ID float	Se target	Se analyzed
1	0.00003		13	0.00030	0.0001
2	0.00030	0.0001	14	0.00003	
3	0		15	0.00030	0.0001
4	0		16	0.00003	
5	0.00003		17	0	
6	0		18	0.00003	
7	0.00030	0.0001	19	0	
8	0		20	0.00030	0.0001
9	0		21	0.00030	0.0002
10	0		22	0	
11	0		23	0	
12	0.00003		24	0	

TABLE VI. STATISTICALLY DESIGNED LOW-EXP. BOROSILICATE GLASS COMPOSITIONS (WEIGHT FRACTION)

Sample ID borosilicate	SiO_2 target	SiO_2 analyzed	B_2O_3 target	B_2O_3 analyzed	Al_2O_3 target	Al_2O_3 analyzed	CaO target	CaO analyzed
1	0.84	0.8390	0.10	0.0990	0.02	0.0221	0	0
2	0.80	0.7880	0.10	0.1030	0.02	0.0218	0.02	0.0191
3	0.75	0.7460	0.10	0.1000	0.02	0.0226	0.02	0.0194
4	0.76	0.7610	0.10	0.1040	0.07	0.0687	0	0
5	0.70	0.6940	0.10	0.1020	0.07	0.0714	0	0
6	0.71	0.7070	0.10	0.1030	0.07	0.0713	0.02	0.0201
7	0.74	0.7370	0.15	0.1490	0.02	0.0218	0	0
8	0.73	0.7330	0.15	0.1490	0.02	0.0221	0	0
9	0.70	0.7130	0.15	0.1500	0.02	0.0225	0.02	0.0194
10	0.70	0.7060	0.15	0.1540	0.07	0.0719	0	0
11	0.67	0.6740	0.15	0.1500	0.07	0.0731	0.02	0.0203
12	0.72	0.7180	0.15	0.1510	0.07	0.0717	0.02	0.0201
13	0.75	0.7740	0.10	0.0899	0.02	0.0214	0	0
14	0.80	0.8040	0.10	0.1020	0.02	0.0208	0	0
15	0.77	0.7670	0.10	0.1050	0.02	0.0213	0.02	0.0190
16	0.77	0.7850	0.10	0.0986	0.07	0.0689	0	0
17	0.74	0.7410	0.10	0.0998	0.07	0.0725	0.02	0.0196
18	0.73	0.7250	0.10	0.0998	0.07	0.0722	0.02	0.0205
19	0.76	0.7600	0.15	0.1520	0.02	0.0224	0	0
20	0.77	0.7650	0.15	0.1500	0.02	0.0230	0.02	0.0175
21	0.71	0.7180	0.15	0.1550	0.02	0.0221	0.02	0.0198
22	0.72	0.7200	0.15	0.1530	0.07	0.0725	0	0
23	0.67	0.6520	0.15	0.1490	0.07	0.0712	0	0
24	0.63	0.6190	0.15	0.1540	0.07	0.0734	0.02	0.0210

TABLE VI CONT'D.

Sample ID borosilicate	Na_2O target	Na_2O analyzed	K_2O target	K_2O analyzed	BaO target	BaO analyzed
1	0.04	0.0374	0	0	0	0
2	0.04	0.0385	0	0	0.02	0.0195
3	0.08	0.0799	0.03	0.0290	0	0
4	0.04	0.0389	0.03	0.0300	0	0
5	0.08	0.0864	0.03	0.0321	0.02	0.0205
6	0.08	0.0868	0	0	0.02	0.0207
7	0.04	0.0411	0.03	0.0296	0.02	0.0205
8	0.08	0.0821	0	0	0.02	0.0208
9	0.08	0.0795	0.03	0.0291	0	0
10	0.08	0.0778	0	0	0	0
11	0.04	0.0410	0.03	0.0302	0.02	0.0203
12	0.04	0.0436	0	0	0	0
13	0.08	0.0769	0.03	0.0272	0.02	0.0195
14	0.08	0.0784	0	0	0	0
15	0.04	0.0410	0.03	0.0300	0.02	0.0196
16	0.04	0.0300	0	0	0.02	0.0169
17	0.04	0.0389	0.03	0.0290	0	0
18	0.08	0.0831	0	0	0	0
19	0.04	0.0397	0.03	0.0303	0	0
20	0.04	0.0405	0	0	0	0
21	0.08	0.0808	0	0	0.02	0.0201
22	0.04	0.0461	0	0	0.02	0.0211
23	0.08	0.0827	0.03	0.0289	0	0.0198
24	0.08	0.0833	0.03	0.0305	0.02	0.0211

TABLE VII. STATISTICALLY DESIGNED TV PANEL GLASS COMPOSITIONS (WEIGHT FRACTION)

Sample ID TV panel	SiO_2 target	SiO_2 analyzed	Al_2O_3 target	Al_2O_3 analyzed	MgO target	MgO analyzed	CaO target	CaO analyzed
1	0.650	0.6475	0.013	0.0138	0	0.0001	0	0.0004
2	0.706	0.6970	0.035	0.0345	0	0	0	0
3	0.711	0.7127	0.035	0.0336	0.015	0.0146	0	0
4	0.707	0.7103	0.035	0.0360	0.015	0.0151	0.035	0.0311
5	0.631	0.6365	0.035	0.0358	0.015	0.0143	0.035	0.0336
6	0.514	0.5339	0.035	0.0363	0.015	0.0154	0.035	0.0326
7	0.527	0.5396	0.013	0.0136	0.015	0.0140	0.035	0.0319
8	0.509	0.5258	0.035	0.0368	0	0	0.035	0.0332
9	0.535	0.5423	0.013	0.0138	0.015	0.0137	0	0.0006
10	0.588	0.5926	0.035	0.0351	0	0	0.035	0.0333
11	0.627	0.6316	0.035	0.0366	0.015	0.0136	0	0.0007
12	0.566	0.5712	0.013	0.0135	0.015	0.0135	0.035	0.0316
13	0.603	0.6061	0.013	0.0132	0	0.0007	0.035	0.0344
14	0.496	0.5068	0.035	0.0355	0	0	0	0

TABLE VII CONT'D.

Sample ID TV panel	SiO_2 target	SiO_2 analyzed	Al_2O_3 target	Al_2O_3 analyzed	MgO target	MgO analyzed	CaO target	CaO analyzed
15	0.567	0.5746	0.035	0.0352	0.015	0.0142	0	0.0005
16	0.663	0.6609	0.013	0.0135	0.015	0.0143	0.035	0.0321
17	0.583	0.5834	0.013	0.0134	0	0.0004	0.035	0.0346
18	0.519	0.5331	0.035	0.0345	0	0	0	0
19	0.649	0.6492	0.013	0.0134	0.015	0.0143	0	0.0008
20	0.679	0.6762	0.035	0.0345	0	0	0.035	0.0333
21	0.599	0.6031	0.013	0.0135	0.015	0.0142	0	0.0005
22	0.606	0.6072	0.013	0.0136	0	0.0004	0.035	0.0338
23	0.703	0.7101	0.013	0.0141	0	0	0	0
24	0.834	0.8062	0.013	0.0120	0	0	0	0

TABLE VII CONT'D.

Sample ID TV panel	SrO target	SrO analyzed	BaO target	BaO analyzed	Li_2O target	Li_2O analyzed	Na_2O target	Na_2O analyzed
1	0.01	0.0093	0.13	0.1270	0		0.09	0.0911
2	0.01	0.0100	0.02	0.0201	0.005	0.0050	0.06	0.0603
3	0.01	0.0106	0.02	0.0192	0	0.0000	0.09	0.0827
4	0.01	0.0103	0.02	0.0193	0	0.0000	0.06	0.0560
5	0.10	0.0967	0.02	0.0202	0		0.06	0.0586
6	0.10	0.1004	0.13	0.1321	0	0.0000	0.06	0.0670
7	0.10	0.0976	0.13	0.1283	0.005	0.0050	0.06	0.0605
8	0.10	0.0992	0.13	0.1330	0.005	0.0054	0.09	0.0945
9	0.10	0.0996	0.13	0.1300	0.005	0.0050	0.09	0.0893
10	0.01	0.0096	0.13	0.1298	0.005	0.0051	0.09	0.0918
11	0.10	0.0986	0.02	0.0206	0.005	0.0050	0.09	0.0890
12	0.01	0.0094	0.13	0.1284	0		0.09	0.0904
13	0.10	0.1020	0.02	0.0208	0.005	0.0051	0.06	0.0613
14	0.10	0.0969	0.13	0.1286	0	0.0000	0.09	0.0934
15	0.01	0.0093	0.13	0.1280	0.005	0.0050	0.06	0.0612
16	0.01	0.0100	0.02	0.0201	0.005	0.0050	0.09	0.0925
17	0.10	0.1009	0.02	0.0207	0		0.09	0.0910
18	0.10	0.0965	0.13	0.1290	0	0.0000	0.06	0.0655
19	0.01	0.0095	0.13	0.1288	0.005	0.0050	0.06	0.0614
20	0.01	0.0099	0.02	0.0198	0.005	0.0050	0.09	0.0924
21	0.10	0.1030	0.02	0.0209	0		0.09	0.0911
22	0.01	0.0095	0.13	0.1280	0		0.06	0.0619
23	0.10	0.1026	0.02	0.0198	0.005	0.0052	0.06	0.0574
24	0.01	0.0102	0.02	0.0189	0	0.0000	0.06	0.0590

TABLE VII CONT'D.

Sample ID TV panel	K_2O target	K_2O analyzed	TiO_2 target	TiO_2 analyzed	CeO_2 target	CeO_2 analyzed	ZrO_2 target	ZrO_2 analyzed
1	0.06	0.0606	0.001	0.0010	0.007	0.0068	0.03	0.0297
2	0.09	0.0892	0.001	0.0008	0	0	0.03	0.0286
3	0.06	0.0608	0.005	0.0057	0	0	0	0
4	0.09	0.0881	0.001	0.0008	0.007	0.0081	0	0
5	0.06	0.0594	0.005	0.0050	0	0.0001	0.03	0.0294
6	0.06	0.0581	0.001	0.0008	0.007	0.0057	0	0
7	0.06	0.0608	0.001	0.0010	0	0.0008	0.03	0.0295
8	0.06	0.0580	0.001	0.0008	0	0	0	0
9	0.09	0.0889	0.001	0.0010	0	0.0007	0	0.0002
10	0.09	0.0903	0.005	0.0052	0	0.0005	0	0
11	0.09	0.0885	0.005	0.0051	0.007	0.0062	0	0
12	0.09	0.0896	0.005	0.0052	0.007	0.0068	0.03	0.0298
13	0.09	0.0907	0.005	0.0052	0.007	0.0065	0.03	0.0305
14	0.06	0.0588	0.005	0.0046	0.007	0.0059	0.03	0.0289
15	0.09	0.0899	0.001	0.0010	0.007	0.0070	0.03	0.0297
16	0.06	0.0601	0.005	0.0051	0	0.0002	0.03	0.0298
17	0.09	0.0894	0.001	0.0010	0.007	0.0066	0	0.0002
18	0.09	0.0882	0.005	0.0048	0	0	0.03	0.0284
19	0.06	0.0608	0.005	0.0051	0.007	0.0070	0	0
20	0.06	0.0604	0.001	0.0009	0.007	0.0060	0.03	0.0288
21	0.09	0.0901	0.001	0.0010	0	0.0003	0.03	0.0308
22	0.09	0.0897	0.005	0.0052	0	0.0008	0	0
23	0.06	0.0598	0.005	0.0056	0.007	0.0077	0	0
24	0.06	0.0636	0.001	0.0009	0	0	0	0

TABLE VII CONT'D.

Sample ID TV panel	PbO target	PbO analyzed	ZnO target	ZnO analyzed	As_2O_3 target	As_2O_3 analyzed	Sb_2O_3 target	Sb_2O_3 analyzed
1	0	0.0001	0	0	0.003	0.0025	0.006	0.0071
2	0.03	0.0284	0	0	0	0	0.006	0.0055
3	0.03	0.0296	0.015	0.0209	0	0	0.002	0.0020
4	0	0	0.015	0.0197	0.003	0.0031	0.002	0.0018
5	0	0.0003	0	0	0.003	0.0027	0.006	0.0067
6	0.03	0.0294	0	0	0	0	0.006	0.0054
7	0	0.0002	0.015	0.0147	0	0.0004	0.002	0.0025
8	0.03	0.0291	0	0	0.003	0.0020	0.002	0.0019
9	0	0.0002	0.015	0.0150	0	0.0003	0.006	0.0069
10	0	0	0	0	0.003	0.0027	0.002	0.0026
11	0	0.0001	0	0	0	0	0.006	0.0070
12	0	0	0	0.0001	0	0	0.002	0.0026
13	0.03	0.0312	0	0	0	0.0003	0.002	0.0025
14	0.03	0.0279	0.015	0.0120	0	0	0.002	0.0019

TABLE VII CONT'D.

Sample ID TV panel	PbO target	PbO analyzed	ZnO target	ZnO analyzed	As_2O_3 target	As_2O_3 analyzed	Sb_2O_3 target	Sb_2O_3 analyzed
15	0.03	0.0283	0.015	0.0147	0.003	0.0028	0.002	0.0025
16	0.03	0.0295	0.015	0.0147	0.003	0.0028	0.006	0.0069
17	0.03	0.0311	0.015	0.0147	0.003	0.0030	0.006	0.0070
18	0	0	0.015	0.0121	0.003	0.0021	0.006	0.0053
19	0.03	0.0293	0	0.0001	0.003	0.0028	0.006	0.0071
20	0	0.0001	0.015	0.0143	0	0	0.006	0.0068
21	0.03	0.0321	0	0.0001	0.003	0.0031	0.002	0.0025
22	0.03	0.0292	0.015	0.0148	0	0.0001	0.006	0.0071
23	0	0	0.015	0.0206	0.003	0.0035	0.002	0.0020
24	0	0	0	0	0	0	0.002	0.0019

TABLE VII CONT'D.

Sample ID TV panel	F target	F analyzed	Fe_2O_3 analyzed	Sample ID TV panel	F target	F analyzed	Fe_2O_3 analyzed
1	0	0.0006		13	0	0.0007	
2	0.007	0.0062	0.0004	14	0	0	0.0002
3	0.007	0.0061	0.0004	15	0	0.0005	
4	0	0	0.0004	16	0	0	
5	0	0		17	0.007	0.0078	
6	0.007	0.0067	0.0002	18	0.007	0.0063	0.0002
7	0.007	0.0076		19	0.007	0.0080	
8	0	0	0.0002	20	0.007	0.0079	
9	0	0		21	0.007	0.0074	
10	0.007	0.0075		22	0	0	
11	0	0.0007		23	0.007	0.0069	0.0004
12	0.007	0.0084		24	0	0	0.0004

TABLE VIII. STATISTICALLY DESIGNED FIBER WOOL GLASS COMPOSITIONS (WEIGHT FRACTION)

Sample ID wool	SiO_2 target	SiO_2 analyzed	B_2O_3 target	B_2O_3 analyzed	Al_2O_3 target	Al_2O_3 analyzed	MgO target	MgO analyzed
1	0.780		0.03		0		0.01	
2	0.712		0.03		0		0.05	
3	0.628		0.03		0		0.05	
4	0.666		0.03		0.06		0.01	
5	0.600	0.5780	0.03	0.0298	0.06	0.0586	0.01	0.0098
6	0.592	0.5540	0.03	0.0340	0.06	0.0644	0.05	0.0496
7	0.652	0.6280	0.09	0.0885	0		0.01	0.0092
8	0.628	0.6050	0.09	0.0888	0		0.01	0.0098
9	0.578		0.09		0		0.05	
10	0.592	0.5290	0.09	0.0931	0.06	0.0621	0.01	0.0111
11	0.554		0.09		0.06		0.05	
12	0.614		0.09		0.06		0.05	

TABLE VIII CONT'D.

Sample ID wool	SiO_2 target	SiO_2 analyzed	B_2O_3 target	B_2O_3 analyzed	Al_2O_3 target	Al_2O_3 analyzed	MgO target	MgO analyzed
13	0.652	0.6620	0.03	0.0313	0		0.01	0.0099
14	0.712		0.03		0		0.01	
15	0.674		0.03		0		0.05	
16	0.688		0.03		0.06		0.01	
17	0.638	0.6200	0.03	0.0300	0.06	0.0635	0.05	0.0454
18	0.614	0.6110	0.03	0.0306	0.06	0.0589	0.05	0.0443
19	0.674		0.09		0		0.01	
20	0.666		0.09		0		0.05	
21	0.600	0.6130	0.09	0.0889	0		0.05	0.0096
22	0.638		0.09		0.06		0.01	
23	0.554		0.09		0.06		0.01	
24	0.486		0.09		0.06		0.05	

TABLE VIII CONT'D.

Sample ID wool	CaO target	CaO analyzed	Na_2O target	Na_2O analyzed	K_2O target	K_2O analyzed	Fe_2O_3 target	Fe_2O_3 analyzed
1	0.05		0.13		0		0	
2	0.05		0.13		0.02		0	
3	0.11		0.17		0		0.006	
4	0.05		0.17		0		0.006	
5	0.11	0.1100	0.17	0.1710	0.02	0.0201	0	
6	0.11	0.1270	0.13	0.1440	0.02	0.0221	0.006	0.0069
7	0.05	0.0480	0.17	0.1730	0.02	0.0201	0.006	0.0062
8	0.11	0.1130	0.13	0.1320	0.02	0.0201	0.006	0.0065
9	0.11		0.17		0		0	
10	0.11	0.1350	0.13	0.1370	0		0	
11	0.05		0.17		0.02		0	
12	0.05		0.13		0		0.006	
13	0.11	0.0536	0.17	0.1830	0.02	0.0211	0	
14	0.11		0.13		0		0.006	
15	0.05		0.17		0.02		0.006	
16	0.05		0.13		0.02		0.006	
17	0.05	0.0528	0.17	0.1780	0		0	
18	0.11	0.1160	0.13	0.1340	0		0	0.0004
19	0.05		0.17		0		0	
20	0.05		0.13		0		0.006	
21	0.11	0.0500	0.13	0.1330	0.02	0.0197	0	
22	0.05		0.13		0.02		0	
23	0.11		0.17		0		0.006	
24	0.11		0.17		0.02		0.006	

TABLE VIII CONT'D.

Sample ID wool	F target	F analyzed	SO_3 target	SO_3 analyzed	Sample ID wool	F target	F analyzed	SO_3 target	SO_3 analyzed
1	0		0		13	0.006	0.0047	0.002	0.0022
2	0.006		0.002		14	0		0.002	
3	0.006		0		15	0		0	
4	0.006		0.002		16	0.006		0	
5	0		0		17	0		0.002	0.0021
6	0	0.0001	0.002		18	0.006		0	
7	0		0.002	0.0017	19	0.006		0	
8	0.006	0.0053	0		20	0.006		0.002	
9	0		0.002		21	0		0	
10	0.006	0.0078	0.002	0.0028	22	0		0.002	
11	0.006		0		23	0		0	
12	0		0		24	0.006		0.002	

GLASS MELTING AND CHEMICAL ANALYSIS

The Corning Engineering Laboratory Services (CELS) group at Corning Incorporated (Corning, NY) was subcontracted to melt and chemically analyze the laboratory-scale melts of E-glass (E), low-expansion borosilicate glass (LO) and color TV panel glass (TV). Approximately four-kilogram quantities of 72 different glasses were prepared and analyzed by CELS. Students at Alfred University prepared laboratory scale melts of the container (CO), float (FL), and fiber wool (WO) glasses (approx. 1 kg each). Glass melting procedures are outlined below in Table IX.

TABLE IX. GLASS MELTING PROCEDURES

sample ID	CO	E	FL	LO	TV	WO
batch mix melt temp. melt time	ball mill 1500 °C 5 hr.	ball mill 1600 °C overnight	ball mill 1500 °C 5 hr.	ball mill 1550 °C 8 hr.	ball mill 1550 °C 4 hr.	ball mill 1450 °C 5 hr.
Pt crucible covered stir	 no no	 yes 15 min. 1 hr. fining	 no no	 yes 30 min. 3.5 hr. fining	 yes no	 no no
anneal temp. batch size	650 °C 1 kg	560 °C 5 kg	no anneal 1 kg	560 °C 5 kg	525 °C 4 kg	no anneal 1 kg

With the exception of 15 float and 15 fiberglass wool compositions not yet analyzed, and 16 container glass compositions analyzed by Owens-Corning's Integrex Testing Services group (Granville, OH), all of the chemical analysis was performed by CELS. The analysis data are shown with the target compositions in Tables III-VIII.

The deviations from the target compositions are summarized below in Tables X-XV as number of compositions above/below target value and the maximum per cent differences relative to the target values (% diff.= [(target value – analyzed value) / target value] * 100). A sample was considered to be 'off-target' when the per cent difference was greater than the accuracy of the analysis (i.e. % diff. > +/- 0.5-1 for levels > 2 wt. %).

Container glass compositions

Of the 24 container glass samples analyzed, 16 were off-target for CaO, 20 for Na_2O, 2 for TiO_2, and 3 for SO_3. The majority of the deviations were below target value, most likely due to losses occurring during melting in an uncovered crucible.

E glass compositions

As Table XI illustrates, the E glass samples exhibited considerable deviation from the target compositions. Two of the samples, E8 and E20, were visibly inhomogeneous with cords and color variations respectively. Half of the compositions were off target for SiO_2, B_2O_3, MgO, and F, and almost all deviated from the target values for Al_2O_3 and CaO. The majority of deviations were below target for B_2O_3, Al_2O_3, and F, and above target for MgO, CaO, Na_2O, and K_2O.

Float glass compositions

Eight of the nine float glass samples analyzed showed poor agreement with target for Al_2O_3 content, particularly with large per cent differences above target for 4 samples. The 6 samples that measured above target for Al_2O_3 content were also below target for MgO and above target for CaO. Selenium loss is presumably due to volatization as the melt crucible was not covered.

Low-expansion borosilicate glass compositions

As shown in Table XIII, measured values for Al_2O_3, B_2O_3 and Na_2O were greater than target values for half or more of the samples tested. In total, Al_2O_3 and Na_2O values were off target in 21 of the 24 samples tested, although no apparent trends were identified.

Inhomogeneities were observed in several of the compositions. Small seeds were widespread in compositions 1, 12, 17, and 22. Composition 4 showed large seeds and cords throughout, and composition 16 was cloudy in appearance, formed into irregular balls rather than poured out as one thick plate. Opalescence was observed uniformly throughout composition 20 indicating that phase separation has occurred. Heat treatment and phase analysis studies are currently being undertaken for compostions 1, 2, 3, 4, 11, 12, 16, 20, and 22 to further investigate phase separation.

TV panel glass compositions

As shown below in Table XIV, samples that measured above target for Na_2O were also below target for CaO and above target for Al_2O_3.

Fiber wool glass compositions

Of the nine wool glass samples measured, eight were above target for Na_2O. Six of those samples were also below target for SiO_2. Overall agreement was poor as the majority of the samples were off target for SiO_2, B_2O_3, Al_2O_3, MgO, CaO, and Na_2O.

TABLE X. DIFFERENCES BETWEEN TARGET AND ANALYZED COMPOSITIONS: CONTAINER GLASS

	below target	max.% diff.	above target	max.% diff.
Ca0	12	5	4	7.14
Na_2O	18	10.67	2	2
TiO_2	1	6	1	8
SO_3	3	63.33	0	

24 compositions analyzed

TABLE XI. DIFFERENCES BETWEEN TARGET AND ANALYZED COMPOSITIONS: E GLASS

	below target	max.% diff.	above target	max.% diff.
SiO_2	6	3.22	6	8.57
B_2O_3	10	5.56	1	3.33
Al_2O_3	19	22.25	3	25
Mg0	3	66	10	298
Ca0	8	15.63	15	8.12
Na_2O	0		7	5
K_2O	1	8	6	12
Fe_2O_3	1	37.5	0	
TiO_2	1	20	1	10
F	11	71.67	0	

24 compositions analyzed

TABLE XII. DIFFERENCES BETWEEN TARGET AND ANALYZED COMPOSITIONS: FLOAT GLASS

	below target	max.% diff.	above target	max.% diff.
SiO_2	3	1.65	1	2.92
Al_2O_3	2	8	6	130
Mg0	9	20	0	
Ca0	1	7.14	7	8.89
Na_2O	4	11.33	1	2.5
K_2O	0		8	18
Fe_2O_3	3	15.33	1	30
Cr_2O_3	1	97.5	0	
Co_3O_4	0		2	33.33
TiO_2	1	6	3	9
Se	6	66.67	0	

9 compositions analyzed

TABLE XIII: DIFFERENCES BETWEEN TARGET AND ANALYZED COMPOSITIONS: BOROSILICATE GLASS

	below target	max.% diff.	above target	max.% diff.
SiO_2	3	2.69	4	3.2
B_2O_3	1	10	11	5
Al_2O_3	2	1.43	19	15
CaO	1	10	0	
Na_2O	8	25	13	15
K_2O	5	10	2	6.67
BaO	1	15	0	

24 compositions analyzed

TABLE XIV: DIFFERENCES BETWEEN TARGET AND ANALYZED COMPOSITIONS: TV PANEL GLASS

	below target	max.% diff.	above target	max.% diff.
SiO_2	2	3.33	7	3.87
Al_2O_3	5	7.69	9	8.46
MgO	6	10	0	
CaO	12	11.14	0	
SrO	7	10	4	10
BaO	8	5	6	5
Li_2O	0		1	8
Na_2O	7	7.78	14	11.67
K_2O	9	3.33	6	6.67
TiO_2	1	8	2	14
CeO_2	6	18.57	2	15.71
ZrO_2	5	6.67	2	3.33
PbO	7	6.67	3	6.67
ZnO	2	20	3	39.33
As_2O_3	2	33.33	0	
Sb_2O_3	3	11.67	15	30
F	3	12.86	7	20

24 compositions analyzed

TABLE XV: DIFFERENCES BETWEEN TARGET AND ANALYZED COMPOSITIONS: FIBER WOOL GLASS

	below target	max.% diff.	above target	max.% diff.
SiO_2	6	10.64	2	2.17
B_2O_3	3	1.11	4	13.33
Al_2O_3	2	1.67	3	6.67
MgO	4	80	1	10
CaO	3	54.55	5	22.73
Na_2O	0		8	10.77
K_2O	0		1	10
Fe_2O_3	0		2	15
F	2	21.67	1	30
SO_3	0		1	40

9 compositions analyzed

Chapter 3

GAS SOLUBILITY AND WATER DIFFUSION IN GLASSES AND MELTS OF COMMERCIAL COMPOSITIONS

James E. Shelby
New York State College of Ceramics at Alfred University
Alfred, New York 14802

INTRODUCTION

Gas solubility in melts plays a major role in the fining of melts [1,2]. Trapping of atmospheric gases during the initial stages of melt formation, retention of gases released from the batch components, and reactions with the furnace gases all contribute to the total gas content of melts and of the glasses formed by cooling these melts. The initial processes of gas trapping and batch decomposition produce large quantities of bubbles which are removed by either rising to the surface of the melt or by absorption into the melt. Exosolution of these dissolved gases can result in the formation bubbles during changes in melt temperature or in the nature of the gas in the atmosphere above the melt.

Gases of concern in glass technology can be divided into 3 categories. First, those gases which are present in the atmosphere and can be trapped during initial melt formation include nitrogen, oxygen, argon, carbon dioxide, and water vapor, with lesser amounts of other gases. Decomposition of batches can release carbon dioxide, sulfur oxides, nitrates, water vapor, halides, etc. The furnace atmosphere will contain all of the atmospheric gases, but will also contain those gases generated through combustion, esp. carbon dioxide and water vapor. In gas-fired furnaces, the furnace atmosphere will consist primarily of carbon dioxide and water vapor, with lesser amounts of the normal atmospheric gases, while the atmospheres in electrically heated furnaces will contain primarily the common atmospheric gases.

This study considers the solubility of helium, neon, argon, carbon dioxide, and water in the base glasses. No measurements were made for the variation compositions due to the very large amount of time required for every measurement. While not commonly found in commercial glasses in significant quantities, helium and neon were included in this study to provide a better understanding of the mechanisms of gas solubility in glasses and melts. The high diffusivity of helium in these materials allowed extension of the measurements to temperatures in or below the glass transformation range. The results will be presented in categories of (a) inert gases, i.e. helium, neon, and argon, (b) carbon dioxide, and (c) water vapor. Measurement techniques for the inert gases and carbon dioxide are similar, while those used for water are very different from those used for the other gases.

EXPERIMENTAL PROCEDURES FOR INERT GASES AND CARBON DIOXIDE

Sample Preparation

Solubility measurements using the mass spectrometer method for inert gases and carbon dioxide could not be made on the as-received commercial compositions due to foaming during remelting under vacuum [3]. Foaming of these melts when heated under vacuum occurs due to the presence of residual carbonate and/or sulfur in these glasses. The residual carbonate arises from decomposition of carbonate raw materials used during the initial glass melting process. Sulfur may be present from use as a fining agent. In order to avoid foaming of the samples during the saturation-outgassing measurements, "mock" commercial glasses were prepared using nitrates and oxides as raw materials as opposed to the carbonates and sulfates used in the as-received glasses. The borosilicate glass, which did not foam on remelting, was the only glass which was used in the as-received form.

Glasses for this portion of the study were prepared using reagent grade alkali nitrates, alkaline earth oxides, and 99.9% pure alumina and silica. Batches were calculated to yield 25 grams of glass. Compositions were dry mixed and melted in platinum crucibles in an electric resistance furnace at times and temperatures ranging from 1 to 4 hours and 1400 to 1600°C depending upon the viscosity of the specific composition being melted. The melts were then water quenched, removed from the crucible, and annealed at their respective glass transformation temperatures (T_g) as determined by differential scanning calorimetry (DSC). Samples in the form of either 1 mm thick plates (for study of glasses) or cubes weighing 2 to 4 grams (for study of melts) were cut from the bulk glasses.

The compositions for the mock glasses correspond to typical compositions used for float, container, TV panel, wool, and E-glass. With the exception of the borosilicate tubing, these compositions are listed in Table I, where each component is given in mole percent. While the borosilicate tubing is known to contain roughly 92 mol% glass forming oxides, i.e. SiO_2 and B_2O_3, the exact composition was not specified by the supplier.

TABLE I. COMPOSITIONS AND PROPERTIES OF COMMERCIAL GLASSES CONSIDERED IN THIS STUDY.

Oxide (mol%)	Container	TV Panel	Float	Wool	E-Glass
SiO_2	74.32	72.23	71.43	62.33	57.21
B_2O_3	---	---	---	4.34	6.28
Al_2O_3	0.75	1.36	0.36	2.37	9.19
Na_2O	13.13	8.52	13.15	15.59	0.51
K_2O	0.17	5.71	0.09	0.64	---
MgO	0.30	---	5.64	5.99	6.20
CaO	11.32	---	9.33	8.62	20.06
SrO	---	6.25	---	---	---
BaO	---	4.30	---	---	---
TiO_2	---	0.37	---	---	0.39
ZrO_2	---	0.84	---	---	---
ZnO	---	0.42	---	---	---
Fe_2O_3	0.01	---	---	0.11	0.16
Density (g/cm^3)	2.47	2.65	2.47	2.53	2.58
Annealing Temp. (°C)	583	543	571	553	687

Saturation Procedure

The apparatus designed by Jewell [1,4] for the saturation of glasses and melts under controlled temperatures and pressures was used throughout this study. A schematic of this apparatus is shown in Figure 1 [4]. The system is constructed of 1.5 inch stainless steel vacuum fittings with copper gasket seals. The saturation chamber is either a vitreous silica or mullite tube, depending on the saturation temperature used. Gas is supplied to the sample through 1/4 inch stainless steel tubing connected by Swagelock® fittings.

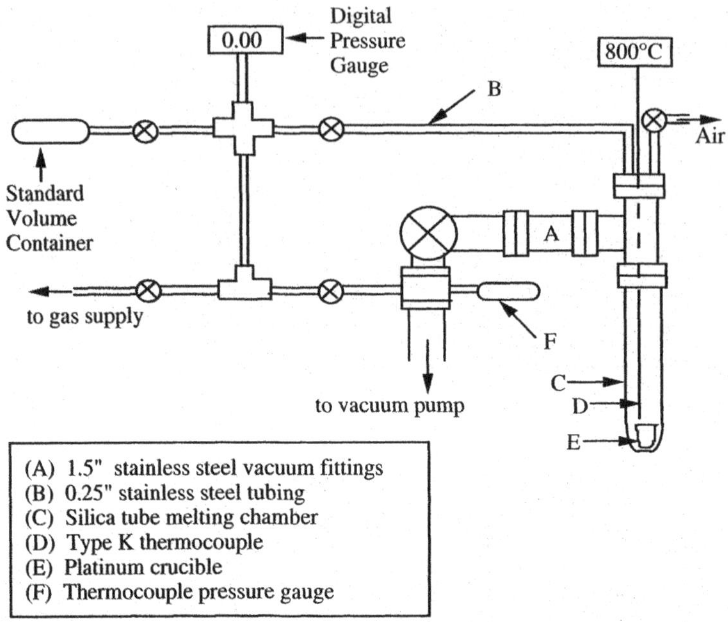

Figure 1. Gas saturation system used for glasses and melts (reference 1)

The saturation chamber is evacuated with the mechanical pump to a pressure of less than 30 millitorr, with the system pressure monitored continually by a thermocouple vacuum gauge. An vertical electric furnace is used to heat the vitreous silica saturation chamber, whereas a stationary electric resistance furnace is used with the mullite tubes used for temperatures above 1000°C due to the poor thermal shock resistance of mullite. A type-K thermocouple placed within 1 inch of the specimen continually monitors the temperature of the glass in the vitreous silica tube. A type-R Pt-alloy thermocouple is used to monitor the temperature in the mullite tube. The thermocouple used in the high temperature apparatus is removed prior to insertion of the sample in order to allow enough room for the crucible to be lowered into the tube. Digital monometers are used to measure the applied gas pressures in both systems.

Treatment times required to reach the equilibrium gas concentration are a function of the diffusivity of a specific gas in the glass or melt under study, as well as the depth of the melt, or thickness of the glass plates. Approximations to Fick's second law for specific geometries, which allow calculations of the approximate time to reach equilibrium with the surrounding gas phase, are found in Kingery, et al. [5]. For plate geometries, the time to equilibrium is estimated from the expression

$$t = \frac{L^2}{2D} \tag{1}$$

where t is the time to reach greater than 98% of equilibrium, L is the half-thickness of the plate, and D is the diffusion coefficient. The use of 1 mm thick plates instead of a several mm deep melt significantly reduces the equilibration time by decreasing the diffusion path length by a factor of 5 to 10 relative to the melts, as well as allowing the gas to enter from two surfaces as opposed to only entering the sample from the top surface as is the case for melts. Equilibration times for the melts in this study varied from 1 to 24 hours, while glass plate samples required from 1 to 120 hours as a function of glass composition and treatment temperature to reach equilibrium with the surrounding atmosphere.

A check of the validity of Equation 1 for estimating equilibration times was performed on random samples by re-treating the samples for longer periods under the same conditions as the initial saturation treatment. A solubility which is independent of the treatment time indicates that the samples were treated long enough to have reached equilibrium with the surrounding gas phase for each measurement.

Slightly different methods are used to produce saturated samples of melts and glasses. Melt samples are saturated by placing a cube of glass weighing 2 to 4 grams into a platinum crucible, which is inserted into a platinum cradle suspended from the top of the saturation chamber by platinum wire. The crucible is then lowered into the mullite tube, which is attached to the saturation system and surrounded by a furnace preheated to approximately 1400°C. Samples are thermally equilibrated at 1400°C while a vacuum is drawn in a series of pressure decreases of 100 torr increments in order to remove the ambient atmosphere in the tube and to remove any residual gases and bubbles in the melt. The pressure must be decreased slowly to avoid foaming of the melt from dissolved gas or bubbles. The furnace temperature is then changed to the desired treatment temperature and the sample allowed to thermally equilibrate at the new temperature. Once thermally equilibrated, the chamber is filled to the desired pressure with the gas of interest. The samples are held at the desired temperature and pressure for sufficient time to reach equilibrium, as estimated from Equation 1 and confirmed as described above, after which the dissolved gas is frozen into the sample by rapidly removing the crucible from the chamber and quenching the bottom of the crucible to a depth above the melt line in water. It is assumed that very little, if any, gas escapes from the samples during quenching, and hence the gas in the sample is indicative of the equilibrium concentration.

Samples of the glass consisting of 2 to 4 grams of 1 mm thick plates are saturated by placing the plates in a small platinum crucible. The crucible containing the sample is placed into a vitreous silica tube, which is then attached using a Wilson seal to the treatment system. A vertical furnace preheated to the desired treatment temperature is raised into place around the sample. The plates are thermally equilibrated at this temperature while a vacuum is drawn in order to remove the ambient atmosphere in the tube. Once thermally equilibrated, the chamber is filled to the desired pressure with the gas of interest. The plates are held at the desired temperature and pressure for sufficient time to reach equilibrium, as estimated from Equation 1 and confirmed as described above, after which the dissolved gas is frozen into the sample by rapidly lowering the furnace from around the chamber and quenching the sample by blowing air from a small fan over the vitreous silica tube. The temperature of the specimens decreases rapidly upon removal of the furnace, dropping from 500 to 800°C to less than 300°C within minutes of removal from the furnace, and reaching room temperature in 5 to 7 minutes. It is assumed that very little, if any, gas escapes from the samples during this time.

Outgassing Procedure

The apparatus, shown schematically in Figure 2, designed by Jewell [1,4] for the outgassing and quantitative analysis of gas content in glasses and melts is used throughout this study. The construction of this apparatus is essentially identical to that of the saturation facilities described above. A mechanical pump is used to "rough" the system, after which the valve to the diffusion pump is opened, producing a pressure of less than 10^{-7} torr. After the desired pressure is reached, a vertical electric resistance furnace is raised into place around the vitreous silica sample chamber. A type-K thermocouple placed within 1 inch of the specimen monitors the temperature of the glass or melt, while the system pressure is monitored by a high vacuum gauge. A thermocouple pressure gauge monitors the pressure of the sample chamber when it is isolated from the rest of the outgassing system (valve 2 is closed).

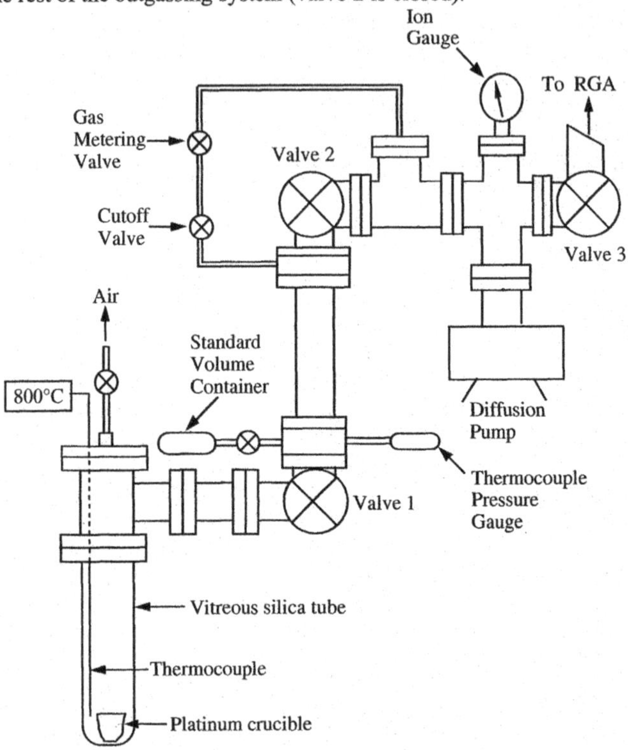

Figure 2. Outgassing system used for glasses and melts (reference 1)

A quadrapole mass spectrometer, i.e. a residual gas analyzer (RGA), is used to detect the gas as it is released from the sample, generating an "outgassing curve". The gases are introduced to the RGA through a metering valve to control the rate of gas flow from the sample to the mass spectrometer. A port also exists for introducing gas from a container of a known volume in order to calibrate the system and determine absolute gas solubilities.

Outgassing of both glasses and melts is carried out by transferring the sample to a different platinum crucible which has not been exposed to the saturating gas. This crucible is placed in the vitreous silica tube, which is then attached to the outgassing apparatus. Prior to each measurement, the vitreous silica tube is degassed by heating under vacuum to ensure that the only gas detected by the RGA comes from the sample. The system is evacuated to approximately 10^{-7} torr to remove the ambient atmosphere. After reaching the desired pressure, the RGA background signal is measured for five minutes before a furnace preheated to the desired outgassing temperature is raised around the vitreous silica tube. Glass plates are normally outgassed at 500°C while melts are outgassed at 950°C. Specimens are held at temperature until all the dissolved gas has diffused out of the sample and into the mass spectrometer system, as indicated by the return of the RGA signal to the background level. Typical outgassing curves for both a high solubility (vitreous silica) and low solubility (sodium silicate) glass, along with the calibration curve obtained from the standard container are shown in Figures 3 and 4 [6].

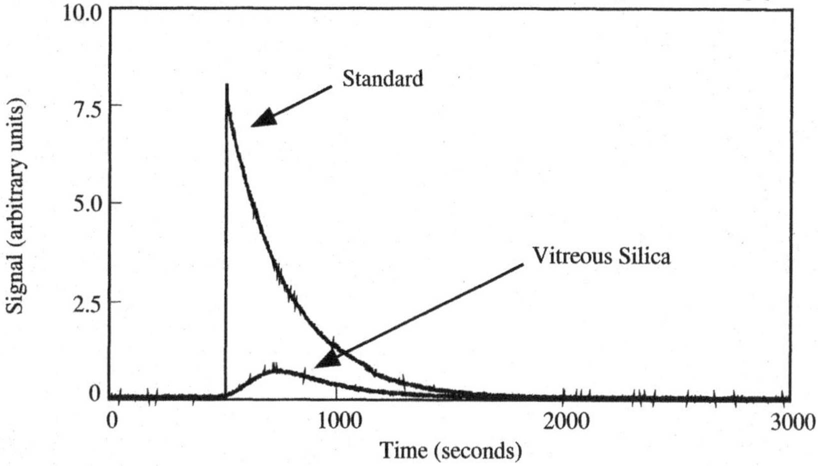

Figure 3. Typical outgassing curves for vitreous silica and for a standard volume container (reference 6)

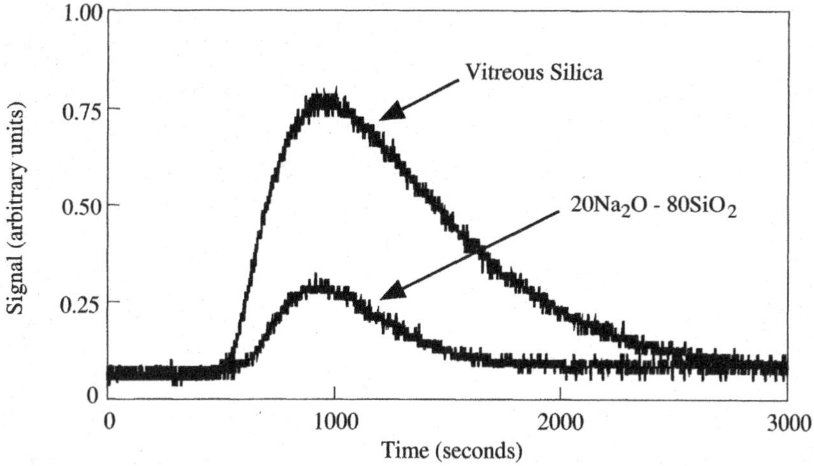

Figure 4. Typical outgassing curves for vitreous silica and a sodium silicate glass (reference 6)

A slightly different procedure from that used for helium or neon was used for melts containing argon or carbon dioxide. The higher diffusivities of helium and neon allow the melt samples to be reused for multiple measurements, since the gas can be driven out of the melt at the measuring temperature of 950°C in a reasonable experimental time. The lower diffusivities of argon and carbon dioxide in these melts require the use of a flux (borax) to produce a more fluid melt which can be completely outgassed in an acceptable time. Quenched samples of the equilibrated melt are fractured and mixed with a few grams of borax before placing the crucible into the outgassing tube. Since the sample is destroyed in this procedure, a fresh sample must be used for each measurement. As a result, far more time is required for argon and carbon dioxide solubility measurements than for helium and neon studies.

Data Analysis

The area under each experimental outgassing curve is directly proportional to the amount of gas dissolved in the sample. Since the RGA measures relative and not absolute values, a calibration curve is obtained after each sample measurement by introducing a known amount of gas from a standard container of known volume filled to a known pressure. The resulting curves for both the sample and standard are imported into a spreadsheet and the solubility of gas in the specimen determined by the expression

$$S = \left(\frac{A_{sample}}{A_{std}}\right)\left[\left(V_{std}\right)\left(\frac{P_{std}}{760torr}\right)\right]\left[\frac{N_A}{\left(M_{sample}\right)\left(22.4*10^3\right)}\right] \tag{2}$$

where the first term in this expression is the ratio of the area under these two curves, the second term is the volume of gas contained in the standard normalized to 760 torr, N_A is Avogadro's number, M is the mass of the sample, and S is the gas solubility expressed as atoms per gram at STP.

The concentration of dissolved gas is given by the ratio of the area under the sample and standard curves. It follows that these areas must increase linearly as a function of increasing pressure or their ratio becomes meaningless. To confirm this relationship, and hence the validity of Equation 2, the standard container was filled with helium to various pressures over the range from 10 to 120 torr. The areas under the resulting outgassing curves were determined. The results are shown as a function of fill pressure in Figure 5. The area under the curve clearly increases linearly with increasing pressure, as is assumed throughout this study.

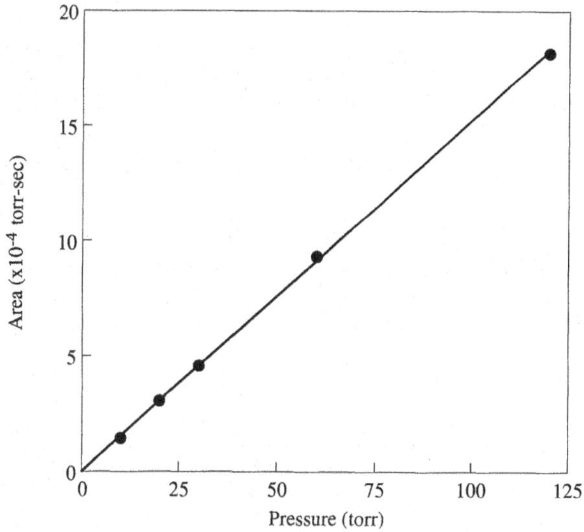

Figure 5. Effect of filling pressure on the area under the curve for helium release from the standard volume container (reference 7).

The average solubility obtained from at least three individual measurements typically represents each of the values reported by Tounour [7], while individual values are reported for the work by Mesko. The range of values reported by Tournour for any given set of measurements varies from a few percent for compositions with high gas solubility (typically greater than 7×10^{16} atoms/gm-atm) to as much as 25% for compositions with low gas solubility (typically less than 3×10^{16} atoms/gm-atm). This increase in the range of values with decreasing gas solubility is expected due to the decrease in the area under the outgassing curve with decreasing gas solubility. The very small areas obtained from samples containing very small amounts of gas are significantly more difficult to distinguish from the background signal of the RGA. Results for carbon dioxide are particularly suspect due to the very low carbon dioxide concentrations in these melts and problems associated with contamination of the samples with organic materials which convert to carbon dioxide during outgassing and result in an erroneously high apparent gas solubility. It is necessary to avoid any contact of the crucible or sample with the fingers of the experimenter during transfer from the saturation to the outgassing apparatus. Touching the crucible with a finger, or any other contact with a source of organic contamination, will void any subsequent measurement of carbon dioxide concentration for these very low concentrations.

EXPERIMENTAL PROCEDURES FOR WATER SOLUBILITY MEASUREMENTS

The water solubility measurements are made using a completely different method. The details of these measurements have been presented in a number of publications [3, 8-11]. A small amount of glass is placed in a Pt/5Au crucible and remelted. Since the glasses studied here do not adhere to these crucibles, they can be easily removed after each heat treatment. The melt is exposed to a known partial pressure of water vapor for a known time, the crucible is removed from the furnace and rapidly cooled it to room temperature, the glass is removed, and the infrared spectrum in the region from 4000 to 2000 cm^{-1} is measured, the thickness of the glass at the point where the infrared measurement was made is measured, and then the sample is replaced in the crucible for further treatment. Treatments are continued until no further change in water concentration can be detected in the infrared spectrum.

Most of the work reported here was carried out at in an atmosphere obtained by passing flowing air above boiling water, i.e. at one atmosphere of water vapor. Lower partial pressures of water vapor are obtained by varying the temperature of the water bath used to provide the water vapor.

After the water concentration reaches equilibrium, each sample is removed from the crucible, annealed, cooled slowly to room temperature, and sliced using a low speed diamond saw parallel to the original top surface of the melt to produce a plate about 1 mm thick. After polishing, the amount of water present as hydroxyl is determined at several points on the sample using the Beer-Lambert Law, i.e.

$$A = \varepsilon c L \quad (3)$$

where A is the absorbance measured at a given wavelength, c is the concentration of water in the sample, L is the sample thickness at the point of measurement, and ε is the extinction coefficient for hydroxyl in that glass.

The absorption data were converted to concentrations of "water" in the sample using the single band method and the extinction coefficients determined for these compositions by Mesko and Shelby [3]. These values are listed in Table II. Values reported here are averages of concentrations determined at several points on each plate. Measurements made on as-received plates of a commercial float glass sample yield a relative deviation of only 0.3% of the mean value, indicating that the infrared measurement itself is highly reproducible. Measurements on our saturated samples exhibit a wider range of variation, which is attributed to greater variation in the thickness of the plate over the sample area for our hand polished samples and to some variation in the actual water content of the sample in different regions of the melt. In the worst case, the precision in the water solubility determination is within than ±5% of the value reported. The absolute value of the solubility will vary with the choice of the extinction coefficient used in the Beer-Lambert equation. As a result, the results of different studies can only be compared if the values of the extinction coefficients used to convert the absorbance data to water concentrations are specified in each report.

TABLE II: EXTINCTION COEFFICIENTS FOR HYDROXYL IN COMMERCIAL GLASSES

Glass	Extinction Coefficient at 3500 cm^{-1} (L/mol-cm)	Source of Value
Float	41.3 ± 1.8	Mesko & Shelby [3]
Container	42.4 ± 1.7	Mesko & Shelby [3]
TV Panel	18.2 ± 1.2	Mesko & Shelby [8]
Wool	46.8 ±1.5	Mesko & Shelby [9]
E glass	49.1 ± 2.6	Nemec & Gotz [13]

RESULTS

Inert Gas Solubilities

The solubilities of helium, neon, and argon found in this study are given in Tables III to VIII. A comparison of the solubilities of these gases in the melts at 1300°C is given in Table IX. In general, the solubilities of helium and neon are similar, while those of argon are much smaller. Helium solubilities are much higher in the borosilicate glasses and melts than in the other compositions, while the helium solubilities are lower in the E-glass melts than in the melts of the other compositions. In general, the isothermal solubility of any inert gas in both glasses and melts decreases with decreasing concentration of glass former oxides in the composition.

Examination of the results presented in Tables III to VIII also reveals that the solubility of helium passes through a minimum with increasing temperature when data cover the temperature range from below T_g to the fluid melt. This behavior appears to be the norm for silicate compositions, since it has also been found for several commercial borosilicate glasses [12] and for simple binary sodium silicate compositions [6]. Tables III-VIII: Inert Gas Solubility in DOE Glasses and Melts as a Function of Temperature (All values are x 10^{16})

TABLE III: FLOAT GLASS

Temperature (°C)	Helium (atoms/gm-atm)	Neon (atoms/gm-atm)	Argon (atoms/gm-atm)
399	1.39		
451	1.23		
483	1.15		
547	1.27		
614	0.77		
619	0.73		
669	1.08		
687	0.72		
734	1.00		
1101	0.45		
1055	1.08		
1151	0.85		
1200	1.14		
1300	1.19		
1375	2.24		
1300*	3.47	2.9	1.5

*Tournour

TABLE IV: CONTAINER GLASS

Temperature (°C)	Helium (atoms/gm-atm)	Neon (atoms/gm-atm)	Argon (atoms/gm-atm)
222	1.16		
294	1.18		
398	1.17		
464	1.01		
499	1.16		
508	0.98		
540	1.71		
541	1.49		
567	1.16		
587	1.13		
645	1.48		
670	1.51		
713	1.70		
762	1.95		
950	2.25		
1000	1.94		
1000	2.33		
1050	2.25		
1075	2.46		
1150	1.50		
1175	1.29		
1175	1.51		
1200	1.42		
1225	1.32		
1250	2.71		
1250	2.14		
1325	2.18		
1350	1.51		
1350	1.79		
1300*		2.5	0.27

*Tournour

TABLE V: WOOL GLASS

Temperature (°C)	Helium (atoms/gm-atm)	Neon (atoms/gm-atm)	Argon (atoms/gm-atm)
595	0.99		
650	0.76		
695	1.25		
1000	1.36		
1100	1.37		
1200	2.04		
1200	2.30		
1300	1.76		
1400	2.05		
1300*	3.7	2.7	0.44

*Tournour

TABLE VI: E-GLASS

Temperature (°C)	Helium (atoms/gm-atm)	Neon (atoms/gm-atm)	Argon (atoms/gm-atm)
601	1.47		
601	1.27		
694	1.28		
741	0.67		
759	1.37		
900	1.56		
1000	0.89		
1100	0.98		
1250	1.12		
1300	0.75		
1375	1.45		
1300*			0.38

*Tournour

TABLE VII: TV PANEL GLASS

Temperature (°C)	Helium (atoms/gm-atm)	Neon (atoms/gm-atm)	Argon (atoms/gm-atm)
291	0.69		
382	0.94		
438	1.22		
472	1.15		
500	0.96		
576	1.02		
588	1.27		
636	1.23		
671	1.30		
711	1.19		
900	1.52		
1000	1.34		
1100	1.83		
1150	1.87		
1200	3.11		
1300	3.56		
1375	3.56		
1300*	4.1	2.8	0.97

*Tournour

TABLE VIII: BOROSILICATE TUBING GLASS

Temperature (°C)	Helium (atoms/gm-atm)	Neon (atoms/gm-atm)	Argon (atoms/gm-atm)
206	9.47		
247	9.73		
317	8.56		
365	7.93		
423	7.97		
473	7.36		
517	7.62		
567	7.48		
616	7.62		
666	7.98		
714	8.73		
732	6.48		
777	7.46		
850	8.36		
900	7.10		
1000	6.80		
1050	9.66		
1100	7.56		
1150	8.99		
1200	8.31		0.14
1250	8.05		0.18
1300	8.56	2.0	0.23
1350	7.17		0.21
1400			0.21
1000*	7.21		
1100*	7.11		
1200*	6.30		
1300*	6.15		0.72
1400*	9.50		

*Tournour

TABLE IX: INERT GAS SOLUBILITY AT 1300°C IN BASE COMMERCIAL MELTS

(Solubility is given in 10^{16} atoms/gm-atm)

Glass Identity	Helium (atoms/gm-atm)	Neon (atoms/gm-atm)	Argon (atoms/gm-atm)
Float	1.7 (Mesko) 3.4 (Tournour)	2.9 (Tournour)	1.5 (Tournour)
Container	2.4 (Mesko)	2.5 (Tournour)	0.27 (Tournour)
Wool	2.0 (Mesko) 3.7 (Tournour)	2.7 (Tournour)	0.44 (Tournour)
E-glass	1.1 (Mekso)	---	0.38 (Tournour)
TV Panel	3.3 (Mesko) 4.1 (Tournour)	2.8 (Tournour)	0.97 (Tournour)
Borosilicate	8.6 (Mesko) 6.2 (Tournour)	2.0 (Mesko)	0.23 (Mesko) 0.72 (Tournour)

Carbon Dioxide Solubilities

Carbon dioxide solubilities for these compositions are listed in Table X as a function of melt temperature. The values are considerably lower than those of the inert gases, with the exception of argon. A few of the points listed in this table appear to be inconsistent with the remainder of the data. In particular, the values for carbon dioxide solubility in the wool melts at 950 and 1100°C appear to be unusually large. These values are the averages of several measurements which were internally consistent. The validity of these particular values is not known. It does appear that carbon dioxide solubility is somewhat higher in the borosilicate melts than in melts of the other compositions.

TABLE X: CARBON DIOXIDE SOLUBILITY IN BASE COMMERCIAL MELTS

(Solubility is given in 10^{16} molecules/gm-atm)

Temp. (°C)	Float	Container	TV Panel	Wool	E-glass	Borosil.
950	xtal	xtal	---	4.5	---	---
975	---	---	0.25	---	---	---
1000	0.24	---	1.02	0.87	---	---
1050	---	0.18	0.27	0.60	---	---
1100	0.31	0.18	0.19	2.3	xtal	xtal
1150	---	---	---	---	1.19	1.98
1200	0.68	0.13	0.24	0.25	1.11	1.97
1250	---	---	---	---	0.61	0.98
1300	0.47	0.28	0.30	---	0.62	1.26
1350	---	---	---	---	---	0.43
1400	---	0.43	0.97	---	---	0.53

Water Solubility

The solubilities of water, expressed as the concentrations in wtppm obtained by melting under one atmosphere of water vapor pressure, are listed in Table XI. With the exception of the TV panel composition [8], these values are roughly the same for all of the glasses. No values can be obtained for the borosilicate melts since no extinction coefficient is available for this glass. Attempts to determine this extinction coefficient during this study were unsuccessful. The values listed for the E-glass melts are suspect due to loss of boron during the long exposures to water vapor necessary to reach equilibrium in the melt [9].

TABLE XI: WATER SOLUBILITY IN WTPPM

Temp. (°C)	Float	Container	TV Panel	Wool	E-glass#	Borosil.
900	---	---	---	1001	---	---
1000	1072	1036	1585	1138	---	---
1050	---	---	---	---	1035	---
1100	1080	1053	1533	1066	---	---
1200	1107	980	1631	1077	1048*	---
1250	1042	1002	---	973	987*	---
1300	1116	1108	1698	1063	---	---
1350	1060	967	---	---	---	---
1400	1136	1080	1677	---	---	---

*Composition change during measurement (boron loss)

#Based on extinction coeff. of Nemec and Gotz (49.1 L/mol-cm), which has been confirmed by the results of this study.

REFERENCES

1. J. E. Shelby, *Handbook of Gas Diffusion in Solids and Melts*, ASM International, Materials Park, OH, 1996.
2. J. E. Shelby, *Introduction to Glass Science and Technology*, Royal Society of Chemistry, Cambridge, 1997.
3. M. G. Mesko, J. E. Shelby, "Water Solubility and Diffusion in Melts of Commercial Silicate Glasses," Glastech. Ber. Glass Sci. Technol. **73**(C2) 13-22 (2000)
4. J. M. Jewell, "Apparatus for the Determination of Gas Solubility in Glass Melts," M. S Thesis, Alfred University, Alfred, NY, 1986.
5. W. D. Kingery, H. K. Bowen, & D. R. Uhlmann, *Introduction to Ceramics*, 2nd *Edition*, pp. 225-6, John Wiley & Sons, New York, 1976.
6. M. G. Mesko, K. Newton, J. E. Shelby, "Helium Solubility in Sodium Silicate Glasses and Melts," Phys. Chem. Glasses **41**(3) 111-16 (2000).
7. C. C. Tournour, "Solubility and Diffusion of Gases in Glasses and Melts," Ph. D. Thesis, Alfred University, Alfred, NY, 2004.
8. M G. Mesko, J. E. Shelby, "Solubility and Diffusion of Water in Melts of a TV Panel Glass, Phys. Chem. Glasses, **42**(1) 17-22 (2001)
9. M. G. Mesko, J. E. Shelby, "Solubility and Diffusion of Water in Melts of E and Wool Glasses," Phys. Chem. Glasses **42**(6) 389-96 (2001)
10. P. B. McGinnis, J. E. Shelby, "Diffusion of Water in Float Glass Melts," J. Non-Cryst. Solids, **177** 381-88 (1994).

11. D. B. Rapp, J. E. Shelby, "Water Diffusion and Solubility in Soda-Lime-Silica Melts," Phys. Chem. Glasses **44**(6) 393-400 (2003)
12. M. G. Mesko, B. E. Kenyon, J. E. Shelby, "Helium Solubility in Commercial Borosilicate Glasses and Melts," Glastech. Ber. Glass Sci. Technol. 73(C2) 33-42 (2000)
13. L. Nemec, J. Gotz, "Infrared Absorption of OH^- in E. Glass", J. Am. Ceram. Soc.- Discussions and Notes, **53** (9) 526 (1970)

WATER DIFFUSION

INTRODUCTION

Water enters the structure of silicate glasses and melts by reacting to form hydroxyl [1,2]. Since replacement of a bridging oxygen by a hydroxyl reduces the connectivity of the network, hydroxyl significantly reduces the glass transformation temperature (T_g) of silicate glasses, with variations of 20 to 30 K for soda-lime-silicate glasses [2] for changes in water concentration of 1000 wtppm. The transformation range viscosity is altered by more than an order of magnitude by varying the hydroxyl concentration over this range [2]. Crystallization rates, which are very sensitive to viscosity, can be altered by orders of magnitude by changes in hydroxyl content [3,4]. Changes of this magnitude are particularly important in annealing and reshaping glasses and in the prevention of crystallization during processing.

Although the product of the reaction of water with oxide melts is hydroxyl, the reaction is the result of exposure to water vapor. Since the reaction rate is diffusion-controlled, this process is usually termed "water diffusion". While the diffusing species is usually assumed to be molecular water, details of the mechanisms involved in the production of hydroxyl have not been established [5]. Use of the term "water" here only indicates that the species in the atmosphere surrounding the melt is molecular water and does not imply any particular diffusion mechanism.

The diffusivity of water has usually been determined by measuring the rate of removal of hydroxyl from a glass or melt, i.e. during dehydroxylation, utilizing either direct measurements of water content or infrared spectral measurements of the hydroxyl concentration [6,7]. Very few studies have measured the diffusivity of water in glasses or melts during the hydroxyl formation process. Since studies of hydroxyl formation and removal in vitreous silica [8] yield different diffusivities, there is some question regarding the validity of comparison of results obtained during hydroxylation with those obtained during dehydroxylation.

Only one study has considered the diffusivity of water in float glass melts [9]. No reports of water diffusion in melts of the other glasses considered here were found. The work reported here deals with both hydroxylation and dehydroxylation of glasses and melts of the base compositions. In some cases, the measurements cover the range from below Tg to 1500°C. The results indicate that there is no change in the activation energy for water diffusion from the solid glass to the fluid melt.

EXPERIMENTAL PROCEDURE

Samples used in this study were produced by remelting the commercial glasses in Pt/5Au crucibles. Details of the measurement technique have been discussed in several other papers [8,9,10-14].

Diffusivities are determined by measuring the rate of change of absorption at the maximum of the infrared band due to hydroxyl, which occurs at approximately 3500 cm^{-1} for these glasses. In general, hydroxyl is added to melts by exposure to an atmosphere enriched in water vapor. Since laboratory air has a lower water vapor partial pressure than that found in the glass tanks where these glasses were manufactured, however, remelting of commercial glasses in air will usually cause the melt to release water vapor. The final hydroxyl content of the glass will be determined by the humidity at the time of the study. If the glass or melt is exposed to vacuum, it is usually assumed that the water vapor partial pressure is zero, so that all of the hydroxyl will be removed from the glass if sufficient time were available. This assumption is necessary for the determination of the diffusivity in glasses at the lower temperatures since the time to reach the final water concentration would be prohibitively large.

Water diffusion coefficients were measured using the method developed by McGinnis and Shelby [9] and subsequently optimized by Mesko and Shelby [10-14]. Three to five grams of glass were remelted in a 30 ml Pt/5Au crucible to produce a base sample. Since these glasses do not adhere to gold-doped platinum crucibles, they are easily removed from the crucible by tapping the bottom of the crucible after cooling. After removal from the crucible, the infrared spectrum of the glass is measured, and the sample returned to the crucible in the same orientation as the original melt. The crucible is then placed in the controlled-atmosphere furnace (Figure 1) at the desired temperature as described below.

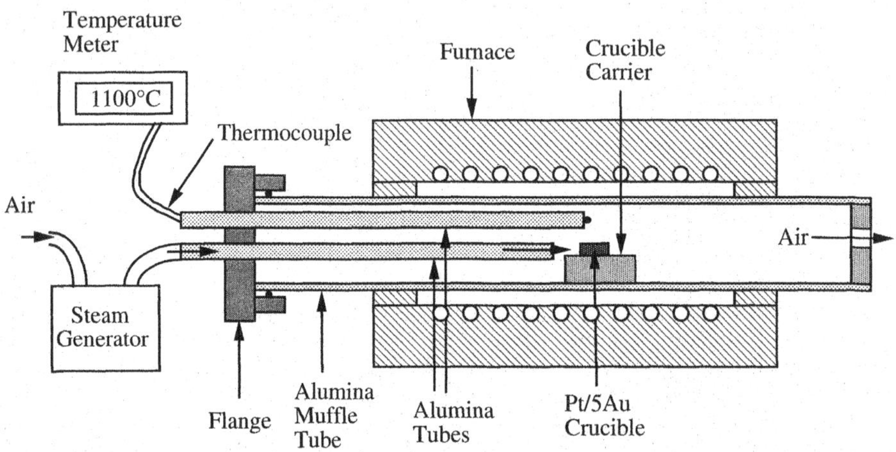

Figure 1. Apparatus for determining the diffusivity of water in melts by exposure to water vapor.

The change in relative hydroxyl concentration from that of the initial condition after each treatment is obtained by subtracting the value for the absorption/mm at the maximum of the selected infrared absorption band for the initial state (C_o) from each of the values obtained during the heat treatments (C_t). Dividing the change in absorption/mm by the value obtained by subtracting the initial melt value from that of the saturated melt yields the fractional change, F, in concentration for each treatment time. The fraction of saturation is given by the expression

$$F = (C_t - C_o)/(C_\infty - C_o). \tag{1}$$

The effective diffusivity for hydroxyl formation, D, into or out of a semi-infinite plate, where diffusion occurs from both surfaces, is given by the expression

$$\frac{M_t}{M_\infty} = \frac{4}{\sqrt{\pi}}\sqrt{\frac{Dt}{L^2}} \tag{2}$$

where M_t is the concentration after treatment time t, M_∞ is the total change in concentration ($M_t/M_\infty = F$), and L is the thickness of the sample. If the data are plotted as F versus $\sqrt{t}/L$, it can be shown that the diffusivity is given by

$$D = (\pi/16)R^2 \tag{3}$$

where R is the slope of the plot of F versus •t/L at short times (the plot is essentially linear up to a value of F = 0.5). This expression is used to obtain D for water vapor diffusion from the plate samples used in the transformation range.

Since Eq. 2 assumes that the sample is exposed to the diffusing species on both surfaces, Eq. 3 must be corrected by a factor of 4 to yield the expression for diffusion into or out of a melt, where the sample is only exposed to the atmosphere on one surface, or

$$D = (\pi/4)R^2. \tag{4}$$

The apparatus used to expose melts to air saturated with water vapor is shown in Figure 1. The furnace is equilibrated at the desired temperature, the water is brought to a boil, and the air flow is initiated. The crucible containing the premelted sample is placed in the ceramic boat, which is then pushed into place in the apparatus. The crucible is located approximately 1 cm from the end of the measurement thermocouple, which in monitored throughout the treatment. The temperature indicated by this thermocouple drops initially on insertion of the crucible, but returns to the desired temperature within 5 minutes. Since most of the treatments begin with a treatment of 30 minutes, the heating time for the crucible is considered negligible. At high diffusivities, the first treatment time is only 15 minutes. Under these conditions, some degree of error is introduced by the thermal lag of the sample during this heating period. Eventually, the upper limit of measurable diffusivities is partially determined by the total time of equilibration needed to reach saturation through the melt depth.

Dehydroxylation of the melt is much easier. In this case, it is only necessary to place the crucible into a preheated furnace with an uncontrolled atmosphere. Since the water vapor partial pressure under these conditions is controlled by the ambient humidity, it has been found that this method is more successful during the winter months than during the summer months. Treatments during the winter of float and container melts typically remove about 2/3 of the original hydroxyl content.

Hydroxylation of plates is carried out in a facility identical to that used for the melts, with the exception that the samples are placed in a ceramic carrier which holds them in the stream of flowing humid air rather than in a crucible. Dehydroxylation of the plates is carried out in a different system (Figure 2), which is capable of evacuation to ≈30 millitorr. The remainder of the procedure is similar to that used for the other work.

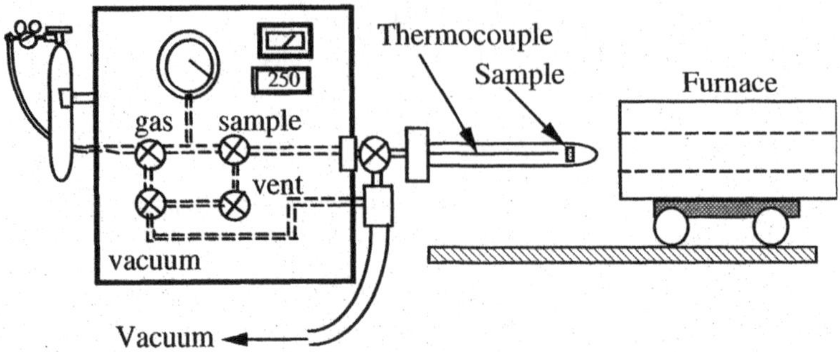

Figure 2. Apparatus for determining the diffusivity of water in glasses by vacuum dehydroxylation.

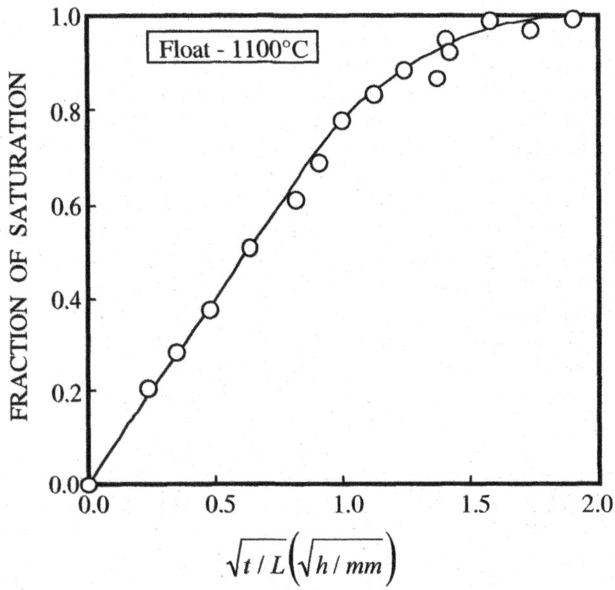

Figure 3. Typical results of a diffusivity measurement.

RESULTS

An example of the determination of the water diffusivity in a melt is shown in Figure 3. The linear region below F = 0.50 is used to determine the value of R. The diffusivity is then calculated using either Eq. 3 (for plates of glass) or Eq. 4 (for melts in a crucible).

The effect of temperature on the diffusivity of water in float, container, TV panel, and glass fiber compositions in glasses and melts is shown in Figures 4 through 7, respectively.

Measurements were made as discussed above. Data used to generate these figures are listed in Tables I through VI.

While a number of studies have separately considered the diffusion of water in the melt and glass transformation ranges [2, 6-9], this is the first report of a study which covers both temperature regions using the identical glasses as samples. The results indicate that there is no significant deviation in the log(D) versus reciprocal temperature plot over a range of 1000 K. In particular, results for float glass convincingly show that no unusual behavior in water diffusivity occurs in the glass transformation range. This result is rather surprising in that other diffusion-related phenomena, e.g. sodium [15] and helium [2] diffusion and ionic conductivity due to alkali migration [15], have been reported to show significant deviations in the glass transformation range from the simple behavior shown here.

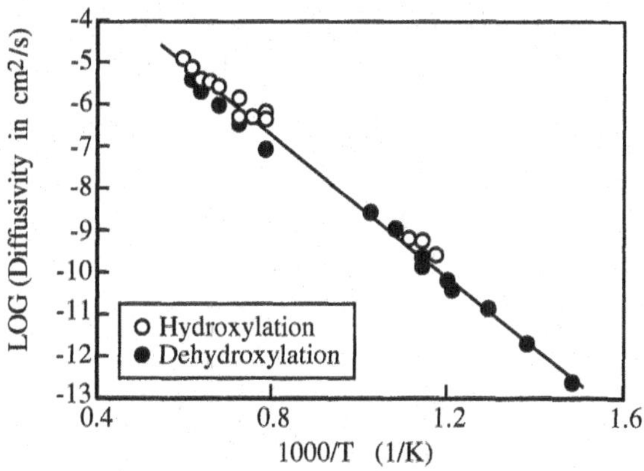

Figure 4. Diffusivity of water in glasses and melts of float composition.

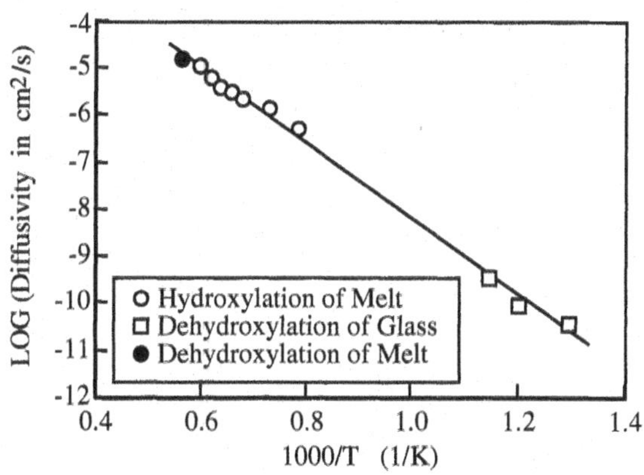

Figure 5. Diffusivity of water in glasses and melts of container composition.

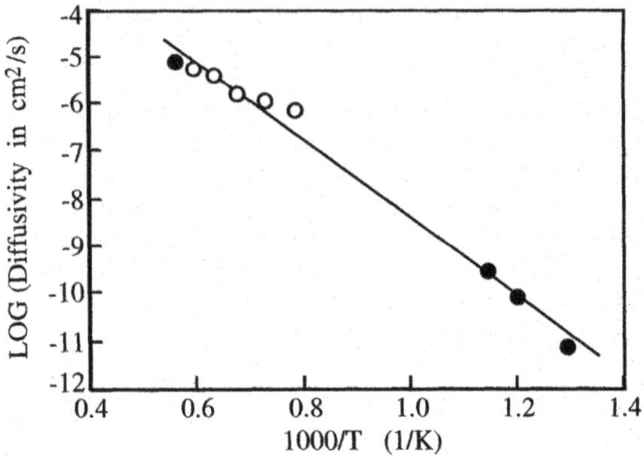

Figure 6. Diffusivity of water in glasses and melts of TV panel composition.

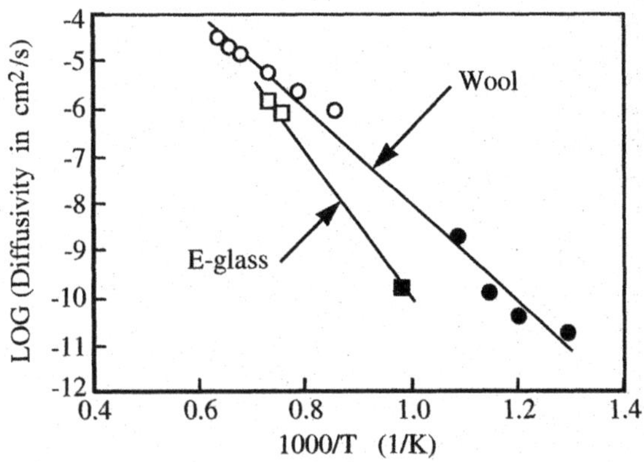

Figure 7. Diffusivity of water in glasses and melts of compositions used in the production of commercial glass fibers (wool and E-glass).

TABLE I. DIFFUSIVITY OF WATER IN FLOAT GLASSES AND MELTS

Temperature (°C)	D (addition) Present Work (cm^2/s)	D (removal) Present Work (cm^2/s)	D (addition) McGinnis (cm^2/s)	D (removal) McGinnis (cm^2/s)
400				2.5×10^{-13}
450				2.0×10^{-12}
500				1.4×10^{-11}
550				4.1×10^{-11}
560		6.3×10^{-11}		
575			2.7×10^{-10}	
600		1.4×10^{-10}	6.1×10^{-10}	2.5×10^{-10}
625			7.0×10^{-10}	
650				1.0×10^{-9}
700				2.8×10^{-9}
1000	6.2×10^{-7}		4.3×10^{-7}	9.1×10^{-8}
1050			5.0×10^{-7}	
1100	1.5×10^{-6}		5.3×10^{-7}	3.6×10^{-7}
1200	2.6×10^{-6}		9.6×10^{-7}	9.6×10^{-7}
1250	3.6×10^{-6}			
1300	4.1×10^{-6}		2.4×10^{-6}	2.2×10^{-6}
1350	7.8×10^{-6}		4.2×10^{-6}	3.9×10^{-6}
1400	1.2×10^{-5}			

TABLE II. DIFFUSIVITY OF WATER IN CONTAINER GLASSES AND MELTS

Temperature (°C)	D (addition) (cm^2/s)	D (removal) (cm^2/s)
500		3.8×10^{-11}
560		8.7×10^{-11}
595		3.5×10^{-10}
650		2.3×10^{-9}
1000	5.8×10^{-7}	
1100	1.6×10^{-6}	
1200	2.4×10^{-6}	
1250	3.2×10^{-6}	
1300	4.0×10^{-6}	
1350	6.5×10^{-6}	
1400	1.4×10^{-5}	

TABLE III. DIFFUSIVITY OF WATER IN TV PANEL GLASSES AND MELTS

Temperature (°C)	D (addition) (cm^2/s)	D (removal) (cm^2/s)
500		7.1×10^{-12}
560		8.5×10^{-11}
595		3.1×10^{-10}
650		2.8×10^{-9}
1000	7.9×10^{-7}	
1100	1.2×10^{-6}	
1200	1.7×10^{-6}	
1300	3.9×10^{-6}	
1400	5.8×10^{-6}	

TABLE IV. DIFFUSIVITY OF WATER IN WOOL GLASSES AND MELTS

Temperature (°C)	D (addition) (cm^2/s)	D (removal) (cm^2/s)
500		1.8×10^{-11}
560		4.1×10^{-11}
595		1.4×10^{-10}
650		2.0×10^{-9}
900	9.1×10^{-7}	
1000	2.4×10^{-6}	
1100	6.0×10^{-6}	
1200	1.5×10^{-5}	
1250	2.1×10^{-5}	
1300	3.4×10^{-5}	

TABLE V. DIFFUSIVITY OF WATER IN E-GLASS GLASSES AND MELTS

Temperature (°C)	D (addition) (cm^2/s)	D (removal) McGinnis (cm^2/s)
???		
1000	crystallized	
1050	8.5×10^{-7}	
1100	1.5×10^{-6}	
1200*	5.2×10^{-6}	
1250*	8.7×10^{-6}	

*These melts lost considerable boron during the measurement. The listed values are considered to be highly suspect.

TABLE VI. DIFFUSIVITY OF WATER IN BOROSILICATE TUBING GLASSES AND MELTS

Temperature (°C)	D (addition) (cm^2/s)	D (removal) (cm^2/s)
600		2.5 x 10-11
1000	crystallized	
1100	crystallized	
1200	approx. 3 x 10-6	
1300	approx. 1.5 x 10-3	

REFERENCES

1. J. E. Shelby, "Glass Melting"; pp. 25-47 in *Introduction to Glass Science and Technology*, by J. E. Shelby, Royal Society of Chemistry, Cambridge, 1997.
2. J. E. Shelby, "Water in Glasses and Melts"; pp. 217-234 in *Handbook of Gas Diffusion in Solids and Melts*, by J. E. Shelby, ASM International, Materials Park, OH, 1996.
3. M. R. Heslin and J. E. Shelby, "Effect of Hydroxyl Content on the Nucleation and Crystallization of Lithium Disilicate Glass, *Ceram. Trans.*, **30** 189-96 (1993).
4. C. J. R. Gonzalez-Oliver, P. S. Johnson and P. F. James, "Influence of Water Content on the Rates of Crystal Nucleation and Growth in Lithia-Silica and Soda-Lime-Silica Glasses," *J. Mater. Sci.*, **14** 1159-69 (1979).
5. R. H. Doremus, "Speciation of Water in Silicate Glasses and Melts"; pp. 129-47 in *Diffusion of Reactive Molecules in Solids and Melts*, by R. H. Doremus, Wiley Interscience, New York, 2002.
6. H Scholze and H. O. Mulfinger, "The Influence of Water in Glasses V. The Diffusion of Water in Glasses at High Temperatures," *Glastech. Ber.*, 32 [9] 381-86 (1959).
7. B. J. Todd, "Outgassing of Glass," *J. Appl. Phys.*, **26** [10] 1238-43 (1955).
8. P. B. McGinnis and J. E. Shelby, "Diffusion of Water in Vitreous Silica," *J. Non-Cryst. Solids*, **179** 185--93 (1994).
9. P. B. McGinnis and J. E. Shelby, "Diffusion of Water in Float Glass Melts," *J. Non-Cryst. Solids*, **177** 381-88 (1994).
10. M. G. Mesko and J. E. Shelby, "Water Solubility and Diffusion in Melts of Commercial Silicate Glasses," *Glastech. Ber. Glass Sci. Technol.*, **73** [C2] 13-22 (2000).
11. M. G. Mesko and J. E. Shelby, "Solubility and Diffusion of Water in Melts of a TV Panel Glass," *Phys. Chem. Glasses*, **42** [1] 17-22 (2001).
12. M. G. Mesko and J. E. Shelby, "Solubility and Diffusion of Water in Melts of E and Wool Glasses," *Phys. Chem. Glasses*, **42** [6] 389-96 (2001).
13. M. G. Mesko and J. E. Shelby, "Water Solubility and Diffusion in Alkali Silicate Melts," *Phys. Chem. Glasses*, **42** [3] 173-8 (2001).
14. M. G. Mesko, P. A. Schader and J. E. Shelby, "Water Solubility and Diffusion in Sodium Silicate Melts," *Phys. Chem. Glasses*, **43** [6] 283-90 (2002).
15. R. H. Doremus, "Electrical Conductivity and Ionic Diffusion," pp. 271-6 *Glass Science, 2nd Ed.*, by R. H. Doremus, Wiley Interscience, New York, 1994.

Chapter 4

SOLUBILITY AND DIFFUSION OF SO_3 IN SODA-LIME-SILICATE GLASS MELTS AT HIGH TEMPERATURES

Oleg A. Prokhorenko
Laboratory of Glass Properties LLC
5a Varshavskaya Street, bldg. 50, 3rd floor
St. Petersburg, 196084
Russian Federation

ABSTRACT

Sulphates remain the main fining agents for melting of many commercial glasses. Correct description of transport of SO_3 between the melt and the gas phase (bubbles and surrounding atmosphere) is a key to successful mathematical modelling of sulphate fining. The main task of the present chapter was to describe exchange by SO_3 between the soda-lime-silicate melt (having matrix similar to float and container glasses) and the gas phase combining mathematical modelling of the diffusion process with experimental studies of SO_3 solubility and diffusion coefficients. The described work has been performed in 1998-2001 in the Laboratory of Glass Properties under support of NSF Industry/University Center for Glass Research (CGR) at Alfred University (Alfred, NY).

INTRODUCTION

For successful modeling of fining of molten glass by sulfates, efficient mathematical procedures describing the processes of transport of sulfur within the melt as well as between the melt and the gas phase (bubbles and surrounding atmosphere) are needed. Dependable data on solubility and diffusion of SO_3 in the molten glasses are required as input parameters for the modeling. It should be noted that the process of transport of sulfur through the interface between molten glass and gas phase is complicated by multi-stage chemical reactions between sulfur in the gas phase and components of the melt. As a result of these reactions, formation of a layer of sodium sulfate is observed. Convection of the layers of the melt having different density and surface tension and some other processes take place as well. Study of the contribution of each of these processes separately is a very difficult task. Bruckner [1, 2] attempted to study transport of sulfur in the system molten glass – melted Na_2SO_4 – gas. In these papers some general patterns of influence of a number of factors (melt and gas composition, redox (reduction-oxidation) conditions, temperature, humidity, gas pressure, etc.) on the final sulfur content in molten glass have been formulated. Unfortunately the lack of experimental data allowed the author to make only a qualitative description of the process in question.

As it was mentioned already, studies of transport of SO_3 between the molten glass and gas atmosphere should include mathematical modeling and experimental parts. Mathematical modeling of sulfur oxide diffusion should be performed under different conditions. We have chosen the following four gas-liquid systems:

System I. Diffusion of SO_3 between the molten glass and the gas atmosphere to the achievement of the state of equilibrium: $[SO_3]^{gas}\Rightarrow[SO_3]^{eq1}$ and $[SO_3]^{melt}\Rightarrow[SO_3]^{eq2}$ (closed system).

System II. Diffusion of SO_3 between the molten glass and the gas atmosphere at $[SO_3]^{gas}$=const; $[SO_3]^{melt}\Rightarrow 0$ (degassing) or $[SO_3]^{melt}\Rightarrow[SO_3]^{eq2}$ (saturation).

System III. Exchange by SO_3 between the molten glass and the gas in the course of forced bubbling by the gases having different content of SO_3.

System IV. Extraction of excess of SO_3 from the melt (supersaturated by SO_3 with respect to the gas atmosphere) into small bubbles spontaneously forming, growing and rising in the molten glass at high temperatures.

Several laboratory techniques for studies of diffusion and solubility of SO_3 in the molten glass have been developed. Preliminary calculations helped us to determine experimental conditions for these studies. The following experimental techniques for the studies of exchange by sulfur oxide between the molten glass and gas phase were developed and used for measurements of diffusion coefficients and solubility of SO_3:

(1) Diffusion of sulfur between high-SO_3 and sulfur-free melts (two-layers technique);

(2) Saturation by SO_3 of thin layers or droplets of molten glass positioned in sulfur containing gas atmosphere;

(3) Exchange by SO_3 between the melt and small bubbles.

MATHEMATICAL MODELING

Below mathematical models describing transport of SO_3 in four different gas-melt systems are considered. Brief descriptions of the mathematical procedures are followed by the results of calculations.

System I. Diffusion of SO_3 through the boundary between molten glass and gas atmosphere to achievement of state of equilibrium in the closed system containing no bubbles. The diffusion of sulfur from the layer of the molten glass to the gas atmosphere (and conversely) can be described by the following equation:

$$\partial_t C = D\partial^2_{xx} C \tag{1}$$

Where: t – is the time, x – is the coordinate directed in perpendicular direction to the melt surface, C – is the concentration of SO_3 in the molten glass, D – is the diffusion coefficient of sulfur in the molten glass.

The concentration of SO_3 in the gas phase was determined by the following equation:

$$\partial_t C_a = -\frac{F}{V} D\partial_x C\big|_{x=\delta} \tag{2}$$

Where: C_a – is the concentration of sulfur in the gas phase, V – is the volume of the gas phase, F – is the surface area of the melt-gas interface through which sulfur leaves the melt, δ – is the thickness of the molten glass layer.

Let us assume that the concentration of sulfur at the molten glass – gas interface obeys the Henry's law:

$$c|_{x=\delta} = hC_a \tag{3}$$

Where: h – is the Henry constant, which is connected with solubility by the following equation:

$$h = \frac{\tilde{R}T}{\mu P_0} S \tag{4}$$

Where: $\tilde{R}$ – is the absolute gas constant, μ – is the molar mass of SO_3, T – is the temperature, P_0 – is the atmospheric pressure.

The following parameters have been used for the calculation of the transport of sulfur in the closed system. Thickness of the molten glass layer – 1 inch, molten glass volume – 1 cubic inch, gas volume to the melt volume ratio – 1/100, initial concentration of sulfur in the glass – 6 kg/m^3, initial concentration of sulfur in the gas – 0 kg/m^3, absolute gas constant – 8.314 J/mol·K, molar mass of SO_3 – 80 g/mol, diffusion coefficients - $logD_{1200°C}$ = - 7.24 cm^2/s; $logD_{1300°C}$ = - 7.08 cm^2/s; $logD_{1400°C}$ = - 6.83 cm^2/s (values were taken from Table I, which can be found below in the text); solubility of SO_3 - $[SO_3]_{1400°C}$ = 0.4 wt.%; $[SO_3]_{1300°C}$ = 0.6 wt.%; $[SO_3]_{1200°C}$ = 0.8 wt.% (temperature-averaged and rounded values were taken from Table II, which can be found below in the text).

The computation was performed for three temperatures: 1200, 1300, and 1400°C.

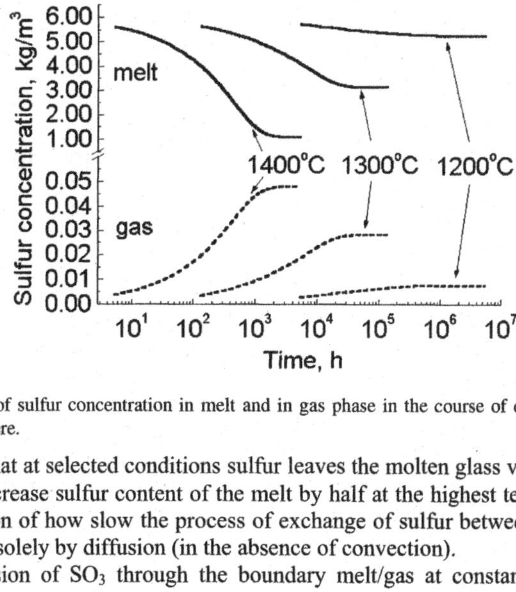

Fig.1. Time dependence of sulfur concentration in melt and in gas phase in the course of diffusion of sulfur into surrounding gas atmosphere.

Fig. 1 shows us that at selected conditions sulfur leaves the molten glass very slowly. It takes over 100 hours to decrease sulfur content of the melt by half at the highest temperature 1400°C. It is a good illustration of how slow the process of exchange of sulfur between the molten glass and the gas proceeds solely by diffusion (in the absence of convection).

System II. Diffusion of SO_3 through the boundary melt/gas at constant concentration of sulfur in gas phase (open system).

The initial conditions of the task are similar to those taken for System I. Instead of the conditions (3), however, let us take:

$$C|_{x=\delta}=0. \tag{5}$$

Let us assume also, that flow of the gas passing over the surface of the molten glass is equal to $\dot{Q}$. Then concentration of SO_3 in the flow can be determined by the following equation:

$$C_a^{out}=\frac{-FD\partial_x C|_{x=\delta}}{\dot{Q}}. \tag{6}$$

The volume of gas flow equal to 1 l/h (61 cubic inch per hour) was taken. The rest of the parameters are the same as were taken for System I.

Below in Fig. 2 one can see the time dependence of concentration of sulfur in both the molten glass and the gas passing over the melt calculated at three temperatures.

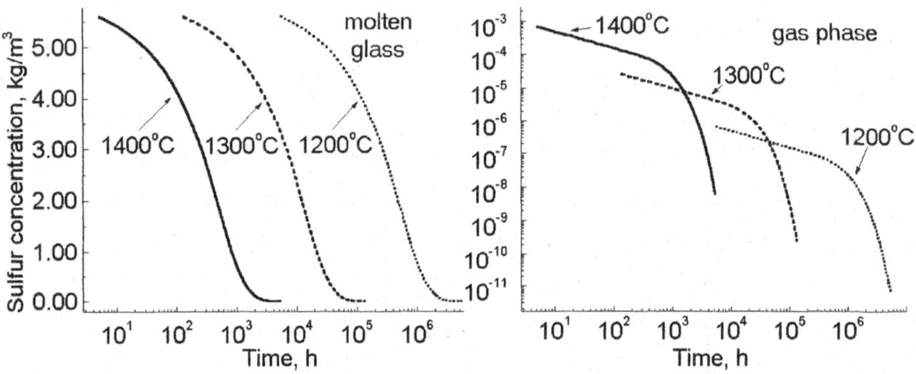

Fig. 2. Time dependence of concentration of SO_3 in molten glass and in flowing gas in the course of its diffusion into surrounding gas atmosphere.

The results of calculation show us the possibility to register a noticeable decrease of sulfur content in molten glass during 10-100 hours of heat treatment in a flow of gas initially containing no sulfur only at 1400°C. At lower temperatures one can observe changes in concentration of SO_3 during very extended heat treatments only.

In Fig. 2 (the right graph) we can see the time dependence of sulfur concentration in effluent gas. This characteristic can be used as analytical signal for the study of degassing of the molten glass at stationary conditions. During 100 hours at 1400°C, the content of SO_3 in gas phase changes by an order of magnitude.

System III. Exchange by SO_3 between molten glass and a gas atmosphere during forced bubbling of the melt by gas (enriched or depleted by SO_3 with respect to molten glass).

Let us take the concentration of SO_3 dissolved into the molten glass as equal to C and that into gas bubbles as equal to C_a correspondingly. In the course of diffusion of SO_3 from the molten glass into the bubbles the whole mass of sulfur remains constant. In this situation the concentrations of SO_3 in the melt and in the gas (inside bubbles) are connected to each other by the following equation:

$$\varepsilon C_a + (1-\varepsilon)C = \text{const} \tag{7}$$

Where: ε - is the volume fraction of bubbles in the melt.

Change of concentration of sulfur inside a bubble can be described by the following known equation:

$$d_t C_a = \frac{3Dh}{R^2}\left(\frac{C}{h} - C_a\right) \tag{8}$$

Where: R – is the bubble radius.

Combining equations (7) and (8), and assuming that the radius of the bubble and the volume fraction of the bubbles at the bubbling change slightly, we obtain the following equation, which describes change of sulfur concentration in the molten glass:

$$d_t C = -\frac{\varepsilon}{1-\varepsilon}\frac{3D}{R^2}(C - hC_a) \tag{9}$$

Integration of equation (9) gives us the following equation:

$$C = C_0 e^{-t/\tau} + hC_a\left(1 - e^{-t/\tau}\right). \tag{10}$$

Where: $\tau = \frac{1-\varepsilon}{\varepsilon}\frac{R^2}{3D}$ - is the characteristic time of degassing of the melt at bubbling.

Note that in Systems I, II characteristic times are equal to $\tau = \frac{\delta^2}{D}$, where: δ - is the molten glass layer thickness.

Below (in Fig. 3) the results of calculation of the process of the forced bubbling are presented. It is seen from Fig. 3 (left graph) that for practical degassing of the studied melt one should provide an intensive forced bubbling (volume fraction of the bubbles equal 0.2) of the melt specimen during 10 hours at 1400°C, 300 hours at 1300°C, and 10000 hours (!) at 1200°C. These results show us higher efficiency of exchange by sulfur in the course of bubbling in comparison with the holding of the melt in gas atmosphere. One should note, however, that in practice the continuous intensive bubbling could not be provided easily. Estimations show that the bubbling of the gas mixture through a single platinum tube having diameter 2 mm does not accelerate the process of degassing (saturation) of the molten glass sufficiently in comparison with the passing of the same gas mixture when the platinum tube is raised over the surface of the melt.

Thus one can complete the practical degassing or saturation of the molten glass by sulfur in the course of bubbling within a realistic time periods at temperatures over 1350-1400°C.

System IV. Extraction of SO_3 from molten glass (supersaturated by sulfur) into small bubbles.

Below one can find a mathematical description of the process of evolution (changes of size, content of sulfur inside and rate of rising) of bubbles spontaneously incipient inside the molten glass. Here we don't describe the process of formation of a bubble from bubble nucleus but start with the stage of existence of a small bubble, which grows due to diffusion of sulfur from the surrounding molten glass.

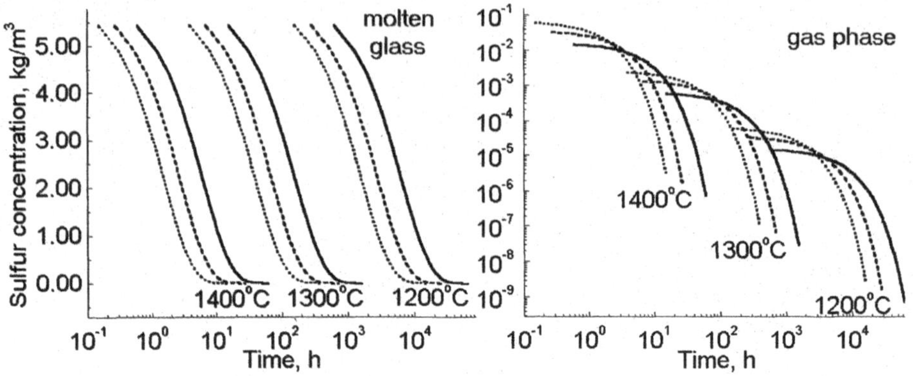

Fig.3. Degassing of molten glass at different fractions ε and radius R of bubbles. Gas flow - 1 l/h (61 cubic inch per hour); melt volume - $\delta = 16.39$ cm^3 (1 cubic inch); $\varepsilon = 0.1$, $R = 1$mm (solid line); $\varepsilon = 0.2$, $R = 1$mm *(dash line)*; $\varepsilon = 0.1$, $R = 0.5$mm (dot line).

The increase of mass of a bubble rising in the molten glass is controlled by the diffusion of SO_3 into the bubble. This process can be described by the following equation:

$$d_t\left(C_a \frac{4}{3}\pi R^3\right) = 4\pi R^2 D \frac{C - hC_a}{R}, \tag{11}$$

Assuming that the concentration of the gas (SO_3) inside the bubble (C_a) does not change. That assumption is quite correct for big enough bubbles (having diameter 0.1 to 1 mm) rising in 25 mm (~1 inch) thick layer of molten glass). We obtain the following law for the growing of the bubble:

$$R^2 = R_0^2 + 2D\frac{C - hC_a}{C_a}t \approx 2D\frac{C - hC_a}{C_a}t, \tag{12}$$

Where: R_0 – is the initial radius of the bubble. At the beginning of the growing the bubble is considerably smaller than at the end (bursting). C_a – the concentration of SO_3 inside the bubble, at which the pressure inside the bubble $P = C_a \frac{\tilde{R}T}{\mu}$, is equal to the gas pressure in the surrounding atmosphere.

In the course of the growing of the bubble it rises under the Archimedes's buoyant force at rate:

$$d_t x = \frac{2}{9}\frac{\rho g R^2}{\eta} \tag{13}$$

Where: ρ - is the density of the molten glass; g - is the acceleration of gravity; η - is the molten glass viscosity.

Substituting the equation for the bubble radius into equation (13) we obtain:

$$d_t x \approx \frac{4}{9}\frac{\rho g}{\eta} D \frac{C - hC_a}{C_a} t \tag{14}$$

From equation (14) the time of rising of the growing bubble can be determined:

$$t_{rise} \approx \sqrt{\frac{9\eta\delta C_a}{2\rho g D(C - hC_a)}} \tag{15}$$

The maximal radius, which the bubble attains at the melt – gas interface can be determined from equations (12), and (15) in the following manner:

$$R_{rise} \approx \sqrt{2D\frac{C - hC_a}{C_a} t_{rise}} \tag{16}$$

Each bubble extracts and removes from the molten glass a small portion of SO_3:

$$m_a = C_a \frac{4}{3}\pi R_{rise}^3 \tag{17}$$

Rate of the ablation of sulfur by all bubbles is determined as follows:

$$d_t M_a = \frac{nm_a}{t_{rise}} = \frac{C_a}{t_{rise}} n \frac{4}{3}\pi R_{rise}^3 = \frac{C_a}{t_{rise}} \varepsilon \delta^3 \tag{18}$$

Where: ε - is the volume fraction of the bubbles in the molten glass.

Thus we obtain the following equation for computation of the concentration of SO_3 in the molten glass:

$$d_t M = \delta^3 d_t C = -d_t M_a = -\frac{C_a}{t_{rise}} \varepsilon \delta^3 \tag{19}$$

We can take the equation for the time of the rising of the bubble (15) and finally obtain:

$$d_t C = -\varepsilon \sqrt{\frac{2\rho g D(C - C_a h) C_a}{9\eta\delta}} \tag{20}$$

Parameters used for the simulation of the extraction of sulfur into the gas phase by rising bubbles were as follows. The melt viscosity – ln (η, Pois) = -4.15 + 10373/T, the content of sulfur in the gas phase - C_a = 0.9 · 10^{-3} kg/m^3, the melt density ρ = 2200 kg/m^3. The remaining parameters are the same as for System I.

As it is seen from Fig. 4 (left graph) the process of degassing of the molten glass by means of spontaneous blistering is about as intensive as that by the intensive forced bubbling. If one, however, doesn't state a problem of complete degassing (saturation) of the molten glass but needs to determine just the beginning part of the time dependence of concentration of sulfur in the melt, the described method can be used successfully as the simplest one. In case of registration of the beginning part of the dependence *C=f(t)* (the first one third of the logarithmic curve) one requires ~2 (at 1400°C), ~40 (at 1300°C), or ~900 hours (at 1200°C); that sounds realistic for the temperatures exceeding 1300°C.

Analysis of the content of sulfur in the effluent gas cannot give us accurate information about the state of the studied system. The curve $C_a=f(t)$ is not informative enough due to a small slope of the curve during down to a practical degassing of the molten glass (see Fig. 4 – right graph).

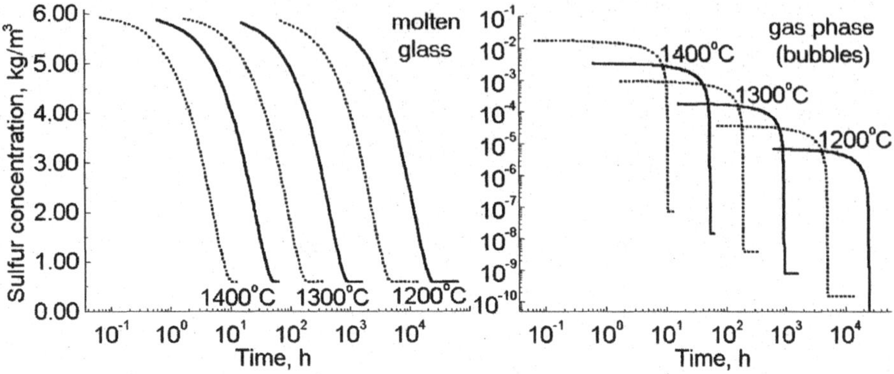

Fig. 4. Degassing of the molten glass into small bubbles at different fractions (ε) of bubbles: $\varepsilon = 0.1$ (solid line) or 0.5 (dot line). Melt volume – $\delta^3 = 16.39$ cm³ (1 cubic inch).

EXPERIMENTAL (DIFFUSION COEFFICIENTS)

Below experimental techniques used for studies of SO_3 solubility and diffusion in soda-lime-silicate glass melts are described. Experimental glasses were melted from pure-grade batch materials in an electrical furnace in platinum crucibles at 1400 - 1450°C during 1.5 hours. Composition of the base glass (in wt. % by chemical analysis) was as follows: SiO_2 – 73.3, Al_2O_3 – 0.1, CaO – 9.1, MgO – 4.0, Na_2O – 13.4, K_2O – 0.1. Base glass with no sulfur added is denoted as glass #1, with 0.65 wt. % of SO_3 added – as glass #2, with 0.97 wt. % of SO_3 added – as glass #3. Distribution of SO_3 in glass specimens was analyzed by an electron microprobe. Average content of SO_3 in glass specimens was determined by chemical analysis.

For studies of diffusion of SO_3 by the two-layers technique a graphite experimental cell was used (see Fig. 5). In order to protect graphite against an intensive oxidation at the high temperatures covered, the outer surfaces were coated by special high-temperature coating.

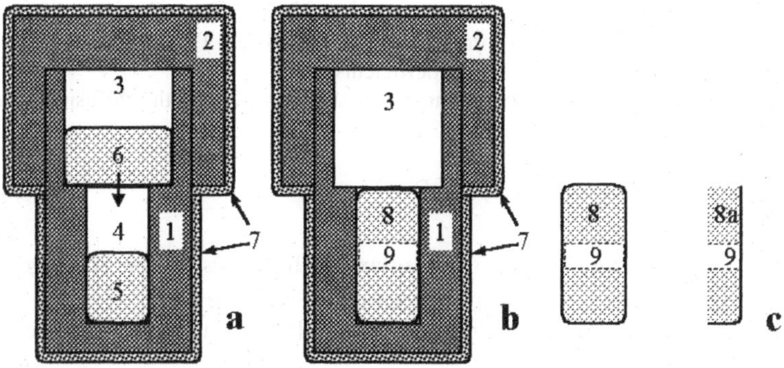

Fig. 5. Graphite cell used for study of diffusion of SO_3. Cutaway view of glass specimens before (a) and after (b) heat time. Preparation of sample for electron microprobe (c). 1 - graphite crucible; 2 - graphite cover; 3 - top cylindrical chamber; 4 - bottom cylindrical chamber; 5 - piece of sulfur-free glass; 6 - piece of sulfur-containing glass; 7 - protective coating on graphite surface; 8 – joint sample (after heat treatment); 8a – half-cylindrical sample ready for electron microprobe analysis; 9 – zone of diffusion of sulfur.

Two cylindrical pieces of glass (5, 6) were positioned into the inner cylindrical chamber of the graphite crucible (1) (see Fig. 5). The piece of sulfur-free glass was positioned at the bottom of the chamber (4). The piece of sulfur containing glass was at the top (3). In the course of heating the pieces of glass (5, 6) melted. The drop of the molten glass at the top flowed into the bottom section under the force of gravity. After the heat treatment and the subsequent cooling, the joined sample (8 in Fig. 5) was taken out from the graphite crucible. One side of the sample was ground and polished in order to obtain the half-cylindrical sample (8a in Fig. 5) used for electron microprobe analysis.

Distribution of sulfur within a zone of diffusion was determined by electron microprobe. Example of the distribution of sulfur measured by electron microprobe is given in Fig. 6.

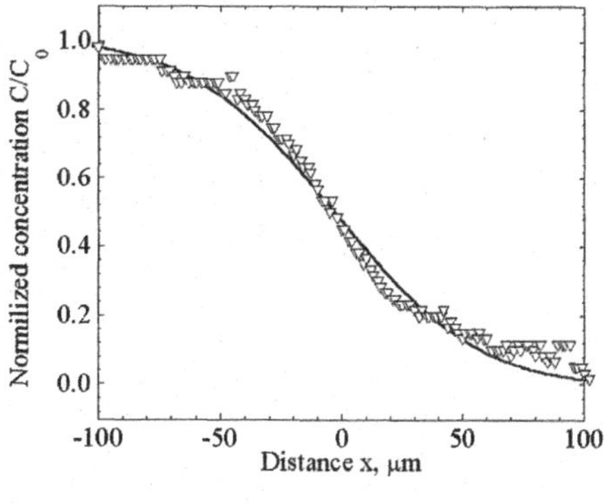

Fig. 6. Distribution of sulfur within zone of diffusion measured by electron microprobe. Experimental points are shown by open triangles. Solid line presents distribution of sulfur calculated as described below in this Section.

The minimal time of the heat treatment was equal to 2 hours (time of diffusion process). The maximal time required for joining two drops was equal to 30 sec. Thus, the time of flowing can be neglected in the calculation of diffusion coefficients.

Let us suppose that the process of mass transfer of sulfur in glass is the diffusion process that obeys to the following law:

$$\vec{j} = -D\nabla C \tag{21}$$

Where: $\vec{j}$ - is the vector of density of mass flux of sulfur; ∇- is the operator of gradient; D - is the coefficient of diffusion; C - is the partial density of SO_3 in the glass.

We assume that the coefficient of diffusion D depends on the temperature of the glass melt only.

To find the values of the diffusion coefficient of sulfur from the diffusion profiles of the kind presented in Fig. 6 we do the following.

Let us consider the evolution of the jump in concentration determined in the initial moment. That problem can be described by the equation of diffusion

$$\partial_t C = D\partial_{xx}^2 C \tag{22}$$

Where: t - is the time; x - is the distance, directed normally to the boundary of the samples. With the following boundary and initial conditions:

$$\begin{aligned} &\text{at}\, x \to -\infty \quad C \to 1, \\ &\text{at}\, x \to \infty \quad C \to 0, \\ &\text{at}\, t = 0 \quad C = \begin{cases} 1 & x < x_0 \\ 0 & x > x_0 \end{cases}. \end{aligned} \tag{23}$$

Where: x_0 - is the location of initial concentration jump.

Note that because the coefficient of diffusion does not depend on the density of the diffusing component its density can be normalized to unity.

By computing the mathematical problem given by Eqs (22) and (23) we determine the concentration C as function of time, coordinate x and of two parameters – coefficient D and location x_0, i.e. the function $C = C(x, t, D, x_0)$. Consider the difference between measured (C_{ex}) and calculated (C) distributions of concentration taken at the same time moment according to the following manner:

$$\Phi(D,x_0) = \sum_k \left[C_{ex}(x_k,t) - C(x_k,t,D,x_0)\right]^2 \tag{24}$$

Thus, the problem of calculation of the diffusion coefficient D is the problem of determination of the point of minimum of function Φ. The minimization of the function was performed by the gradient method.

It should be noted that the error of calculation of diffusion coefficients did not exceed 0.01 unit of *log D*.

TABLE I. VALUES OF DIFFUSION COEFFICIENTS OF SO_3 IN THE STUDIED MOLTEN GLASS OBTAINED BY USING THE TWO-LAYERS TECHNIQUE.

Number of individual specimens	T, °C	Time, hrs.	Mean diffusion coefficients and their random errors, *logD*, cm^2/s
12	1200	8	-7.24±0.08
12	1300	4	-7.08±0.12
12	1400	2	-6.83±0.14

EXPERIMENTAL (SOLUBILITY)

For determination of solubility of sulfur in molten glass, one should ensure uniform distribution of the studied substance within the liquid phase as well as equilibrium between the liquid phase and sulfur containing gas atmosphere. This state will be denoted below as "saturation". Let us formulate quantitative criteria for saturation of the melt by SO_3.

Below the major factors influencing the saturation process are analyzed in order to determine favorable conditions of experimental work.

Let us consider a process of dissolution of sulfur in a stationary layer of the molten glass (having thickness equal to δ) lying in a plane surface. Sulfur penetrates into the molten glass and proceeds through its top surface by diffusion. The time required for the achievement of the ratio between concentrations of sulfur at the bottom interface to that at the top equal to 0.9 (the diffusion time) can be calculated according to the following equation:

$$t_D = \frac{\delta^2}{D} \tag{25}$$

Where: δ - is the thickness of the glass layer, D - is the diffusion coefficient.

To achieve uniform distribution of SO_3 within the layer via the diffusion process the time t_{sat}, which is much longer than the diffusion time t_D, is required. Let us, however, formulate a clear practical criterion of saturation of the molten glass. At the moment the diffusion process begins, the concentration of sulfur within the layer of the molten glass is equal to zero.

$$C_S(x = 0, t) = 0 \tag{26}$$

Where: x - is the coordinate, directed into the bulk of the layer of the molten glass perpendicular to the surface of the layer; t - is the time.

Assume also that at the surface of the molten glass the concentration of sulfur is kept constant in the course of the whole process and equal to the equilibrium concentration of sulfur in the melt at the given content of sulfur in the gas atmosphere.

$$C_S^{equil}\left(x = 0,\ t\right) = 1 \tag{27}$$

The process of the saturation of the layer of the molten glass by sulfur will be defined by the following parameters:

Average concentration of sulfur within the layer of the molten glass.

$$\langle C\rangle_S(t) = \frac{1}{\delta}\int_0^{\delta} C(x, t)\, dx \tag{28}$$

Concentration of sulfur at the bottom interface of the molten glass layer.

$$C_S^{bottom}(t) = C(x = \delta,\ t) \tag{29}$$

Time dependencies of $\langle C\rangle_S(t)$ and $C_S^{bottom}(t)$ are presented in Fig. 7.

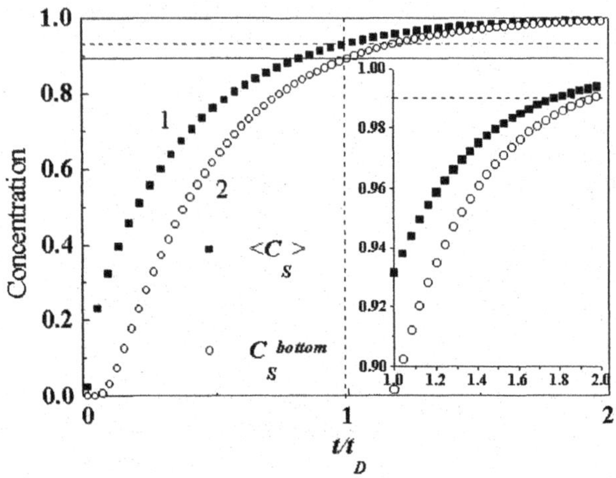

Fig. 7. Dependence of average concentration ($\langle C\rangle_S$) and concentration at the bottom (C_S^{bottom}) of diffusing compound on time to diffusion time ratio.

If $\langle C\rangle_S \approx C_S^{bottom} = 0.99\ C_S^{equil}$, one can formulate a prerequisite to the attainment of saturation in the form of equation (30). (This prerequisite seems to be reasonable from the practical point of view because difference between C_S^{bottom} and C_S^{equil} equal to 1% is less than the error of a majority of analytical methods.) This equation provides good accuracy of determination of time of practical saturation.

$$t_{sat} \geq 2 \cdot t_D = \frac{2 \cdot \delta^2}{D} \tag{30}$$

From equation (30) it appears that the time of saturation at constant temperature can be decreased only by a decrease of the thickness of the molten glass layer. In Fig. 8, the dependence of time of saturation of the studied molten glass on thickness of the glass layer at different temperatures is presented. t_{sat} depends on the thickness of the layer squared. Thus, a thinning by half of the layer leads to a decrease of the time needed for the saturation by a factor of four.

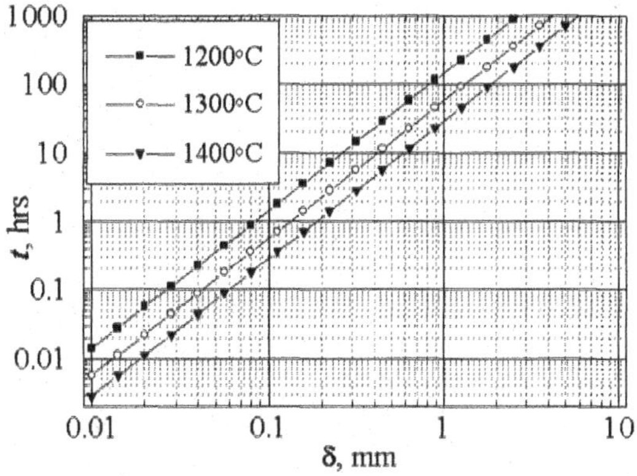

Fig.8. Dependencies of t_{sat} on the thickness of the molten glass layer and the temperature.

The thinning can be achieved by decrease of either a physical or an effective thickness of the layer of the molten glass. In order to decrease the effective thickness of the molten glass layer and the time needed for its saturation we applied two techniques. The first one was swinging of a drop of the molten glass on an inclined surface. The second one was exchange by SO_3 between the molten glass and a large number of small bubbles. In order to find optimal conditions of these experiments we used mathematical modeling.

Let us make a drop of a molten glass moving along an inclined plane surface. The drop is surrounded by SO_3 containing gas. The molten glass can wet or not wet the surface. In both cases (overturning or sliding), the effective thickness of the drop is half less than its physical thickness.

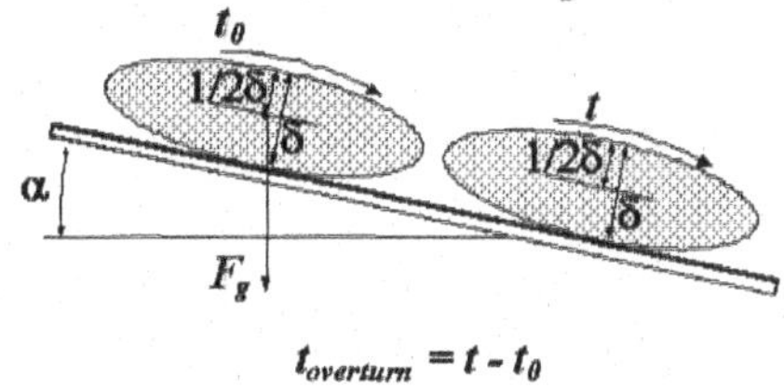

Fig. 9. Schematic presentation of process of moving of drop of molten glass along inclined plane surface.

Let us assume that the drop catches on the surface by only its bottom part (see Fig. 9). Under the action of the inclining the drop starts overturning, which leads to permanent exchange of the surface of the molten glass being exposed to the surrounding atmosphere. Swinging (moving forth and back) the drop creates conditions for faster saturation of the melt by SO_3.

Let us consider a small movement of the drop along the inclined plane surface. Then the rate of the movement (U), can be determined by the following equation.

$$U = \frac{\rho g \delta^2}{2\eta} \sin(\alpha) \tag{31}$$

Where: ρ - is the density of the molten glass; g - is the acceleration of free fall; δ - is the thickness of a molten glass layer (equal to the height of the drop); η - is the viscosity of the molten glass; α - is the angle of inclining of the plane to the horizon.

From equation (32) the time of the drop overturn can be estimated:

$$t_{overturn} = \frac{\delta}{U} = \frac{2\eta}{\rho g \delta \sin(\alpha)} \tag{32}$$

The equation (32) gives the time of the complete overturn of the drop. If the drop is positioned on the swinging surface then $t_{overturn}$ determines the minimal time required for keeping the plane surface inclined at an angle α. The time $t_{overturn}$ is proportional to the viscosity of the melt. If the time of the saturation is longer than the time of overturn ($t_{sat} >> t_{overturn}$) sulfur penetrates into the melt through the whole outer surface of the drop. In this situation, the effective thickness of the melt is equal to half of the drop height ($1/2\times\delta$), and the time of saturation decreases by a factor of four in comparison with that calculated for a stationary layer having thickness δ.

In Fig. 10 calculated dependencies of t_{sat} and $t_{overturn}$ on the height of the drop are presented.

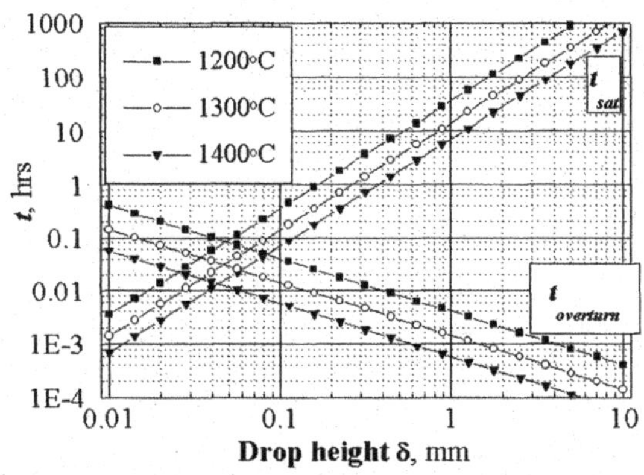

Fig.10. Dependence of t_{sat} and $t_{overturn}$ on the height of the molten glass drop and the temperature.

The parameters of the experiment on the drop swinging can be found from Fig. 10. For the drop of molten glass having the height equal to 1 mm, the times of the drop overturn (period of swinging of the plane), are equal to 7×10^{-4} (1200°C), 2×10^{-3} (1300°C), and 5×10^{-3} (1400°C) hours (~3, ~7, and ~18 seconds correspondingly). The times of saturation are equal to ~30 hours at 1200°C, ~12 hours at 1300°C and ~7 hours at 1400°C correspondingly.

Perhaps, the quickest way to achieve equilibrium between the molten glass and the gas phase is to exchange sulfur between the molten glass and a large number of small bubbles. The bubbles can be generated by heating the glass powder in sulfur containing gas atmosphere to the temperature corresponding to closing the pores. Let us assume that the pores are filled with sulfur containing gas mixture at room temperature. After heating to the high temperatures, the pores are closed and the bubbles filled with the sulfur containing gas are formed. The time required for achievement of the equilibrium between the molten glass and the gas atmosphere inside the bubbles can be calculated by the following equation:

$$t_D = \frac{l^2}{4D} \tag{33}$$

Where: l - is one half of the distance between the neighboring bubbles.

If the concentration of bubbles in the molten glass is high enough, a distance between neighboring bubbles is equal to their diameter ($l \approx d$). Let us take the diameter of bubbles ten times smaller than the thickness of the molten glass layer. The equation (33) allows comparing the time of the saturation of the layer of the bubble-containing molten glass with the bubble-free one. The bubble-containing layer is saturated a hundred times faster than the bubble-free one. Saturation of the molten glass can be achieved only if the bubbles do not rise to the inner surface of the molten glass too quickly. To evaluate the time of the rise of the bubbles let us take into account that Archimedes buoyant force is equilibrated by the force of drag resistance of Stokes. Characteristic rate of the rise U_A can be calculated by the following equation:

$$U_A = \frac{\rho g d^2}{18\eta} \tag{34}$$

Where: ρ - is the density of the bubble-containing molten glass: $\rho \approx (1-\alpha_\pi) \times \rho_{glass}$; g - is the acceleration of free fall; d - is the bubble diameter; η - is the molten glass viscosity.

Characteristic time of the rise of the bubbles having diameter d can be calculated by the following equation (35):

$$t_{rise} = \frac{\Delta}{U_A} = \frac{18\eta\Delta}{\rho g d^2} \tag{35}$$

Where: Δ - is the thickness of the layer of the molten glass.

Thus, for effective realization of the method, one needs to satisfy the following conditions:

$$2t_D \le t_{sat} < t_{rise} = \frac{\Delta}{U_A} = \frac{18\eta\Delta}{\rho g d^2}$$
$$t_D = \frac{d^2}{4D} \tag{36}$$

In Fig. 11 the dependence of the values of the diffusion time (t_{sat}) and the time of the rise of the bubbles (t_{rise}) on the thickness of the layer of the studied molten glass is presented.

Calculation was performed at temperatures (1200; 1300; and 1400°C) for the volume fraction of bubbles $\alpha_p = 0.3$.

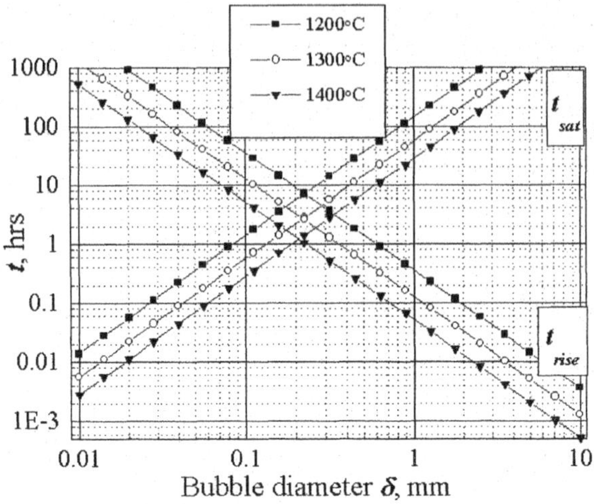

Fig. 11. Dependence of times t_{sat} and t_{rise} on size of bubbles rising into 10 mm thick layer of molten glass, and temperature.

According to equations (36) saturation of the molten glass is going effectively if the diffusion of sulfur from the bubbles into the molten glass proceeds faster than the rise of the bubbles. Thus from the practical standpoint it is necessary to have the bubbles smaller in diameter d than some maximal value d_{max} for which the second equation (36) becomes true. It is seen from Fig. 11 that for saturation of the studied molten glass by sulfur from the gas bubbles their diameter should be equal to or smaller than ~0.2 mm.

It can be seen from Fig. 9 that a 0.1 - 0.3-mm-thick layer of the studied molten glass can be saturated by sulfur in 1 - 10 hours. In order to obtain a thin layer of glass (melt) the following procedure was used. Pieces of glasses #1 and #2 were granulated in a ball-tube mill. Small grains of the studied glasses having diameter 0.3 - 0.7 mm were selected by using a sieve. The grains were washed from dust by ethyl alcohol and dried in air. A portion of the grains was put into a container (10×10×1 mm) made from 0.2 mm thick platinum foil (1 in Fig. 12). The container was heated at 900 - 950°C during ~1 hour. Thin layer of glass (2) forms at the bottom of the container. The container was positioned inside a quartz tube (3). The tube was positioned inside a high temperature furnace supplied with Pt-Rh heating element. The containers were heated at 1200, 1300, or 1400°C during 10, 5, and 2 hours correspondingly. A slow stream of air containing 1 vol. % of SO_2 passed through the tube during the heat treatment. The tube was taken out of the furnace and cooled down to room temperature. Sulfur content in the glass layer was determined by chemical analysis.

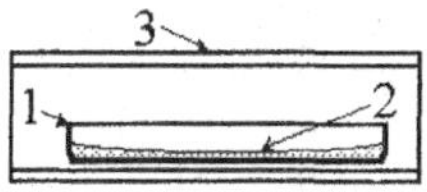

Fig. 12. Platinum container used for saturation of thin layer of molten glass by sulfur. 1 – platinum container; 2 – molten glass; 3 – quartz tube (fragment).

Results of determination of sulfur content in thin layers of the glass on platinum foil treated in air containing 1 vol. % of SO_2 are presented in Table 2.

TABLE II. VALUES OF CONCENTRATION OF SULFUR [SO_3] IN THIN LAYERS OF GLASS ON PLATINUM FOIL TREATED IN ATMOSPHERE OF AIR CONTAINING 1 VOL. % OF SO_2 AT HIGH TEMPERATURES.

Glass #	Number of individual specimens	T, °C	Time, hrs.	Average content of sulfur [SO_3], wt. %
1	3	1200	10	0.78±0.02
2	3	1200	10	0.82±0.02
1	3	1300	5	0.65±0.02
2	3	1300	5	0.63±0.02
1	3	1400	2	0.41±0.02
2	3	1400	2	0.34±0.02

For the saturation of the molten glass by sulfur using swinging droplet technique an experimental device was built. A scheme of the device is given in Fig. 13. The droplet of the molten glass (5) was positioned on a platinum plate (6) inside of a quartz tube (4). The tube (4) was connected with an inlet tube (2). The tubes (2) and (4) were made from silica glass. These tubes were used for three to five heat treatments. For each experiment new quartz tubes were taken.

To make the molten glass droplet swinging, the tubes (2, and 4) and the plate (6) were rotated through ~60 deg. ($\beta \approx$ 120 deg.) to the right and to the left (relative to its initial position). Periodical rotation was provided by a crank drive (9) that was moved by an electrical motor (8). Period of rotation (swinging) of the cell can be adjusted in the range from 0.5 to 10 sec. Amplitude of rotation of the cell is between point A and point B (see Fig. 13).

The slow stream of the gas mixture containing SO_2 (the flow of gas is equal to 20 cm^3 per minute) through the tube (2) into the tube (4) was supplied by a Teflon pump. In order to pump the corrosive gas mixture out of the furnace the outlet tube (3) was connected with a glass water-jet pump. For neutralization of the gaseous SO_2 and SO_3, which were presented in the outgoing exhaust gas, an absorbing flask filled with 5% water solution of Na_2CO_3 was positioned between an outer end of the outlet tube and the water-jet pump. Temperature inside the furnace was controlled by a Pt-PtRh thermocouple (10).

The experimental procedure of the saturation of the molten glass by SO_3 consisted of the following steps. A piece of glass (or a portion of glass powder) was positioned on the platinum plate (6) within the tube (4) at room temperature. The tube was inserted into the furnace. The furnace was heated up to the required temperature. At high temperatures, the piece of the glass on the platinum plate (6) was transformed into a drop of the molten glass. The crank-drive (9) started moving almost simultaneously with the pump (to start pumping of the gas mixture through the system). Swinging of the drop of the molten glass took place during the required period. After that the crank drive and the pumping of the gas mixture were stopped. The outlet tube (3) was taken out of the furnace. After cooling down, the platinum plate with the droplet was taken out of the tube (4). Sulfur content was determined by chemical analysis.

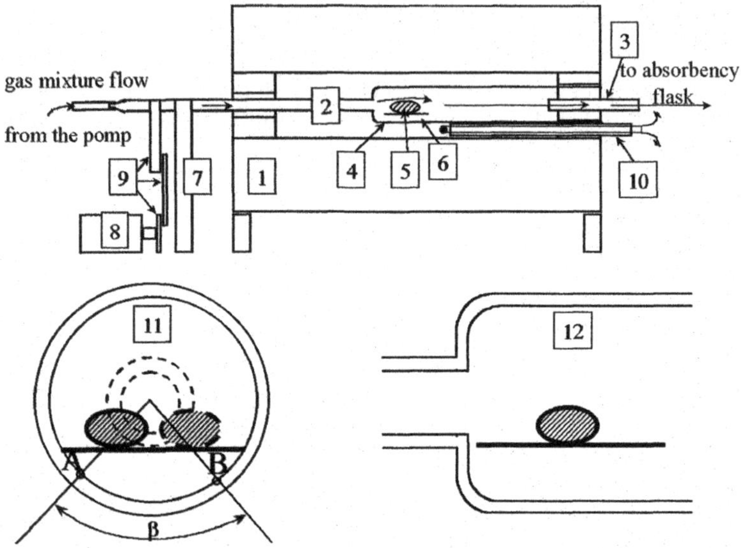

Fig. 13. Experimental device used for saturation of swinging droplet from gas phase. 1 - high-temperature furnace; 2 - inlet tube; 3 - outlet tube; 4 – tube for inserting the samples 5 – platinum plate; 6 - drop of molten glass; 7 - inlet tube support; 8 - electrical motor supplied by reducing gear; 9 - crank drive; 10 - thermocouple; 11 - cross-section; 12 - longitudinal section.

In Table 3 conditions of experiments on saturation of the molten glass #1 by sulfur using swinging drop method and contents of sulfur in the samples after heat treatment are presented.

TABLE III. VALUES OF CONCENTRATION OF SULFUR [SO_3] IN SAMPLES OF GLASS #1 TREATED IN AIR CONTAINING 1 VOL. % OF SO_2 AT HIGH TEMPERATURES BY USING THE SWINGING DROP TECHNIQUE.

Number of specimens	Height of droplets	T, °C	Time, hrs.	Average content of sulfur [SO_3], wt. %
2	~1 mm	1200	30	0.74±0.02
2	~ 1 mm	1300	15	0.59±0.02
2	~ 1 mm	1400	8	0.31±0.02

For formation of small bubbles within the molten glass volume, a powder of the studied glass was used. Small crucibles filled with the powder were positioned inside a vacuum dessicator connected with a rotation vacuum pump for evacuation of air. After evacuation to residual pressure of 0.005 *atm.* the pumping was stopped, and the dessicator was slowly filled with air containing 1 vol. % of SO_2. The crucibles were heated up in air to the temperature T_{close} corresponding to a moment of closing of pores. At further heating, the closed pores were transformed to the bubbles.

Having an average size of particles of the powder d_p one can estimate the average size of the bubbles d_b. If, for example, $d_p = 0.05$ mm, one can suppose that the size of the closed pores was about the same. After closing the pores, the gas mixture with which they were filled started expanding. In the course of heating of the molten glass from T_{close} up to the temperature of heat treatment the volume of the bubbles increased by a factor of $k = T_{treat}(\mathrm{K})/T_{close}(\mathrm{K})$. For a window-type glass T_{close} is equal to ~1000K. For $T_{treat} \approx 1700$K we have $k = 1.7$. Thus, the average size of the bubbles in the volume of the molten glass at 1400°C is equal to ~0.08 mm.

For the height of the molten glass layer equal to 1cm values of t_{rise} are equal to 70h (at 1200°C), 30h (at 1300°C), and 10h (at 1400°C) (see Fig. 11). These values exceed those of t_{sat} at the same temperatures. Thus, the equilibrium between sulfur in the gas phase located inside the bubbles and the surrounding molten glass can be achieved well before the time when all the bubbles rise to the top glass interface.

As it was shown already solubility of sulfur in the studied glass within the temperature range from 1100 to 1500°C varies within a range ~0.3 - ~0.8 wt. %. A simple estimation shows that at the volume fraction of the bubbles (filled with pure SO_3) inside the volume of sulfur free molten glass equal to 0.3 the concentration of SO_3 equal to maximum ~0.05 wt % can be reached. Thus, this process can be completed only if the initial glass is almost saturated or even supersaturated.

For saturation by rising bubbles method, glass #2 (containing 0.65 wt. % of SO_3) and glass #3 (containing 0.97 wt. % of SO_3) were taken. Glass #3 was prepared by filling the powder of glass #2 by pure SO_2 and heating the powder at 1200°C during 3 hours.

The studied glasses were powdered in a ball mill. Fractions of the powders with an average size of the particles equal to 0.05 mm were obtained by using a sieve. Portions of the powder weighing about two grams each were positioned into small crucibles made from high purity Al_2O_3. Three crucibles were positioned into a cassette made from porous SiO_2. The cassette was positioned inside a quartz tube. The tube was positioned inside a high temperature furnace. A slow stream of air containing 1 vol. % of SO_2 through the tube was provided during the whole heat treatment period.

Heat treatment of the glass samples was continued for 3-10 hours depending on the temperature (see Table 4). After taking the cassette out of the tube and cooling it down to room temperature, the glass was taken out of crucibles and analyzed for SO_3 content. Experimental conditions and results of the analysis are presented in the Table IV.

TABLE IV. RESULTS OF THE EXPERIMENTS ON EXCHANGE BY SO_3 BETWEEN THE STUDIED MOLTEN GLASS AND SMALL BUBBLES. BEFORE HEAT TREATMENT, GLASS POWDER WAS "SATURATED" BY SO_3 AT ROOM TEMPERATURE (SEE ABOVE IN THE TEXT). HEAT TREATMENT WAS HELD IN ATMOSPHERE OF AIR CONTAINING 1 VOL. % OF SO_2.

Glass #	Number of specimens	T, °C	Time, hrs.	Average content of sulfur [SO_3], wt. %
2	9	1200	10	0.84±0.02
3	9	1200	10	0.95±0.02
2	9	1300	5	0.69±0.02
3	9	1300	5	0.73±0.02
2	9	1400	3	0.42±0.02
3	9	1400	3	0.51±0.02

CONCLUSION

Mathematical modeling of exchange by SO_3 between soda-lime-silicate glass melt and gas phase at high temperatures can be used as an instrument for the studies of the process in different melt-gas systems having practical interest:

(I) melt layer - gas $[SO_3]^{gas} \Rightarrow [SO_3]^{eq1}$ $[SO_3]^{melt} \Rightarrow [SO_3]^{eq2}$.

(II) melt layer - gas $[SO_3]^{gas}$=const; $[SO_3]^{melt} \Rightarrow 0$ (degassing) or $[SO_3]^{melt} \Rightarrow [SO_3]^{eq2}$ (saturation).

(III) melt - gas flow (as forced bubbling) $[SO_3]^{gas} \neq$const.

(IV) melt - gas outflow (inside rising small bubbles).

In order to obtain data on diffusion coefficients and solubility of SO_3 in soda-lime-silicate melt at high temperatures several experimental techniques were developed. Preliminary calculations of SO_3 transport processes helped in finding optimal experimental conditions required for obtaining dependable experimental results. Values of SO_3 solubility obtained using different experimental techniques were in a good agreement with each other.

REFERENCES

[1] R. Bruckner, *Glastcch. Ber.*, **34** [11] 515-528 (1961)
[2] R. Bruckner, *Glastcch. Ber.*, **35** [2] 93-105 (1962)E

Chapter 5

RADIATIVE THERMAL CONDUCTIVITY OF MELTS

Oleg A. Prokhorenko
Laboratory of Glass Properties LLC
5a Varshavskaya Street, bldg. 50, 3rd floor
St. Petersburg, 196084
Russian Federation

ABSTRACT

Computer modeling has a fair chance to become a powerful tool for optimization of glass melting, and processing. Calculation of heat transfer by radiation in molten glasses is a key part of modeling programs. For a correct description of transport of radiation, dependable data on absorption spectra of a molten glass are required. In 1994 - 1997 in the Laboratory of Glass Properties under support of the NSF Industry/University Center for Glass Research (CGR) at Alfred University (Alfred, NY) an original technique for measurement of absorption spectra of molten glasses has been developed. Later (in 1998 – 2002) the devices and methods were improved to study commercial glasses predisposed to quick irreversible changes at high temperatures. In this chapter different aspects of development of reliable experimental technique for measurements of absorption spectra of the molten glasses are presented. Major sources of experimental errors are analyzed and measures of their prevention or compensation are described. The results of the study of high-temperature spectra and radiative thermal conductivity characteristics of six types of commercial glasses (container, float, E- and wool-types fiber, color TV panel, low-expansion borosilicate) are presented.

INTRODUCTION

Currently, thermo-physical computations become especially significant in connection with the development of computer models of the basic technological operations in glass manufacturing. At glass melting, forming, annealing, and tempering, heat transfer by radiation contributes greatly in heat exchange processes. Therefore, the absorption spectra in the wavelength range 0.6 - 3.8 μm at temperatures from ~500 to 1600°C are important physical properties of glass-forming melts, required for modeling heat transfer by radiation in these semi-transparent media. The high-temperature spectrophotometry also represents a considerable interest in studying the influence of temperature on the shift of chemical equilibrium involving variable-valence ions and hydroxyl groups having absorption bands in the near-IR.

Multi-component commercial glass melts are not conventional objects for absorption spectrophotometry. The absorption spectra and radiative thermal conductivity (more correctly - the radiative component of heat transfer) of glass-forming melts depend on both the temperature and features of the absorption of a glass-forming melt, which are governed not only by the basic composition of glass but also, to a greater extent, by coloring impurities and melting conditions. Consequently, the dependence $\boldsymbol{K}=f(\lambda, T)$ (where $\boldsymbol{K}$ is the absorption coefficient in cm^{-1}, λ is the wavelength in μm, and T is the temperature in °C) should be experimentally determined for each new type of commercial glass. Moreover, the structure of all commercial glasses is frozen at the time of forming when redox (reduction-oxidation) reactions are often far from the state of equilibrium. Heating to the melting temperatures reactivates red-ox processes in the molten glass, which can change its absorption spectra significantly.

To measure the absorption spectra of a melt, a cell made of heat-resistant materials (platinum alloy or ceramic body, quartz glass or sapphire windows) are used. The melt absorption is determined from the absorption of the cell after a number of corrections for single and multiple light reflections from the interfaces melt-window, window-air, melt-air, etc. If only one specimen is used, corrections are calculated by formulas of classical optics and are often used over the entire temperature range up to 1450°C. By neglecting changes of all reflecting interfaces with the temperature one risks to generate significant systematic error. Using two specimens having considerably different thickness provides compensation of light reflection from all interfaces but only in case of identical design and positions of both cells. The better way, offered in [20] is to determine cell constants within each of three temperature ranges. This technique is also described below in the present chapter.

A comparatively small number of works concerning experimental studies of the absorption spectra of glass-forming melts at high temperatures have been published in the last forty-five years [1-16]. Instruments used in these works were described very briefly, the experimental procedures were only outlined, and specific time-temperature regimes were often not given at all. The exception was the work by Genzel [6], who described in sufficient detail an experimental setup. However, this description can be useful only for the general perception of the problem, because the technique described in [6] for measuring the high-temperature spectra of melts was extremely complicated, and the experimental results were affected by a large number of virtually uncontrollable factors. In most of the published works, the authors presented only one spectrum measured at a given temperature. None of them compared the results of at least two replicate measurements for a glass of the same composition or gave error bars. In a few works the errors of measurement of the absorption spectra at high temperatures were mentioned.

A recent paper by Nijnatten and Efimov [16] describes measurements of high-temperature absorption spectra of TV-panel glass using the most complicated version of two-thickness method. A massive experimental cell is filled with up to 0.5 kg of glass. It took 1 hour to measure spectrum at one temperature. The authors measured the absorption spectra at 1400°C repeatedly for a period of 6 hours and didn't find any changes in the measurement results. However, they have presented neither the results of parallel measurements nor comparison between spectra measured before and after high-temperature exposition. Spectra of two different specimens of the same glass were not presented in this paper.

In three works [1-3], the absorption spectra of glass-forming melts were measured prior to and after the heating to the maximum temperatures. Fig. 1 demonstrates the absorption spectra obtained in [1-3]. A considerable decrease in the absorption over the entire spectral range clearly indicates the oxidation of Fe^{+2} to Fe^{+3}, and partial loss of structural water. These irreversible transformations occurred in iron-containing melts upon its high-temperature treatment and substantially reduce the value of the results obtained.

An attempt to analyze the errors in measurement of the absorption spectra at high temperatures was made by Sergeev and Men' [4]. They considered a number of instrumental errors of a spectrophotometer. The problem of calculating the coefficients of single and multiple reflections of a light beam from the glass-air interface was discussed in sufficient detail. This analysis is possibly proved to be correct in measuring the absorption spectra of crystals below melting temperature and glasses below T_g. However, it ignores the temperature effect on the reflection at the glass-air and melt-transparent window interfaces. No influence of various irreversible changes taking place in a melt at high temperatures was considered as well.

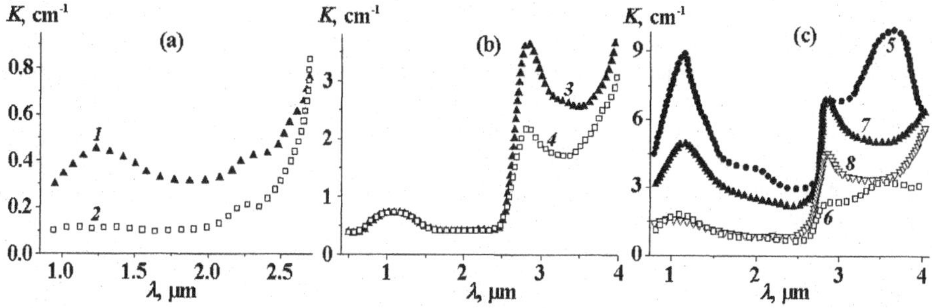

Fig. 1. Absorption spectra of sodium-calcium silicate glass melts containing (a) 0.13 [1], (b) 0.2 [2], and (c) 0.86 wt. % (***FeO*** + ***Fe₂O₃***) [3] at different temperatures °C (1) 600, (2, 3, 4) 800, (5, 6) room temperature, (7) 1070, and (8) 1080. Spectra registered before (solid symbols) and after (open symbols) exposure at the highest temperature in the experiment.

Consequently, the data available in the literature should be used with a great caution even for an approximate estimation of the radiative thermal conductivity of glass forming melts.

Below a number of problems associated with the correct determination of the absorption coefficient for glass-forming melts are considered.

A measuring cell used for recording the absorption spectra of melts at high temperatures is one of the most important elements of an experimental setup. Platinum or platinum-rhodium frames with embedded sapphire or silica glass windows are assigned to cells proposed for the first time by Grove and Jellyman [3]. Later, Coenen [9], Wedding [10], and Blazek [14] applied similar cells without any significant changes.

Neuroth used a cell consisting of a platinum mirror immersed in a platinum crucible [1]. The light beam enters the cell at some angle, reflects from the plane mirror, and leaves the cell. Raising and lowering the crucible, while the mirror position is fixed, change the thickness of the molten glass layer. Franz [9] offered a cell, which was an open sapphire crucible with a polished bottom. The light beam passed through the melt layer in the vertical direction.

The cell used in [16] has a unique combination of features. The cell is assembled inside a large 150mm x Ø85mm crucible, made of alumina ceramics. The crucible has a sapphire window in its bottom part. It holds from 0.1 to 0.5 kg of molten glass. Thickness of the melt layer can vary. A miniature oxygen sensor provides continuous monitoring of red-ox potential of the melt.

There are a number of essential shortcomings connected with the cell's design that are responsible for the non-repeatability of the obtained results.

The cells made of platinum-rhodium alloys have long service life at high temperatures. However, the presence of active catalysts of red-ox reactions can increase many-fold the rate of oxidation of transition metal ions (Fe^{+2}, Cr^{+3}, Mn^{+2}, etc.) by oxygen in air and decolorizing or re-coloring the melts. Redox reactions often involve water molecules, and are accompanied by degassing of the melts.

Open design of some cells results in the volatilization of glass components such as boron oxide, alkali oxides, fluorine and water. The elimination of hydroxyl groups leads to an appreciable decrease in the absorption in the range 2.5 - 3.8 μm. Evaporation of B_2O_3 and Na_2O, can cause surface crystallization followed by devitrification of a molten glass layer.

According to the authors the cell used in [16] provides control of the red-ox state and ensures low volatilization of the melt components. However, its large size doesn't allow heating and stabilizing the specimen quickly enough to measure absorption spectra at the temperatures above 1200 - 1250°C before irreversible processes change the molten glass absorption spectrum. The method described in [16] is applicable only for glasses of little sensitivity to long high-temperature exposure.

In the development of a new experimental technique one should shorten the time of holding glass forming melts within temperature ranges where the melt undergoes different irreversible changes (the ranges with a high rates of nucleation and crystallization, intensive release of dissolved gases, high rates of redox reactions, and some other processes). This goal can be achieved by combining a rapid-scanning spectrophotometer, low-inertia powerful furnace, and the optimal temperature-time schedule for the spectra recording. It seems clear that each new commercial molten glass having different composition or/and melted differently should have its own temperature ranges of "instability". A measuring cell should not cause an additional thermal lag, accelerate irreversible changes, or give rise to new changes in a melt at high temperatures. The original cell developed by the authors for measuring the light absorption in melts was described in [17, 18].

Let us formulate the general requirements to a laboratory technique that can be used for quantitative studies of near-IR absorption spectra of glasses and glass forming melts over a wide temperature range.

(1) In order to obtain dependable data on near-IR absorption spectra of glasses and melts, it is necessary to design a spectrophotometer providing the required accuracy, stability of recording the absorption spectra at room and high temperatures, and flexibility to vary temperature regimes and scanning modes. It should be noted that at the present time reference samples are available for calibration at room temperature only. Reasonable steps should be taken to develop high-temperature standards of optical absorption in the near future.

(2) A measuring cell used for holding a melt specimen should survive numerous heating-cooling cycles without noticeable worsening of its optical and mechanical characteristics, and should not provoke irreversible changes in a melt at high temperatures.

(3) The light reflection at the specimen surface depends on many factors, including temperature. Calculating the reflection loss may lead to noticeable errors. Recording the differential spectra at high temperatures is direct but the most complicated method for determination of the melt volume absorption. It is preferable to divide the whole temperature range for several subranges and determine cell constants, which includes components connected with reflection losses, for each subrange.

(4) A special technique should be developed to measure the spectra at high temperatures. This technique can be based on the accurate determination of absolute absorption coefficients at

room temperature [$K_{room} = f(\lambda)$] and their changes [$\Delta K = f(\lambda, \Delta T)$] upon heating specimens in a certain temperature range. The reproducibility of spectral absorption coefficients from (heating-cooling) cycle to cycle and from experiment to experiment can be considered as a criterion for the correctness of recording temperature changes in the absorption spectra of glasses and melts.

METHODOLOGY

Due to the fact that no high-temperature spectrophotometer is produced commercially available, we have developed and built a series of devices (SF-2-LGP, SF-3-LGP, and SF-4-LGP). These devices allow registering high-temperature spectra of glasses and glass forming melts at temperature ranges from room temperature up to 1500°C. For assembling the spectrophotometers we used commercial parts as well as custom ones. Common features of the devices are as follows. For obtaining monochromatic radiation a single-grating monochromator with a set of glass filters (fixed in a turret) is used. Scanning within wavelength range from 0.6 to 3.8 μm is provided by diffraction gratings having 300 strokes per mm. Moving parts of the spectrophotometer are supplied by step-by-step electrical motors. As a source of light, a commercial quartz halogen lamp is used. As a detector of near infrared radiation, a detector FYO-612-8 (with PbSe sensitive element) is used. The sensitive element is cooled down to -10°C by built-in thermo-battery. The main control unit controls all movable parts, and reads signals of detectors and thermocouples. A separate temperature processor PTR-105 controls the furnace temperature.

Specific features of the devices are briefly described below. SF-2-LGP device allows for measuring spectra of 8 specimens at room temperature or 2 specimens at high temperatures in fully automated mode. Spectrophotometer SF-3-LGP provides fast spectra scanning (under 1 min within the range 0.6 - 3.8 μm), and rapid specimen heating and cooling (at rates up to 30K/min). SF-4-LGP device is additionally supplied with the second light source made of silicon carbide rod, the second detector (bolometer). The monochromator of SF-4-LGP device is equipped with interchangeable diffraction gratings. Two gratings having 150 and 75 strokes per mm are used for registering spectra in the middle infrared range.

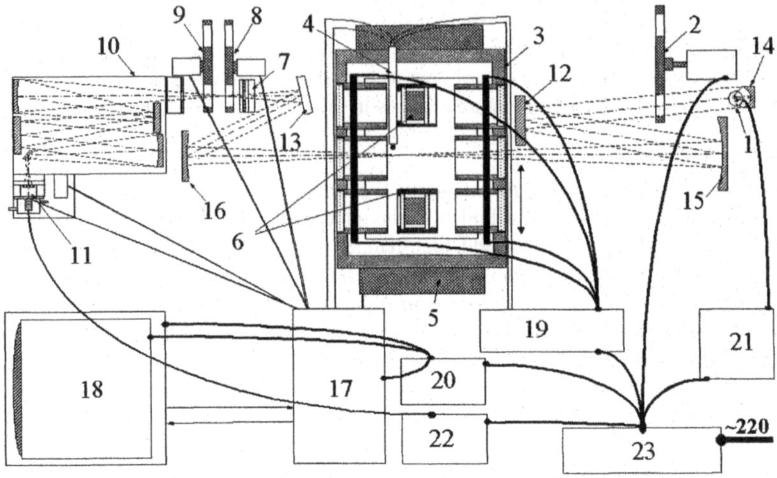

Fig. 2. Typical scheme of high-temperature spectrophotometer. 1 - light source; 2 - light interrupter; 3 - furnace; 4 - thermocouple; 5 - movable table; 6 - cell; 7 – specimen holder; 8 - six specimens holder; 9 – filters holder; 10 - monochriomator; 11 – detector; 12-16 – mirrors; 17 – main control unit; 18 – PC; 19 – furnace control unit; 20-23 – power supplies.

In Fig. 2 a typical scheme of spectrophotometer is given. Light of the lamp (1) is collected by a parabolic short-focus mirror (14), and diverges at a small angle. Before the light source the mechanical light interrupter (2) is fixed. Frequency of interruption of the light beam is 300 Hz (for SF-2-LGP and SF-3-LGP) or 10 Hz for SF-4-LGP.

Then the light beam is reflected by the flat mirror (12) and focused by the parabolic mirror (15) into the high-temperature furnace (3), where one (SF-3-LGP, SF-4-LGP) or two (SF-2-LGP) ceramic cells (6) are positioned. Six holes in the furnace front and back walls are closed by removable ceramic plugs with sapphire windows. The furnace is fixed on the movable table (5). The light beam outgoing from the furnace goes through the system of mirrors (13) and (16) and is focused onto an input slot of the monochromator. On the output slot of the monochromator the detector of radiation (11) is fixed.

All devices were designed as one-beam spectrophotometers. Intensities of light I (passing through the object), and I_0 (reference signal) are registered one after another. Inserting (removal) of the measured specimen in the light beam is made as follows. At the registration of single spectrum at room temperature the specimen is fixed in the holder (7) and taken out of it manually. In the ring holder (8) seven individual specimens can be fixed simultaneously. The specimens are introduced in the light beam one after another automatically. The eighth position of the ring mount is left empty in order to have possibility to register the reference signal. In order to measure spectra above the room temperature specimen(s) fixed in ceramic cells are positioned inside the high-temperature furnace. Reference signal is measured by introduction (removal) of the specimens in the light beam by moving the furnace (3) perpendicular to the optical axes of the device.

As mentioned above, the specimens in ceramic cells are placed in the high-temperature furnace. Fig. 3 depicts two variants of cells designed for the recording of the absorption spectra of glasses and melts. The simplest variant of cell (Fig. 3a) is used for measuring the absorption spectra of glasses at temperatures from room temperature to those corresponding to very high viscosities of the studied melt ($\eta \geq \sim 10^{10}$ P). In this cell, specimen (5) is placed in cylindrical frame (1) and is fixed with a ceramic lute. The cell displayed in Fig. 3b is intended for measuring the spectra of melts in the viscosity range from $\sim 10^{11}$ to $\sim 10^{2}$ P. The cell is equipped with two transparent windows (2) and (3) and ring (4). The window surfaces and also face ring surfaces are plane-parallel to within ±0.01 mm. Glass specimen (5) of required size is placed inside ring (4) between windows. Prior to the assembling, cell parts (2 - 5) are carefully washed and dried in air. Then, the cell parts are placed in frame (1) and outer window (3) is fixed with a special ceramic lute. The mounted cell is dried in air for 8 - 12 h. Parts (2 - 4) produce a gap for a melt, formed upon heating of glass disc (5) to a high temperature.

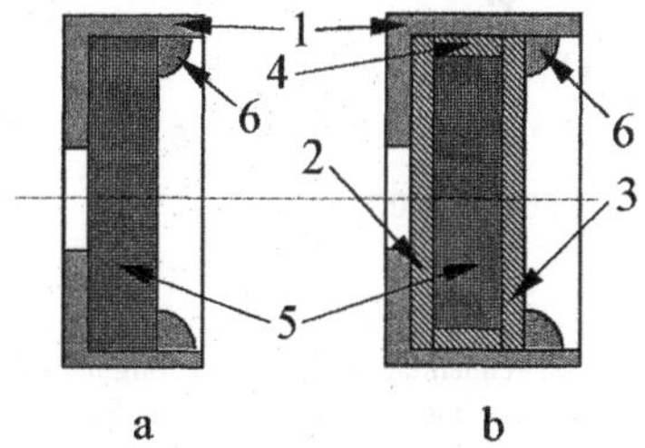

Fig. 3. Cells for measurements of spectra of glasses and melts (a) $\eta > 10^{10}$ P and (b) $10^{11} > \eta > 10^{2}$ P. 1 - ceramic frame; 2, 3 - leucosapphire windows; 4 - leucosapphire ring; 5 – glass (a) or melt (b) specimen; 6 – ceramic lute.

Windows (2) and (3) are made of high-purity leucosapphire, which has a very high melting temperature (~2200°C) and weakly absorbs in the studied spectral range at temperatures up to 1600°C. A 2-mm thick polished leucosapphire disk has optical density (at 3.4 μm) equal to ~0.07 at room temperature and ~0.08 at 1450°C.

Illustrative examples of applicability of the described cell for measurements of spectra of iron-containing melts synthesized under reducing conditions are presented in Figs. 4a and b. Fig. 4a depicts the temperature dependence of the absorption coefficient for the melt of window glass containing 0.4 wt.% of ***FeO*** measured in air. The measurements were carried out at wavelength of 1.1 μm (the absorption at this wavelength is usually attributed to the presence of divalent iron ions in a glass). As can be seen from Fig. 4, a decrease in the absorption coefficient of the melt is observed only after the heat treatment of the cell for 5 hours at temperature of 1100 - 1400°C. The Fe^{2+} ions are oxidized to Fe^{3+} ions by air oxygen diffusing into the melt from the surrounding atmosphere. Under the considered conditions the influence of the convection in a relatively small uniformly heated specimen, to a first approximation, can be ignored. Taking the diffusion coefficient for oxygen $D_{O_2} = \sim 10^{-10}$ m²/s at 1250°C, it is possible to evaluate the time it takes for oxygen to reach the central part of the specimen (to a depth of about 5 mm), which is equal to ~10 h. Another example of long-term stability of the red-ox state of the melt in the cell under consideration is provided by the measurement of the absorption spectra of a melt of multicomponent glass close in composition to window glass, which was synthesized under strongly reducing conditions and contained about 3 wt. % ***FeO***. The absorption spectra recorded at a temperature of 1510°C are displayed in Fig. 4b. The heat treatment of the melt at high temperature for 1.5 h does not bring about a considerable change in its absorption coefficient over the entire studied range of wavelengths.

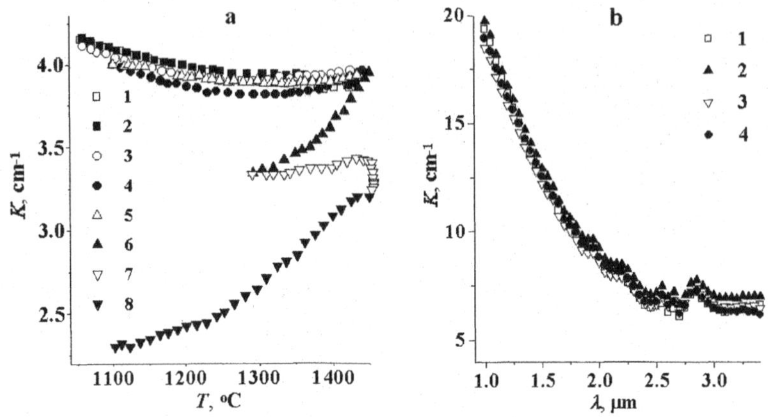

Fig. 4. Temperature dependence of the absorption coefficient for window glass containing 0.6 wt. % of iron oxides (70% ***FeO***) at a wavelength 1.1 μm upon sequential (1, 3, 5, 7) heating and (2, 4, 6, 8) cooling of the melt in a cell with leucosapphire windows. Heating and cooling rates were 5 K/min. Each tenth of a point is shown in the plot (a). Sequentially measured absorption spectra of the melt containing 3 wt. % of Fe_2O_3 and synthesized under reducing conditions (b).

The absorption coefficients of glasses are calculated from the equation.

$$K = D_V/(\log e \cdot h) = 2.303\, D_V/h \qquad (1)$$

Where: D_V is the volume optical density of a glass (melt), K is the absorption coefficient of a glass (melt), h is the specimen thickness, and ***log e*** is the common logarithm of the "***e***" number, which serves in (1) for the conversion to the natural absorption coefficient K.

The volume optical density of a glass is determined from the following relationship.

$$D_V = D - D_S \qquad (2)$$

Where: D is the optical density of a glass specimen, and D_S is the radiation attenuation due to the reflection from two glass-air interfaces.

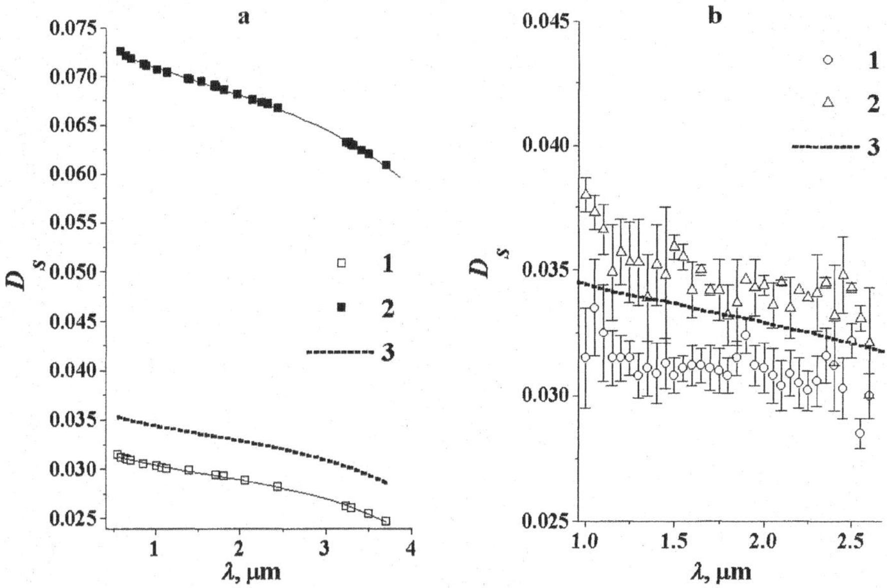

Fig. 5. Wavelength dependence of the surface optical losses $\boldsymbol{D_s}$ for (1) silica glass, (2) leucosapphire, and (3) sheet window glass. Curves 1 and 2 are calculated from equation (4) using data on refractive index from [2], and curve 3 is obtained by the conversion of curve 1 (see text) (a). Wavelength dependence of the surface optical losses $\boldsymbol{D_s}$ for sheet window glasses (1) 1 and (2) 2 (obtained experimentally using differential or two-thickness method) and (3) fragment of curve 3 from Fig. 5a. The confidence intervals for $\boldsymbol{D_s}$ are calculated at a fiducial probability of 95% (b).

The estimations indicate that the multiple reflections can be ignored in calculations of losses by reflection from the surface of glasses with $1.4 < \boldsymbol{n_1} < 1.6$. Therefore, for a plane-parallel specimen with a sufficiently small absorption and an ideal polishing, the total reflection losses are almost completely determined by single reflection from two surfaces.

The $\boldsymbol{D_S}$ density can be calculated using the refractive index of glass $\boldsymbol{n_1(\lambda)}$ at the glass-air interface from the Fresnel equation, in which the quadratic term is omitted due to its smallness.

$$D_s = -\log(1 - 2R) \tag{3}$$

Where: $R = \{[n_1(\lambda) - 1]/[n_1(\lambda) + 2]\}^2$

In order to correctly employ the calculation technique for determining the $\boldsymbol{D_S}$ density, it is necessary to know the dependence of the refractive index on the wavelength for a glass in the studied wavelength range (0.6 - 3.8 μm).

For the majority of glasses, the values of $\boldsymbol{n(\lambda)}$ in the near-IR can be obtained experimentally, which is not a simple procedure. We have used a different approach based on the literature data on refractive indices of pure silica glass and sapphire. Wavelength dependence of the reflection losses at the silica glass and leucosapphire surfaces calculated from equation (4) using the data from [2] are shown in Fig. 5a by solid and open boxes. Despite the fact that these two materials have different structure (crystalline vs. amorphous) and level of reflection loss (~0.03 vs. ~0.07), curves 1 and 2 in Fig. 5a have similar shapes. This implies that dependence $\boldsymbol{D_S} = \boldsymbol{f(\lambda)}$ for window glass can be generated using the corresponding dependence for silica glass, in which all the $n(\lambda)$ values are changed by the difference between the refractive indices n_D of these glasses for the yellow line of sodium (1.52 and 1.46 for window and silica glasses, respectively). The dependence obtained is depicted in Fig. 5a by the dashed line.

To check the correctness of this approximation, the absorption spectra of two commercial window glasses 1 and 2 with close compositions were measured repeatedly. The D_V and D_S densities were calculated from the experimental data for specimens with different thickness.

$$D_{V1} = (D_1 - D_2) \cdot h_1/(h_1 - h_2) \text{ and } D_{V2} = (D_1 - D_2) \cdot h_2/(h_1 - h_2) \qquad (4)$$

Where: D_1, D_2, D_{V1} and D_{V2} are the optical density and the volume optical density for thick and thin specimens, respectively, and h_1 and h_2 are the thickness of these specimens.

The losses at the surface were calculated as

$$D_S = D_1 - D_{V1} = D_2 - D_{V2} \qquad (5)$$

Fig. 5b displays the dependence $D_S = f(\lambda)$ determined by the described technique for window glasses 1 and 2 differing in the absorption level. A fragment of curve 3 from Fig. 5a is given as well. The confidence intervals for D_S were calculated at a fiducial probability of 95%.

One can see from Fig. 5b that the experimental and calculated D_S values differ by no more than 0.005. Consequently, the error in the calculation of reflection losses from equation (3) can be taken to be equal to ±0.005. In the cases when the optical density of a specimen considerably exceeds (more than by one and a half orders of magnitude) this error, the absorption coefficients can be determined with the use of the calculated density D_S. The specimens differing appreciably in thickness should be taken for the reliable determination of the absorption coefficients.

When measuring the absorption spectra in the temperature range corresponding to the viscosities of a glass-forming melt in the range 10^{11} - 10^2 P, the melt is placed in the cell between two leucosapphire windows (Fig. 3b). The determination of the absolute absorption coefficients of a melt requires the knowledge of the cell constant. If the inner surfaces of windows are wetted by the melt, the cell constant (or, more precisely, its component associated with the losses by reflection from two outer surfaces of leucosapphire windows and two leucosapphire-melt interfaces) can be calculated by the equation

$$D_{S1} = -log\left[1-2\left(\frac{n_1(\lambda)-1}{n_1(\lambda)+1}\right)^2\right] - log\left[1-2\left(\frac{n_1(\lambda)-n_2(\lambda)-1}{n_1(\lambda)-n_2(\lambda)+1}\right)^2\right] \qquad (6)$$

Where: $n_1(\lambda)$ and $n_2(\lambda)$ are the refractive indices of leucosapphire and the studied melt, respectively.

The refractive index of leucosapphire in the wavelength range under consideration virtually does not change with an increase in the temperature up to 1500°C. At the same time, the refractive index of the melt can vary in a complicated way over wide ranges of temperatures and wavelengths. The problem of precise measurement of the absorption of a melt can be solved by using the melt layers having different thickness, and under identical conditions. However, this would lead to a substantial complication of the measuring procedure and a rise in the cost of its performance.

In this respect, we preferred to use one cell and to experimentally determine its constant. To accomplish this, the absorption spectra of the studied melt at the temperature T_{11} (corresponding to its viscosity $\eta = \sim 10^{11}$ P) are recorded in two ways. One glass specimen is heated from room temperature to T_{11} in the cell shown in Fig. 3a. Another specimen placed in the cell demonstrated in Fig. 3b is cooled from T_{11} to a temperature, at which the gap between the leucosapphire windows is filled. (The gap between leucosapphire windows is usually filled with a melt at a temperature corresponding to the melt viscosity $\eta = \sim 10^{7}$ P.) The spectra of both specimens are shown in Fig. 6. Spectrum 1 (measured with the cell depicted in Fig. 3a) is the reference spectrum. The difference between spectra 1 and 2 is taken into account in the determination of the temperature dependence of the absorption spectra upon subsequent heating of the melt in a cell (shown in Fig. 3b) from the T_{11} temperature.

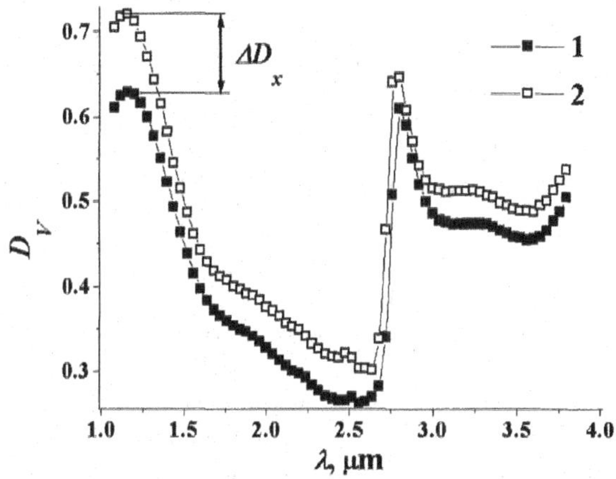

Fig. 6. Volume absorption spectra D_v of iron-containing glass close in composition to the window glass. Measurements are performed at 595°C in the ceramic cell shown in (1) Fig. 3a, and (2) Fig. 3b. In the determination of D_v, the surface losses D_s for spectra 1 and 2 are calculated from equations (3) and (6), respectively.

Calibration of spectrophotometers is performed on a regular basis using two reference samples that considerably differ in the transmittance in the near IR spectral range. The reference samples were examined on a commercial Perkin-Elmer Lambda-9 instrument (certified as the standard instrument at the All-Russia Research Institute of Metrology). Table I presents the mean optical densities (and their confidence intervals) obtained for eight replicate measurements of reference samples #1 and #2 using standard an SF-2-LGP instruments.

It can be seen from Table I that the mean optical densities obtained on the SF-2-LGP instrument differ from those measured on the Perkin-Elmer Lambda-9 instrument by no more than 1.5%. The random error in measurements of the optical density on the SF-2-LGP instrument is no more than that in the measurement on the standard instrument. Calibration of SF-3-LGP and SF-4-LGP devices was performed similarly.

TABLE I. MEAN OPTICAL DENSITIES OF REFERENCE SAMPLES 1 AND 2 AND THEIR CONFIDENCE INTERVALS.

λ, μm	D							
	Reference sample 1				Reference sample 2			
	HP Lambda-9		SF-2-LGP		HP Lambda-9		SF-2-LGP	
	D	ΔD_2^*	D	ΔD_3^*	D	ΔD_2^*	D	ΔD_3^*
1.1	0.1340	0.0007	0.1335	0.0007	1.0515	0.0025	1.0540	0.0005
1.3	0.1180	0.0005	0.1166	0.0007	0.9625	0.0016	0.9490	0.0006
1.5	0.0904	0.0005	0.0908	0.0004	0.6944	0.0014	0.6860	0.0005
1.7	0.0777	0.0005	0.0778	0.0003	0.5393	0.0016	0.5370	0.0006
1.9	0.0769	0.0004	0.0777	0.0004	0.5163	0.0013	0.5120	0.0010
2.1	0.0745	0.0004	0.0740	0.0005	0.4580	0.0016	0.4540	0.0008
2.3	0.0715	0.0004	0.0716	0.0004	0.3983	0.0013	0.3970	0.0008
2.5	0.0710	0.0004	0.0718	0.0009	0.3728	0.0013	0.3720	0.0009
ΔD_{mean}		0.0005		0.0005		0.0016		0.0007

The absolute error in the measurement of the optical density on the instrument under consideration can be determined from the following formula:

$$\Delta D_I = \Delta D_I^* + \sqrt{\Delta D_2^{*2} + \Delta D_3^{*2}} \qquad (7)$$

Where: ΔD_1^* is the absolute error of the standard instrument, ΔD_2^* is the random error in the measurement of the optical density of the reference sample on the standard instrument, and ΔD_3^* is the random error in the measurement of tile optical density of the reference sample on the tested instrument.

The calculated errors ΔD_I^* for various optical densities measured are listed in Table II. The data obtained allow us to make the following conclusions: the absolute error in the measurement of the optical density is reasonable in the optical density range from 0.1 to 1.2, and the absolute error of the standard instrument predominantly contributes to the absolute error in the measurement.

The technique for measuring the temperature dependence of the absorption spectra for melts was developed on the basis of the well-known relative method involving measurements of the absolute value of the property at room temperature $K_r = f(\lambda)$ and the changes in this property $\Delta K_x = f(\lambda, T)$ in the studied temperature range. The values of the property $K_x = f(\lambda, T)$ are determined by the summation of K_r and ΔK_x.

The temperature dependence of the spectra is measured in four steps.

Step I. Measurement of the absorption spectra of the studied glass at room temperature. No less than three specimens in the form of polished discs are used for the determination of reliable mean values. The measurements are carried out using the specimen holders (7) or (8) mounted near the monochromator entrance slit (see Fig. 2).

TABLE II. ABSOLUTE ERRORS OF MEASUREMENTS OF OPTICAL DENSITY ON PE LAMBDA-9 STANDARD INSTRUMENT (ΔD^*_1) AND SF-2-LGP INSTRUMENT (ΔD_1) ACCORDING TO EQUATION (1).

D	ΔD^*_1	$\Delta D^*_1/D$, %	ΔD_1	$\Delta D_1/D$, %
0.10	0.0022	2.2	0.0028	2.8
0.20	0.0028	1.4	0.0034	1.7
0.50	0.0055	1.1	0.0071	1.4
0.70	0.0087	1.2	0.0103	1.5
1.00	0.0170	1.7	0.0186	1.9
1.20	0.0270	2.3	0.0286	2.4

Step II. Recording of changes in the absorption spectrum of glass upon heating specimen from room temperature to T_{11}. The absorption spectra of the specimen in the cell without windows (Fig. 3a) in the furnace are measured, first, at room temperature and then at T_{11}, and, as a result, the change of the absorption coefficients $\Delta K_{r\text{-}11}$ connected with the temperature change by (T_{11} - 20) K. The values of the absorption coefficients K_{11} at the temperature T_{11} are obtained as the sum:

$$K_{11} = K_r + \Delta K_{r\text{-}11} \tag{8}$$

Step III. Recording of changes in the absorption spectrum of the glass forming melt upon heating the specimen in the cell with windows (Fig. 3b) in the temperature range T_{11} - T_{rev}. The T_{rev} is a conditional quantity, which is determined in the course of preliminary experiments. This is the temperature limit, below which a melt can be treated in the cell for a prolonged time without irreversible changes in the optical absorption. To put it differently, the rates of crystal growth, release of dissolved gas bubbles, and chemical corrosion of the leucosapphire windows by the melt in the range from T_{11} to T_{rev} are too small to substantially affect the spectra during several sequential cooling and heating of the cell.

The cell with a specimen is heated up to the T_{11} temperature at a rate of 10-12 K/min. The heating above T_{11} is performed at a rate of 10-12 K/min, and the change in the optical density of the cell is examined at a wavelength of 2 μm. When the viscosity of a melt decreases to $\eta = 10^7$ P, the melt fills the space between the leucosapphire windows, which results in a decrease in the optical density. Thereafter, the temperature is decreased down to T_{11} at a rate of 12 K/min, and the spectrum is measured under isothermal conditions. Then the cell is heated and cooled in the range T_{11} - T_{rev}, and the spectra are recorded. This procedure makes it possible to determine the changes in the absorption coefficient $\Delta K_{11\text{-}x}$ due to the heating of the cell from the T_{11} temperature to T_x temperatures lying in the range T_{11} and T_{rev}.

The absolute absorption coefficients at the T_x temperature are obtained according to the equation:

$$K_x = K_{11} + K_{11-x} \tag{9}$$

Step IV. Recording of changes in the absorption spectrum of glass-forming melts upon its heating from T_{rev} to 1500°C.

At the T_y temperatures ($T_{rev} < T_y < 1500^\circ C$), the absorption of a melt can be irreversibly changed owing to the following factors: partial dissolution of leucosapphire in the melt at the inner surfaces of the windows followed by the formation of crystal layers, the release of bubbles of gases dissolved in a melt, the shift in the red-ox equilibriums, etc. However, we found that the reproducible dependence $\Delta K_{rev\text{-}y} = f(\lambda, T)$ can be obtained in the temperature range T_{rev} - T_y by heating the cell from room temperature to 1500°C at a rate 25 - 35 K/min and measuring the absorption spectra of the cell in the range T_{rev}-1500°C upon cooling and subsequent heating at the same high rate. This procedure allows one to shorten the residence time of a glass-forming melt at temperatures corresponding to the nucleation (within the glass transition range) and other high-rate irreversible transformations (1300 - 1500°C). The absolute absorption coefficients K_y of a melt at temperatures T_y lying in the range T_{rev} - 1500°C are determined from the equation:

$$K_y = K_{rev} + \Delta K_{rev-y} \tag{10}$$

The measuring technique described above was used to investigate the temperature dependences of the absorption spectra for a number of silicate glasses [21 - 26].

As mentioned above, the absorption coefficients for glasses can be determined by measuring specimens with identical or different thickness. The confidence interval for the absorption coefficients of a glass-forming melt is given by:

$$\Delta K^*_{tot} = \frac{\Delta D_1}{h \times lg\, e} + \sqrt{\Delta K_4^2 + \Delta K_5^2 + \Delta K_6^2 + \Delta K_7^2 + \Delta K_8^2} \tag{11}$$

and

$$\Delta K^*_{tot} = \frac{\Delta D_1}{h \times lg\, e} + \sqrt{\Delta K_{41}^2 + \Delta K_{42}^2 + \Delta K_{51}^2 + \Delta K_{52}^2 + \Delta K_6^2 + \Delta K_7^2} \tag{12}$$

Where ΔK_{tot} is the total confidence interval for the absorption coefficient of a melt; ΔD is the absolute error in the determination of the optical density at room temperature; ΔK_4 is the confidence interval for the random error in the measurement of the absorption coefficient of a glass specimen at room temperature; ΔK_5 is the confidence interval for the random error in the measurement of the absorption coefficient of a glass specimen at the T_{11} temperature; ΔK_6 is the confidence interval for the random error in the measurement of the absorption coefficient of a melt specimen at the T_x temperature (ΔK_6 corresponding to the T_{rev} temperature is used in the calculation of ΔK_{tot} at temperatures above T_{rev}); ΔK_7 is the confidence interval for the random error in the determination of the optical density of a melt specimen at the T_y temperature (from T_{rev} to 1500°C); ΔK_8 is the error in the calculation of the surface optical density from relationship (3); ΔK_{41}, ΔK_{42}, ΔK_{51}, and ΔK_{52} are the quantities similar to ΔK_4 and ΔK_5 in relationship (11) for thin (ΔK_{i1}) and thick (ΔK_{i2}) specimens, respectively; and h is the thickness of glass layer. The calculated ΔD_1 values for different optical densities are given in Table II.

Relationships (5) and (6) do not include terms, which are connected with errors in the measurement of the temperature, due to the following reasoning. The temperature coefficients dK/dT for the studied glasses do not exceed 0.003 cm^{-1}/K. The absolute error in the measurement of the temperature is equal to ±3 K. This leads to the error of measurements $\Delta K^{T1} = \pm 0.009$ cm^{-1}, which falls in the range of the total error in the measurement of the absorption coefficients. The errors caused by the lag of the specimen temperature from the thermo-junction temperature are opposite in sign upon heating and cooling of a specimen and are approximately equal in magnitude ($\Delta K^{T+} = -\Delta K^{T-}$). These errors can be treated as the contributions to the corresponding random error (see above).

MEASUREMENT RESULTS

Data on high-temperature absorption spectra measured for six typical commercial glasses (wool fiber glass, E-glass, white container glass, color TV panel glass, low-expansion borosilicate glass, and float glass) are presented in Figs. 7 - 12. These data were obtained using experimental procedure described above in Section "Methodology". Error bars shown on temperature dependence of absorption coefficients determined at 1.1, 2.5, 2.8 and 3.4 μm allow observation of total errors of measurements of absorption coefficients of glasses having different levels of absorption in near IR, and different tendency to irreversible changes.

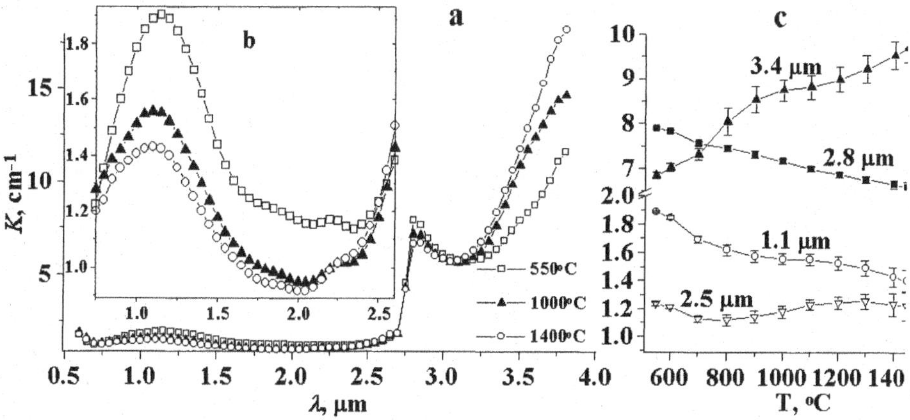

Fig. 7. Absorption spectra of wool fiber glass measured at 550, 1000 and 1400°C (a, b), and temperature dependences of absorption coefficient measured at 1.1, 2.5, 2.8 and 3.4 μm (c).

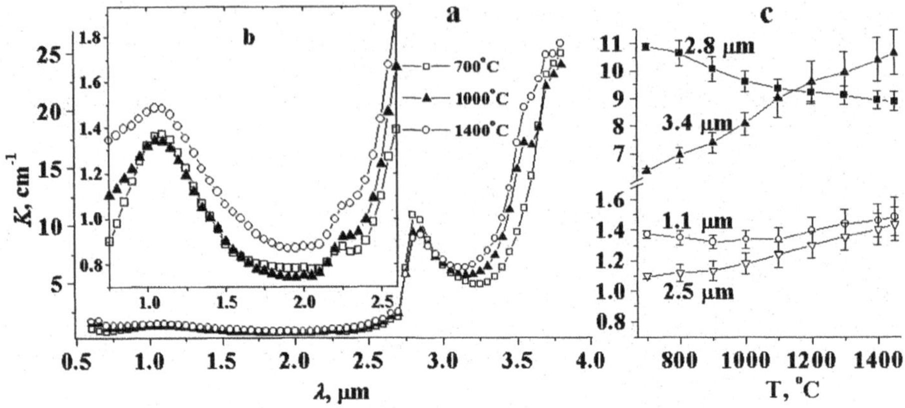

Fig. 8. Absorption spectra of E-glass measured at 700, 1000 and 1400°C (a, b), and temperature dependences of absorption coefficient measured at 1.1, 2.5, 2.8 and 3.4 μm (c).

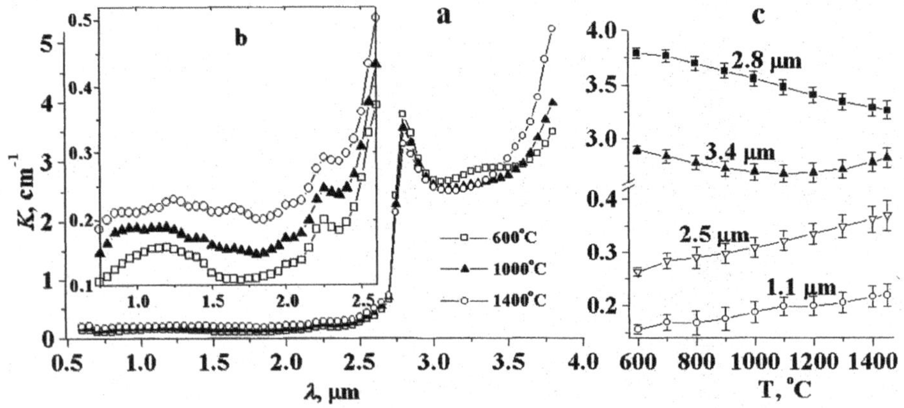

Fig. 9. Absorption spectra of white container glass measured at 600, 1000 and 1400°C (a, b), and temperature dependences of absorption coefficient measured at 1.1, 2.5, 2.8 and 3.4 μm (c).

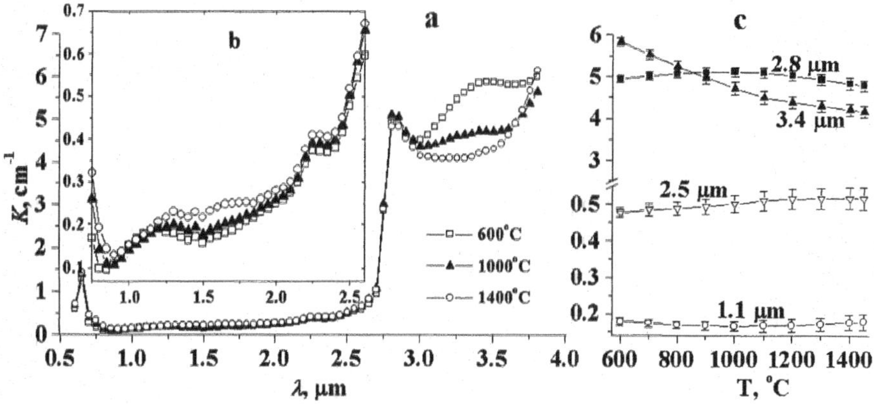

Fig. 10. Absorption spectra of color TV panel glass measured at 600, 1000 and 1400°C (a, b), and temperature dependences of absorption coefficient measured at 1.1, 2.5, 2.8 and 3.4 μm (c).

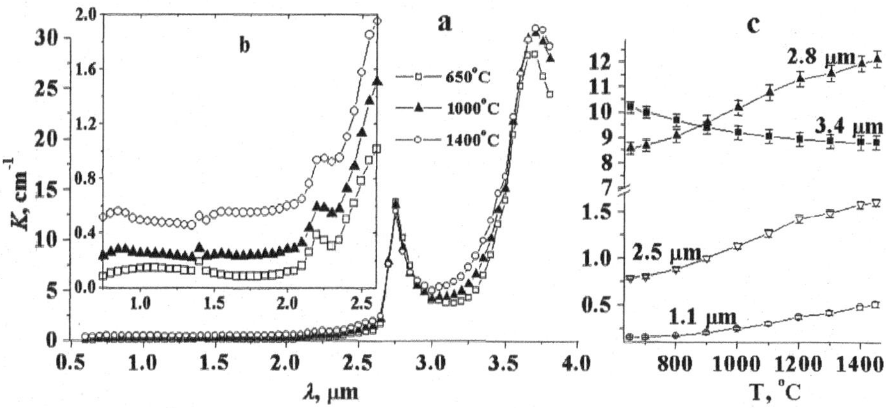

Fig. 11. Absorption spectra of low-expansion borosilicate glass measured at 650, 1000 and 1400°C (a, b), and temperature dependences of absorption coefficient measured at 1.1, 2.5, 2.8 and 3.4 μm (c).

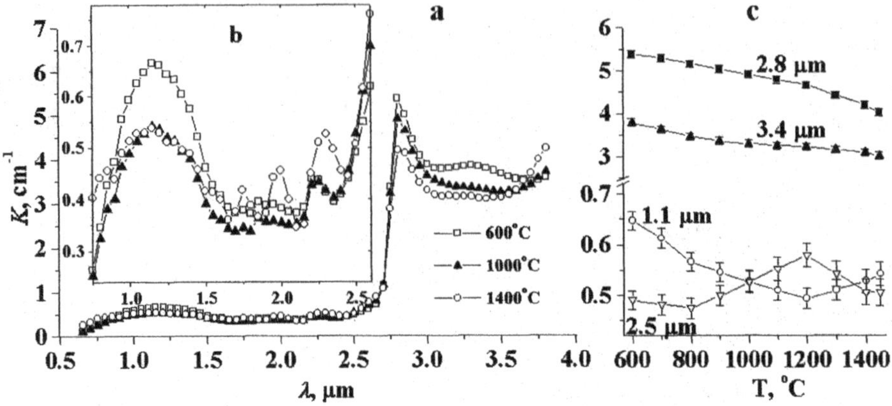

Fig. 12. Absorption spectra of float glass measured at 600, 1000 and 1400°C (a, b), and temperature dependences of absorption coefficient measured at 1.1, 2.5, 2.8 and 3.4 μm (c).

CALCULATIONS OF RADIATIVE THERMAL CONDUCTIVITY

The characteristics of radiative heat transfer (radiative thermal conductivity λ_r and free path length of radiation l) were calculated by using a so-called Radiative thermal conductivity approximation or Rösseland approximation. This approximation is briefly described below.

Let us suppose that for energies (frequencies) of photons taking part in the process of heat transfer in the studied medium the following inequity is fulfilled:

$$l_v/L << 1 \quad (13)$$

Where:

l_v – is the free path length of radiation (for selected frequency v),

L – is the linear extent of the task.

Considering that $l_v = 1/K_v$, inequity (13) can be written as follows

$L \times K_v >> 1$, (14)

Where:

K_v – is the spectral absorption coefficient of the studied media.

As it follows from inequity (13), a spectral free path length of radiation l_v is much less than the linear extent of the task L. Then for a spectral density of radiation U_v the following approximate equation is fulfilled.

$$U_v \cong U_{ve}, \quad (15)$$

Where:

U_{ve} – is the spectral density of equilibrium radiation.

According to Plank

$$U_{\nu e} = \frac{8\pi h\nu^3}{c^3} \frac{1}{e^{h\nu/kT} - 1} \tag{16}$$

Where:
h – is the constant of Plank;
c – is the speed of light in vacuum;
k – is the Boltzmann constant;
T – is the temperature of the medium.

If inequity (13) is true, a spectral flux of energy of radiation is connected with the density of equilibrium radiation in accordance to the following formula:

$$\vec{W}_\nu \cong -\frac{cl_\nu}{3} \nabla U_{\nu e}. \tag{17}$$

Since equilibrium radiation depends on dimensional coordinates only through a temperature of medium, the equation (9) can be rewritten as follows:

$$\vec{W}_\nu \cong -\frac{cl_\nu}{3} \frac{dU_{\nu e}}{dT} \nabla T \tag{18}$$

Integral flux of energy of radiation can be calculated by integration of equation (16) by a whole spectrum of frequencies:

$$\vec{W} = \int_0^\infty \vec{W}_\nu d\nu \cong -\frac{c}{3}\left(\int_0^\infty l_\nu \frac{dU_{\nu e}}{dT} d\nu\right) \nabla T \tag{19}$$

Let's enter into consideration a so-called free path length by Rösseland:

$$l = \frac{\int_0^\infty l_\nu \frac{dU_{\nu e}}{dT} d\nu}{\int_0^\infty \frac{dU_{\nu e}}{dT} d\nu} = \frac{\int_0^\infty \frac{1}{\kappa_\nu} \frac{dU_{\nu e}}{dT} d\nu}{\int_0^\infty \frac{dU_{\nu e}}{dT} d\nu} \tag{20}$$

The Rösseland free path length of radiation l is an averaged by the whole spectrum spectral free path length of radiation with weight $\frac{dU_{\nu e}}{dT}$. Then from equation (17) the following equation can be obtained:

$$\vec{W} = \int_0^\infty \vec{W}_\nu d\nu = -\frac{cl}{3}\nabla\left(\int_0^\infty U_{\nu e} d\nu\right) \tag{21}$$

Taking into account that:

$$\int_0^\infty U_{\nu e} d\nu = \frac{4\sigma T^4}{c} \tag{22}$$

Where:
σ - is the constant of Stephan-Boltzmann;
We obtain the final equation for the flux of radiation:

$$\vec{W} = -\frac{16\sigma l T^3}{3}\nabla T \tag{23}$$

It should be emphasized that the Radiative conductivity approximation (Rösseland approximation) can be applied only if the inequity (13) is fulfilled. In this (and only in this) situation a spectral dependence of absorption of radiation in a medium can be reduced to a single value - free path length by Rösseland.

In Fig. 13 temperature dependence of radiative thermal conductivity and free path length calculated for six commercial glasses by Rösseland formulas using their absorption spectra measured within the temperature range 600 - 1450°C are presented. One can see that the studied glasses can be roughly divided into three groups based on their radiative thermal conductivity at 1450°C. Fiber-glasses (E-type and wool-type) having lowest values of λ_r (~25 W/(m·K), low-expansion borosilicate and float glass with medium values of λ_r (40 - 60 W/(m·K), and white container and color TV panel glass having the highest values of λ_r (~110 W/(m·K). Due to the above mentioned limitations of Rösseland method this classification is very relative and can't be used for any *a priori* judgments on radiative heat transfer in any particular type of glass. One can conclude, however, that the ability of different glasses to absorb infrared radiation and conduct heat by radiation can differ dramatically even within quite small group of commercial glasses. Thus, one should consider measurements of absorption spectra of every new glass forming melt as another complex study and pay significant attention to many specific features of its behavior at high temperatures.

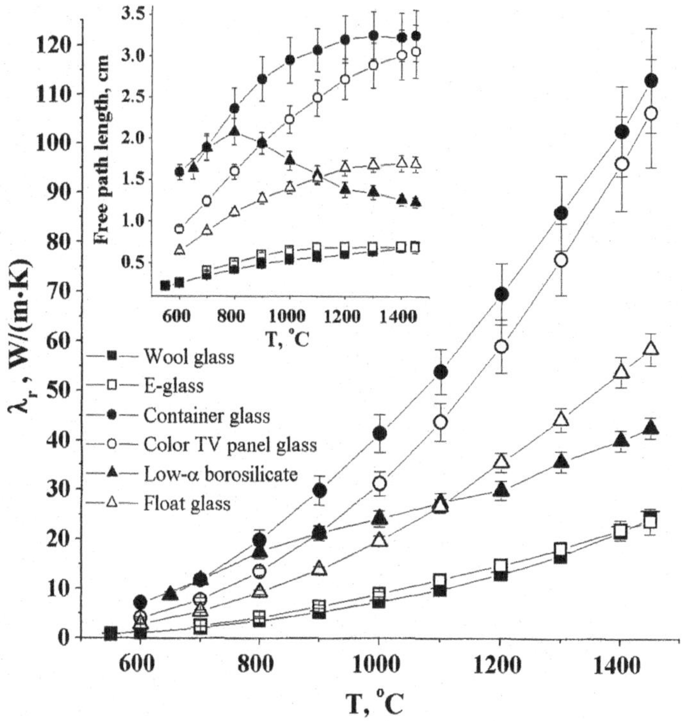

Fig. 13. Radiative thermal conductivity and free path length calculated by Rosseland formulas for six commercial glasses using their absorption spectra measured at high temperatures.

CONCLUSION

Repeatability of consecutive and parallel measurements of glass properties is an essential part of a standard practice of routine check of the quality of testing protocols in glass material science. From the beginning of the experimental work the authors paid special attention to repeatability of the experimental data. Reliability of the method was confirmed by numerous data obtained on at minimum three parallel specimens of the same glass. Absorption spectra of each individual specimen were registered many times at constant temperatures (20°C, T_{11}, T_{rev}) or in the course of many repeated heating-cooling cycles. Thus the method described in this chapter can be considered as the most dependable one at the present time. Unfortunately, the absence of data on reproducibility of high-temperature spectra published by many other authors prevents evaluation of the reliability of those data.

In order to obtain the most accurate results one should perform preliminary study and find a proper time-temperature schedule of the measurements. Since 1994 this simple and obvious approach allowed us to obtain dependable data on absorption spectra of over 100 different glass compositions. Although many glasses required time-consuming studies just to develop time-temperature regimes, we have found no limitations with respect to studies of absorption spectra of commercial silicate glasses.

Reproducibility of experimental results within the limits, which are formulated by practical needs of material science and technology, is an absolutely necessary but not sufficient condition of any effective measurement technique. Calibration of the experimental device(s), careful determination of cell constants, and compensation of irrelevant influences helped us to minimize systematic errors of the measurements of high-temperature spectra of glass forming melts. However, a measurement technique can be considered as error-free only when high-temperature standards of optical absorption in near IR become available.

Due to a high interest from the glass industry to optimize energy consumption during glass melting, the importance of data on near IR absorption and thermal conductivity will only grow. Thus, the work on the further improvement of the measurement technique should continue.

REFERENCES

[1]N. Neuroth, "Der Einfluß der Ttemperatur auf die spektrale Absorption von Gläsern im Ultraroten," *Glastcch. Ber.*, **25** [8] 242-249 (1952).

[2]J. Endrys, F. Geotti-Bianchini and L. De Riu, "Study of the High-Temperature Spectral Behavior of Container Glass," *Glass Sci. Technol.*, (Frankfurt/Main), **70** [5] 126-136 (1997).

[3]F.J. Grove and M.A. Jellyman, "I. The Infrared Transmission of Glass in the Range from Room Temperature to 1400°C," *Soc. Glass Technol*, **39** [186] 3T-15T (1955).

[4]O.A. Sergeev and A.A. Men', "Teplofizicheskie svoistva poluprozrachnykh materialov" (Thermal Properties of Semitransparent Materials), Izd. Standartov, Moscow, (1977).

[5]N. Neuroth, "Der Einfluß der Temperatur auf die spektrale Absorption von Gläsern im Ultraroten: II," *Gtastech. Ber*, **26** [3] 66-70 (1953).

[6]L. Genzel, "Messung der Ultrarot-Absorption von Glaszwischen 20 und 1360°C," *Glastech. Ber.* **24** [3] 55-63 (1951).

[7]F.J. Grove, "Spectral Transmission of Glass at High Temperatures and Its Application to Heat-Trasfer Problem", *J. Am. Ceram. Soc.* **44** [7] 317-320 (1961).

[8]H. Franz, "Infrared Absorption of Soda-Lime-Silica Glasses Containing Transition Metal Oxides," *in Proc. IX Int. Congress on Glass*, Versailles, **AI.2** 243-269 (1971).

[9]M. Coenen, "Durchstahlung des Glasbades bei Farbglasern," *Glastech. Ber.* **41** [l] 1-10 , (1968).

[10]B. Wedding, "Measurements of High-Temperature Absorption Coefficients of Glasses," *J. Am. Ceram. Soc.*, **58** [3-4] 102-105 (1975).

[11]G.A. Digilina and B.V. Tarasov, "An Investigation into the Effect of Temperature on the Near-Infrared Absorption of Commercial Silicate Glasses Colored with Cobalt and Chromium Oxides," *Fiz. Khim. Stekla,* **8** [6] 704-708 (1982).

[12]G.A. Digilina, B.V. Tarasov and B.G. Varshal, "The Effect of Temperature on the Near-Infrared Absorption of Colorless and Colored Lead Silicate Glasses," *Fiz. Khim. Stekla,* **11** [6] 672-676, (1985).

[13]G.A. Digilina, "An Instrument for Determination of Spectral Absorption in Glasses in the Range 293-1723 K," *in Issledovanie stekla i silikatnykh materialov (Investigation of Glasses and Silicate Materials),* Verkhne-Volzhskoe knizhnoe izd., Yaroslavl, 82-88 (1978).

[14]A. Blazek, J. Endrys, J. Kada and J. Stanek, "Strahlungsärmeleitfähigkeit von Glas-Einfluß der Glaszusammensetzung auf seine Wärmedurchlässigkelt," Glastech. Ber., 49 [4] 75-81 (1976).

[15]A. Blazek, J. Endrys and J. Kada, "Einfluss von Farboxiden auf die Absorptionseigenschaften der Glauchmelze bei Honen Temperaturen," *in Proc. XI Int. Congress of Glass*, Prague, **3** 107-116 (1977).

[16]A.M. Efimov and P. A. van Nijnatten, "High temperature optical properties of glass melts in the 0·95–4·5 μm range: an example of television glass melt," *in Proc. XIX Int. Congr. Glass, Edinburgh, 2001, Glass Technol.*, **43C** 15–24 (2002).

[17]O.A. Prokhorenko, O.V. Mazurin and V.M. Zolotarev, "High-Temperature Spectrophotometry: Problems and Ways of Their Solution, " *Opt. Zh.* **65** [10] 67-71 (1998).

[18]O.A. Prokhorenko, "Experimental Cell for Measurement of Temperature Dependence of Absorption Spectra of Molten Glass at Temperatures up to 1550°C", *in Proc. Colloquium on Modeling of Glass Forming Processes*, Valenciennes, 35-41 (1998).

[19]O.A. Prokhorenko, O.V. Mazurin, "Problems of Spectrophotometry of Glass-Forming Melts. I. The Current State of the Problem of Reliable Experimental Data Acquisition," *Glass Physics and Chemistry*, **25** [2] 159-162 (1999).

[20]O.A. Prokhorenko, O.V. Mazurin, M.V. Chistokolova, S.V. Tarakanov, Ju.E. Reznik and I.N. Anfimova, "Problems of Spectrophotometry of Glass-Forming Melts. II. The Method of Measurement of Absorption Spectra of Glasses and Melts within Red and Near Infrared Spectral Regions in the Temperature Range from 20 to 1500°C," *Glass Physics and Chemistry*, **26** [2] 187-198 (2000).

[21]V.K. Leko, M.V. Chistokolova, O.A. Prokhorenko and O.V. Mazurin, "Influence of Composition and Temperature on the Absorption Band Intensity for Water in Alkali Silicate Glasses", *Glass Physics and Chemistry* **23**, [3] 214 – 224 (1997).

[22] O.A. Prokhorenko, O.V. Mazurin, M.V. Chistokolova and I.N. Anfimova, "Study of the Influence of Thermal History on the Absorption of Sodium-Calcium Silicate Glass at a Wavelength of 3.4 μm in the Temperature Range 20 – 630°C," *Glass Physics and Chemistry*, **24** [5] 452 – 459 (1998).

[23] O.A. Prokhorenko, "Some Problems Related to Correct Measurement of Spectra of Highly Transparent Molten Glasses," *in Proc. ICG Annual meeting 2000 (on CD)*, Amsterdam, Session T3-6 (2000).

[24] O.A. Prokhorenko, "Absorption Spectra and Radiative Thermal Conductivity of E-glass at Temperatures up to 1500°C," *in Proc. 6th International Conference Advances in Fusion and Processing of Glass*, Ulm, 33-44 (2000).

[25]O.A. Prokhorenko and O.V. Mazurin, "High-Temperature Radiative Conductivity of Commercial Iron-Containing Glasses," *in Advances in Glass Fusion and Processing of Glass: II Ceramic Trans.*, **82** 323-333 (1998).

[26]O.A. Prokhorenko, "About the Influence of Structural Relaxation on Temperature Dependence of Optical Absorption of Glass Melts," *in Proc. XVIII Int. Congress on Glass*, San Francisco, 64-71 (1998).

Chapter 6

THE MEASUREMENT OF THE DENSITY AND SURFACE TENSION OF GLASS MELTS USING THE SESSILE DROP METHOD

Alexis G. Clare, Ahmet Kucuk, Douglas R. Wing, and Linda E. Jones
New York State College of Ceramics at Alfred University
Alfred, New York 14802 USA

INTRODUCTION

The density of glass melts is a fundamental parameter that affects both surface tension and viscosity (many calculations of the latter properties involve a value or, at least an estimate of the former), and it is a crucial parameter in the modeling of glass flow in tanks. Both density and surface tension affect glass manufacturing through the homogeneity of glass melts (changes in surface tension have led to elimination of cords [1]), fining (decreases in surface tension help fining [3]), refractory corrosion [2], and rupture of the melt.

While the density of glass melts should not be affected by atmosphere, surface tension will most certainly be affected by the gas and any substrate in contact with the glass. Hence it is imperative that surface tension be understood in terms of the atmospheric effects. In fiberglass manufacture for example, drawing parameters and fiber strength change substantially with changes in the drawing atmosphere [4]. While the data presented here are for one atmosphere, argon, we have carried out a number of experiments as a function of atmosphere which have been published elsewhere [5-11].

What we have developed is a method for measuring the density and surface tension of glass melts at temperatures up to 1450°C under different atmospheric conditions. While we have used this method to generate several data sets that impart information not only about glass technology but also basic glass science, the true value of this method lies in the fundamental steps that we have taken to reduce errors by both experimental controls and methods of calculating and interpreting the data. These controls will be discussed in this chapter along with limitations inherent in these types of experiments.

DENSITY

The most popular method of measuring the density of glass melts is probably the Archimedes method based upon buoyancy [12] which involves comparing the weight of a platinum bob immersed in a molten glass to that in air. This method works over a reasonable range of low viscosities (10 to 10^2 Pa.s) with a reported 1-5% accuracy. However, errors associated with surface tension effects on the hangdown wire require knowledge of the surface tension of the liquid to correct. This experiment can be modified by using a two-bob method [13].

The maximum bubble pressure method [14] requires measurements of the maximum pressure of bubbles at the precise moment they are released from a capillary, at two separate depths. This method is, again, suitable for low viscosities ($10\text{-}10^4$ Pa.s) but requires substantial corrections to account for thermal expansion of the crucible and tube. The accuracy is of the order of 1% . The Pycnometric method [15] (10 to 10^5 Pa.s) involves determining the volume of molten glass at a given temperature by measuring the height of the melt in a calibrated crucible and then weighing the glass to obtain the mass. This method has a reported accuracy of 0.5% but this does not take into account errors resulting from expansion of the instrument parts and the problems associated with temperature uniformity within the melt.

The sessile drop method was originally developed by the metallurgical industry to determine the wetting behavior and hence the surface tension of molten metals on a variety of substrates [16, 17] and was quickly adapted to the glass industry to observe the wetting behavior of glass melts [18-20]. It has also been employed to determine the density at high temperature in glass melts [21], which involved fitting the profile of the drop and, after Bashforth and Adams [22], calculating a volume,

$$V = \frac{\pi b^2 x^2}{\beta}\left(\frac{2}{b} - \frac{2\sin\theta}{x} + \frac{\beta z}{b}\right) \quad (1)$$

where:

x is the distance from the center of the contact surface to the contact point,

z is the distance from the apex of the drop to the glass/substrate interface,

θ is the wetting angle,

b is a unit of length related to the radius of curvature at the drop apex (a tabulated scale factor),

and β is the shape factor $\beta = \frac{\rho g b^2}{\gamma}$,where ρ is the density and γ is the surface tension.

Clearly, the appearance of the density and surface tension (the parameters being measured) in the shape factor makes this calculation unsatisfactory. In addition, very small changes in the parameters x and θ measured from the drop profile lead to large errors in density. This brings into question the usefulness of this method to the glass industry.

With the advent of digital video recording, it has been possible to greatly simplify the calculation of volume of the glass at temperature:

$$v = \sum_i \pi\left(\frac{d}{2}\right)^2 h_i \quad (2)$$

where the drop is split up into cylinders one pixel in height and integrated.

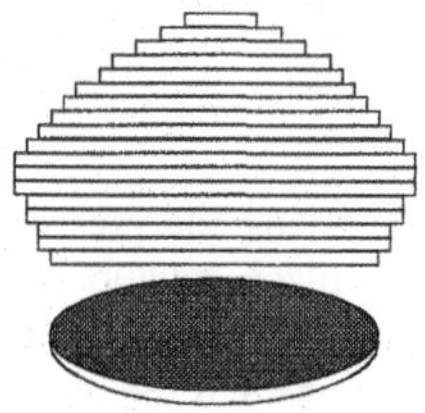

Figure 1. Integrated volume method for determining the volume of a sessile drop

This assumes that the drop is being viewed at exactly 90° to its axis so that the surface of the substrate is parallel to the line of sight (checked by tilting the image until the substrate surface just disappears), and that the drop is axially symmetric (only a problem if the substrate has oriented striations).

In order to calculate the density it is necessary to know the mass of the drop. If a volatile glass is being examined this can lead to fairly serious errors in the mass. We therefore conducted a study [5] of varied compositions, from highly volatile (a commercial glass used for TV funnels) to lower volatility (a high silica soda lime silicate). Both sessile and pendant drop experiments were carried out so that the effect of atmosphere on the volatility could be examined. In a sessile drop experiment the drop size varies, hence to really assess the weight loss accurately the normalized percentage weight loss was determined.

$$\text{NPWL} = \frac{((\text{initial wt/initial surface area}) - (\text{final wt./final surface area})) \times 100}{(\text{initial weight/initial surface area})} \tag{3}$$

The size of the pendant drop is always governed by the ring size hence the calculation of a dynamic mass from the weight loss is very straightforward. It was determined that the mass loss for most commercial glasses could be minimized by reducing the dwell time of the drop in the hot zone. For volatile glasses the normalized percentage weight loss made it possible to determine a dynamic mass and hence an accurate density. None of the base glasses in this project fell under the latter category.

SURFACE TENSION

The pendant drop method is a variant of the drop weight method in which a gob of glass forms from a sized orifice in a platinum crucible [23]. The previous method for calculating the surface tension from the pendant contained several approximations which led to errors. Our approach was to fit a Young-Laplace equation to the drop profile, thereby removing the approximation.

Surface tension can also be measured by looking at the extension of a freely suspended fiber [24] under its own weight, but it is not clear that a fiber with its large surface-area-to-volume ratio is representative of a bulk glass composition. The maximum pull method [25] measures the force needed to pull a hollow cylinder of glass away from contact with a molten glass, but it is a method fraught with cumulative error in both calibration and measurement. The maximum bubble pressure method that was mentioned as a measurement for density can also be used to obtain the surface tension.

The method we use here equates the hydrostatic force of the liquid to the excess pressure that results from the curvature of a drop.

$$\Delta P = \gamma\left(\frac{1}{R_1} + \frac{1}{R_2}\right) \tag{4}$$

The left hand side of the equation represents the hydrostatic force involving the pressure across the interface The right hand side represents the solutions to the Laplace equation for capillary pressure. R_1 and R_2 are the principal radii of curvature of the drop and γ is the interfacial surface tension. Incorporation of the appropriate boundary conditions and the use of the Rotenberg algorithms [26] reduces the problem to a set of simultaneous differential equations, which can be solved given the appropriate boundary conditions,

$$\begin{aligned}\frac{dx}{ds} &= \cos\phi \\ \frac{dz}{ds} &= \sin\phi \\ \frac{d\phi}{ds} &= \frac{2}{R_0} + \frac{(\Delta\rho)g}{\gamma}z - \frac{\sin\phi}{x}\end{aligned} \tag{5}$$

where s is the arc length measured from the apex of the drop and ø is the angle of the principal axes to the drop axis and R_0 is the radius of curvature at the origin.

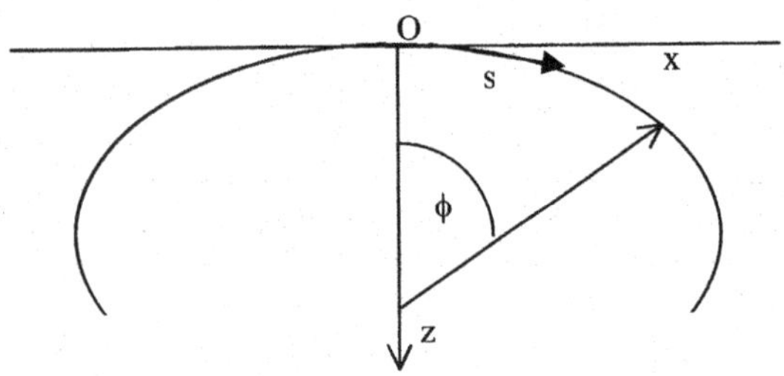

Figure 2. The coordinate system of Rotenberg

This method may be used when dealing with either the sessile drop or a pendant drop. Note that the density is needed to extract the surface tension from this equation. While the density does not change with atmosphere, it is necessary to make a sessile drop measurement to obtain the density because of uncertainty in the drop volume for the pendant drop. For any atmosphere other than inert, however, the pendant drop is a more reliable method for obtaining surface tension. Comparisons of the data obtained from pendant and sessile drop under identical conditions are the same well within the experimental error, which for the surface tension measurement is typically about 2%. In this chapter we have only used the sessile drop since the surface tension for the base glasses was measured in argon.

EXPERIMENTAL PROCEDURE

The experimental arrangement is shown in figure 3. The main differences between this and other previous measurements was the use of the high resolution camera and video, and this situation only improves with better and better imaging systems available. A more recent addition for the measurement of atmosphere is an oxygen sensor. The substrate we use for the sessile drop is graphite polished with 1μm diamond paste. The reason for this is that the interface between the drop and the substrate must be accurately defined to obtain the density, which requires a non-wetting situation. On a non-wetting substrate, surface tension is more controlled by the gas-liquid interface and therefore much more reliable surface tension values can be obtained. The difficulty however is that despite a rather extensive search, graphite is the only feasible non-wetting substrate we have found and hence limits us to inert or reducing conditions.

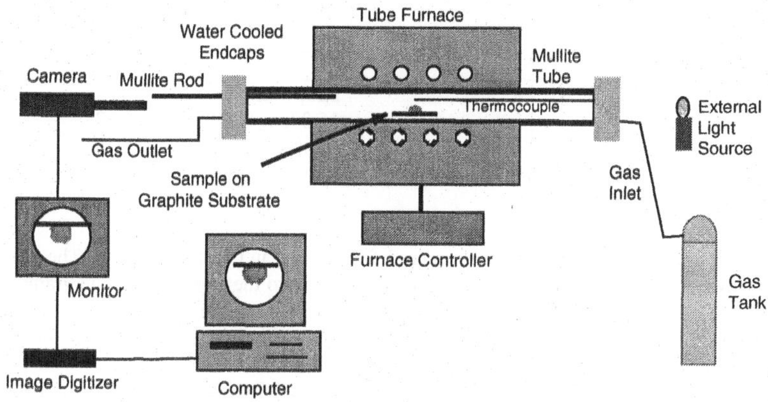

Figure 3. Experimental arrangement for the sessile drop method

Whether conducting a sessile drop or a pendant drop experiment, the specimen is kept in the cold zone (near the cooled windows) until measurement. For the sessile drop the sample is fed in to the furnace hot zone and agitated off of the platinum ring to drop onto the graphite substrate. For a pendant drop measurement the sample remains on the ring.

The drop images, recorded with the time and temperature, are first digitized and converted into a series of still pictures with a 640 x 640 pixel resolution. The images are processed using image software that enhances the contrast between the drop and the surroundings. The computer is then able to digitize the image outline rather simply by applying a decision based upon the gray scale number.

How Good is the Method?

Most of the results of our experiments to date using this method have been reported in the following papers [5, 7-11,27-29]. Nevertheless it may be useful just to compare some high temperature soda lime silica results that have been obtained before with those obtained using our methods. The results are given below.

TABLE I. DATA FOR SODA LIME SILICA MELTS AT 1400°C

Reference	Density (g cm-3)	Surface Tension mN/m	Conditions
Current	2.30±0.01	306	Sessile Drop Atmosphere, $15Na_2O$-,Ar$10CaO$-$75SiO_2$ (mol%)
Parmelee[30]		304	max bubble pressure
Keppler [31]		299	max bubble pressure $13.69Na_2O$-$8.43CaO$-$76.39SiO_2$ (mol%)
Badger [32]		302	Max bubble pressure $17.49Na_2O$-$10.1CaO$-$72.5SiO_2$ (mol%)
William [33]		301	Max bubble pressure $16.649Na_2O$-$10.08CaO$-$72.36SiO_2$ (mol%)
Washburn [34]	2.275	161	dipping cylinder $15.96Na_2O$-$10.69CaO$-$73.35SiO_2$ (mol%)
Din et al. [35]	2.3	312	
Kayalova [36]			Drop weight $15.23Na_2O$-$9.09CaO$-$75.67SiO_2$ (mol%)

Our measurements seem to fall in with the data except for Washburn.

RESULTS

The Base glasses for this study were measured for density and surface tension. The results are given below in tables which contain the errors appropriate to the experiment. The graphs are to show only the trends and the lines are guides to the eye. To scrutinize the data for a single glass composition it is recommended that the data be plotted with the error bars of two standard deviations, to which a linear function can typically be fitted.

As expected the TV glass has the highest density and the LO glass the lowest. The FL and the E-glass have the most change with temperature; dramatic decrease in density with temperature which implies a higher expansion coefficient. All of the glasses have a positive temperature coefficient of surface tension which is what would normally be expected, except the wool glass which is basically temperature independent. This is because competing structural processes occurring in the wool glass prevent the normal increase in surface energy at higher temperature.

TABLE II. DENSITY AND SURFACE TENSION: CONTAINER GLASS

Temperature (°C)	Average Density (g/ cm^3)	Density Standard Deviation	Temperature (°C)	Average Surface Tension (ergs/cm^2)	Surface Tension Standard Deviation
1003	2.3954	0.0020	1100	298.2	2.9
1050	2.3916	0.0052	1216	306.7	1.8
1100	2.3890	0.0023	1303	315.0	6.9
1216	2.3320	0.0063	1400	328.2	6.7
1400	2.3220	0.0250			

TABLE III. DENSITY AND SURFACE TENSION: E-GLASS

Temperature (°C)	Average Density (g/ cm^3)	Density Standard Deviation	Temperature (°C)	Average Surface Tension (ergs/cm^2)	Surface Tension Standard Deviation
1200	2.447	0.0062	1126	315.9	4.4
1250	2.426	0.0113	1230	315.9	2.0
1300	2.400	0.0120	1318	320.1	7.8
1350	2.301	0.0381	1403	328.9	6.5
1400	2.150	0.0385			

TABLE IV. DENSITY: FLOAT GLASS

Temperature (°C)	Average Density (g/ cm^3)	Density Standard Deviation
1200	2.340	0.0087
1250	2.315	0.0061
1300	2.279	0.0053
1350	2.240	0.0139
1400	2.157	0.0111

TABLE V. DENSITY AND SURFACE TENSION: LOW-EXPANSION BOROSILICATE GLASS

Temperature (°C)	Average Density (g/ cm^3)	Density Standard Deviation	Average Surface Tension (ergs/cm^2)	Surface Tension Standard Deviation
1200	2.163	0.0067	371.44	7.19
1250	2.168	0.0026	365.14	6.63
1300	2.138	0.0032	371.88	6.62
1350	2.114	0.0052	375.10	3.63
1400	2.083	0.0042	376.74	3.55

TABLE VI. DENSITY AND SURFACE TENSION: TV PANEL GLASS

Temperature (°C)	Average Density (g/ cm^3)	Density Standard Deviation	Temperature (°C)	Average Surface Tension (ergs/cm^2)	Surface Tension Standard Deviation
909	2.5659	0.0044	1100	260.0	2.7
1003	2.5503	0.0077	1210	266.6	1.7
1100	2.5361	0.0050	1303	277.0	3.4
1303	2.5143	0.0043	1400	281.8	3.5
1400	2.4707	0.0175			

TABLE VII. DENSITY AND SURFACE TENSION: FIBER WOOL GLASS

Temperature (°C)	Average Density (g/ cm^3)	Density Standard Deviation	Average Surface Tension (ergs/cm^2)	Surface Tension Standard Deviation
965	2.3377	0.0007	298.9	3.0
1023	2.3104	0.0051	287.0	2.4
1090	2.3131	0.0053	292.3	1.4
1220	2.2819	0.0072	302.1	3.2
1264	2.2830	0.0060	279.0	2.0
1330	2.2614	0.0080	288.2	2.7
1400	2.2646	0.0073	290.7	6.2

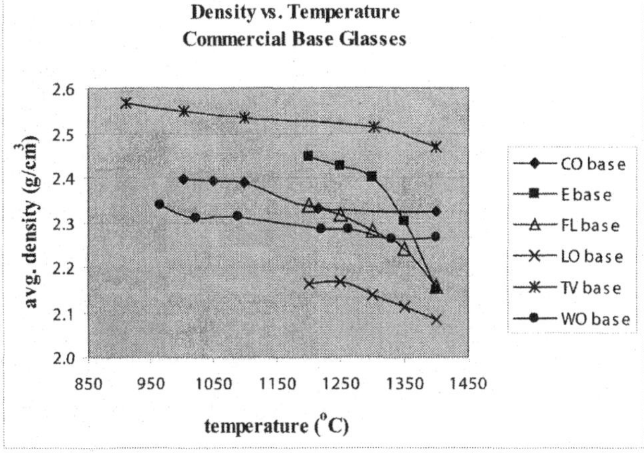

Figure 4. Density (g/cm^3) vs. temperature for container, E, float, low-exp. borosilicate, TV panel, and fiber wool base glasses

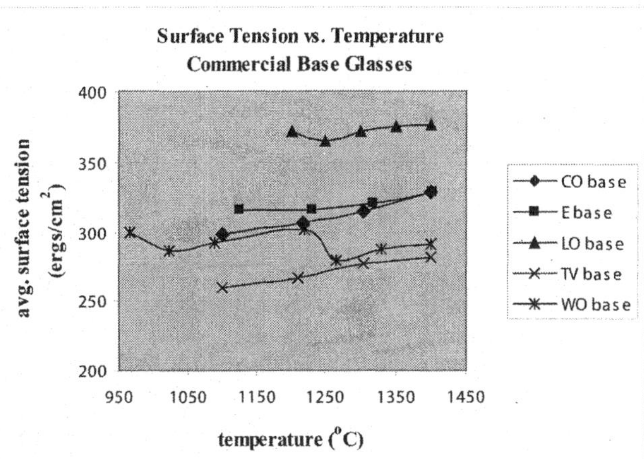

Figure 5. Surface tension (ergs/cm^2) vs. temperature for container, E, low-exp. borosilicate, TV panel, and fiber wool base glasses

LIMITATIONS OF THE TECHNIQUE

The error bars especially at high temperatures are quite substantial for all measurements, but there is a high confidence limit for the data because multiple frames are examined to calculate the density and surface tension. Comparisons with other data in the literature [28] and through our participation in Round Robin tests with the International Commission on Glass suggest that our measurements are reasonable: a Round Robin test on the density of a soda lime silicate glass had our data within one standard deviation of the mean lower than the mean . This method is not always the best choice depending upon the glass melt. Volatilization can result not only in mass errors but also in composition errors, and the sessile drop is prone to volatilization error due to the large surface-area-to-volume ratio. Our data in the Round Robin testing of the density of a TV panel glass came out higher than most of the other laboratories, however the data shown above for a TV panel glass were more consistent with the other Round Robin data . For these reasons, the maximum bubble pressure method, is more suitable for highly volatile samples.

The sessile drop experiment is limited by atmosphere; the requirement for a non-wetting substrate restricts the choice to graphite due to the need to be able to define the drop/substrate interface and to have enough coordinates to get a reasonable fit to the Young Laplace equation. It is also important that the melt and substrate do not interact: bubbles will give large errors in the density. A substrate with oriented structure such as scratches can cause a non-symmetric drop which may be wetting in one direction but not the other. However, the data collected for density and surface tension do help to give a rather better estimate for modeling despite some of the drawbacks and indeed such measurements have been solicited by many in the glass industry.

REFERENCES

1. Silverman, W. B. : Surface Tension of Glass and its Effect on Cords. J. Am. Ceram. Soc. **25** (1942) no. 6, p. 168-73.

2 Busby, T. S.; Eccles, J. A.: A study of the Solution of Single Crystals of Corundum in Molten Glass. Glass. Technol. **5** (1964) no. 3, p. 155-123.

3. Sharp, D. E.; Badger, A. E.: Physical Properties of Glasses. In: Tooley, F.V. (ed.): Handbook of Glass Data. New York: Ogden Pub., 1953. p. 35-39.

4. Members of the NSF Industry-University Center for Glass Research at Alfred: Private Communication (1998).

5. Kucuk, A.; Clare, A. G.; Jones L. E.: Differences Between Surface and Bulk of Glass Melts I: Compositional Differences and Influence of Volatilization on Composition and the Physical Properties. J. Non-Cryst. Solids **261**(1) 2000 28-38

6. Wing D. Çlare A.G. Data from TC18 of the International Commission on Glass Round Robin Tests for High Temperature Density

7. Kucuk, A.; Clare, A. G.; Jones, L. E.: Effect of Humidity in Melting Atmoshere on the Surface Tension of Silicate Glass Melts. In: 18th Int. Cong. on Glass. San Francisco: Am. Ceram. Soc., 1998 (CD ROM).

8. Kucuk A., Clare A.G., Jones L.E. Differences Between Surface and Bulk of Glass Melts II: Influence of Redox Ratio on Surface Properties of Silicate Melts Physics and Chemistry of Glasses **41** (2000) 75-80

9. Kucuk A., Clare A.G., Jones L.E. The Influence of various Atmospheres on the Surface Properties of Silicate Melts Glastech. Berichte Glass Sci. Technol. **73** (5) 2000 1-7

10. Kucuk A., Clare A.G., Jones L.E., Cormack A. Cao Y. The Bulk and Surface Structure of

Potassium and Lithium Silicate Melts at 1700K: A Comparison of Data from Sessile and Pendant Drop Measurements and Molecular Dynamics Simulations. Glastech. Berichte Glass Sci. Technol. **73** C2 2000 pp. 430-440

11. Clare A.G. Wing D Kucuk A. Jones L.E. Density and Surface Tension of Borate Containing Glass Melts Glass Technol. **44** (2) (2003) 59-62
12. Day A.L., Sosman R.B., Hostetter J.C. Determination of Mineral and Rock Densities at High Temperatures. Am. J. Sci. **37** (1914) p. 1-39.
13 Bockris J.O'M Tomlinson J.W., White J.L. "The Stucture of Liquid Silicates Partial Molar Volumes and Expansivities"Trans. Faraday Soc. **52** (1956) 229-310
14. Gaskell D.R., .McLean A., Ward R.G., Density and Structure of Ternary Silicate Melts Trans Faraday Soc. **65** (1965) p. 1498-1508.
15. Johnston J., Adams L.H., The Density of Solid Substances with Special Reference to Permanent Changes Induced by High Pressures. J. Am. Chem. Soc. **34** (1912) p. 563-584.
16. Kingery W.D., Humenik Jr. M., Surface Tension at Elevated Temperatures I. Furnace Method for use of the Sessile Drop Methgod, Surface Tension of Silicon, Iron and Nickel J. Phys. Chem. **57** 359-363 (1953)
17. Kozakevitch P., Urbain G. Surface Tension of Pure Liquid Iron, Cobalt and Nickel at 1550C J. Iron and Steel Inst. London 16 167-173 (1957)
18. Maze G., and Burnet G., A Non-Linear Regression Method for Calculating Surface Tension and Contact Angle from the Shape of a Sessile Drop Surface Sci. 13 451-470 (1969)
19. Butler J.N., A Curve Fitting Method for Calculating Interfacial Tension from the Shape of a Sessile Drop Surface Sci. 4 1-17 (1966)
20. Sangiori R., Caracciolo G., Passerone A., Factors Limiting the Accuracy of Measurements of Surface Tension by the Sessile Drop Method J. Mater. Sci 17 2895-2901 (1982)
21. Ellefson B., Taylor N.W., Surface Properties of Fused Salts and Glasses I. Sessile Drop Method for Determining Surface Tension and Density of Viscous Liquids at High Temperatures J. Am Ceram Soc. 21 193-213 (1938)
22. Bashforth F., Adams J.C. An Attempt to Test the Theories of Capillary Action by Comparing the Theoretical and Measured forms of Drops of Fluids
23. Morey G.W, The Properties of Glass pp. 197-211 Reinhold NY 1938
24. Parikh N.M. The Effect of Atmosphere on Surface Tension of Glass J. Am. Ceram Soc. 41 18-22 (1958)
25. Sharstis L., Spinner S., Capps W. Density, Expansivity, and Viscosity of Molten Alkali Silicates J. Am. Ceram. Soc. 35 155-160 (1952)
26. Rotenberg Y., Borukova L., Neumann A.W. Determination of Surface Tension and Contact Angle from the Shapes of Axisymmetric Fluid Interfaces J. Colloid Interface Sci. 93 169-183 (1983)
27. Kucuk A. Structure and the Physicochemical Properties of Glasses and Glass Melts Ph.D. Thesis Alfred University (1999) Appendix 1 pp. 173-178
28. Kucuk, A.; Clare, A. G.; Jones; L. E.: Density and Surface Tension of Glass Melts as a Function of Composition at 1400 C. Ceram. Trans. **82** (1998) p. 287-97.
29. Kucuk A., Clare A.G., Jones L.E. Density and Surface Tension of Commercial Glass Melts Advances in the Fusion and Processing of Glass II Vol 82, Transactions of the American Ceramic Society (1998)

30. Parmelee C.W., Harman C.G. The Effect of Alumina on the Surface Tension of Molten Glasses J. Am. Ceram. Soc 20 224-230 (1937)
31. Keppler G., The Surface Tension of Glass J. Soc. Glass Technology 21 53-60 (1937
32. Badger A.E., Parmelee C.W., Williams A.E., Surface Tension of Various Molten Glasses J. Am. Ceram Soc 20 325-329 (1937)
33. Williams R.M., Simpson H.E. A Note on the Effect of Lithium Oxide upon the Surface Tension of Several Silicate Melts J. Am. Ceram. Soc. 34 280 (1951)
34. Washburn E.W., Shelton G.R., Libman E.E., The Viscosities and Surface Tensions of Soda Lime Silica Glasses at High Temperatures Bull. Ill Univ. Eng. Exp. Stn. 21 51-71 (1924)
35. Din A. Hennicke H.W., Zum Verhalten Der Magnesiumionen in Silicatglaesern Glasstech. Ber. 47 14-16 (1974)
36. Kayalova S.S., Issedovanie Poverkhnostnogo Natyazheniya Silikatnukh Rasplavov Thesis Leningrad 1963, as cited in O.V. Mazurin, M.V. Streltsina and T. Shavaiko-Shavaikovskaya, Sci Glass Database program. Scivision Lexington MA (1996)

Chapter 7

VISCOSITY OF COMMERCIAL GLASSES

Alexander Fluegel and Arun K. Varshneya
New York State College of Ceramics at Alfred University
Alfred, New York 14802

James E. Shelby
New York State College of Ceramics at Alfred University
Alfred, New York 14802

P. Hrma, C. A. See, O. P. Lam, and K. B. C. Minister
Pacific Northwest National Laboratory
Richland, WA 99352

INTRODUCTION

In this study, viscosity was measured for six types of commercial glasses: container, fiberglass E and wool, float, low-expansion-borosilicate, and TV panel. Each type of glass was represented by 25 compositions, a commercial 'base' glass and 24 statistically-designed variations. Chapter 2 provides a detailed discussion of the commercial and statistically-designed glass compositions and the methods used for glass melting and analysis.

In order to measure viscosity over the entire range of interest (10^0-10^{12} Pa·s), three different techniques were employed. Viscosity data for temperatures between ~900°C and 1550°C (1 Pa·s to 750 Pa·s) were obtained using a rotating spindle viscometer. To measure viscosity in the glass softening range (10^6-10^9 Pa·s), the parallel plate technique was employed. The beam-bending method was used to for viscosity measurements in the transformation range (10^{10}-10^{12} Pa·s). Each of the measurement techniques and the data obtained is reported separately in this chapter.

The high-temperature viscosity measurements were made at Pacific Northwest National Laboratory under the direction of Pavel Hrma*. Measurements by the parallel plate and beam-bending methods were performed at New York State College of Ceramics under the direction of Arun Varshneya (parallel plate) and James Shelby (beam-bending).

*Funding for these measurements was provided by the US Department of Energy through the Glass Manufacturing Industry Council / National Laboratories G+ Project (sponsor: Schott Glass Technologies, Inc.).

HIGH-TEMPERATURE VISCOSITY OF COMMERCIAL GLASSES

P. Hrma, C. A. See, O. P. Lam, and K. B. C. Minister
Pacific Northwest National Laboratory
Richland, WA 99352

ABSTRACT

Viscosity was measured for six types of commercial glasses: low-expansion-borosilicate glasses, fiberglass E and wool glasses, TV panel glasses, container glasses, and float glasses. Viscosity data were obtained with rotating spindle viscometers within the temperature range between 900°C and 1550°C; the viscosity varied from 1 Pa·s to 750 Pa·s. Arrhenius coefficients were calculated for individual glasses.

INTRODUCTION

The melting behavior of glasses depends on their viscosity-temperature relationships. Glassmelting furnace design and operation are largely determined by the temperature span within which melt viscosity allows successful melting, fining, and conditioning of the glass. Therefore, the knowledge of glass viscosity as a function of temperature and glass composition is crucial for efficient glassmaking.

The most commonly used viscosity-temperature relationship for a wide span of viscosities (over 12 orders of magnitude) is the Vogel-Fulcher-Tammann (VTF) equation,

$$\ln\eta = A + \frac{B}{T - T_0} \tag{1}$$

where η is the viscosity, T is the absolute temperature, and A, B, and T_0 are temperature-independent coefficients [1]. However, during the melting process, viscosity of molten glass spans only 3 orders of magnitude (1 to 10^3 Pa·s). For this relatively narrow range, the viscosity-temperature relationship is sufficiently well represented by the Arrhenius equation [2,3],

$$\ln\eta = A + \frac{B}{T} \tag{2}$$

Table I lists composition regions of six types of commercial glasses tested in this study: low-expansion-borosilicate glass (LE), TV panel glass, E glass, fiberglass wool (FW), float glass (F), and container glass (C). The composition matrices of glasses that cover these six regions are presented in Chapter 2. Out of the total number of 150 different compositions (25 compositions—baseline plus 24 variations—for each of the six groups), 147, supplied by the Center for Glass Research at Alfred University, were tested in Pacific Northwest National Laboratory (PNNL), where viscosity was measured with rotating spindle viscometers. Arrhenius coefficients were obtained for the viscosity-temperature functions of individual glasses as slopes and intercepts of least-square lines through data plotted as ln η against $1/T$. The VTF coefficients, defined by Equation (1), were also obtained using the combined high and low viscosity database. The statistical modeling results are reported in Chapter 9 (see also a paper by Fluegel et al. [4]).

EXPERIMENTAL

The viscosity of glasses was measured using two rotating spindle viscometers: a manual Brookfield Digital Viscometer Model LVTD and a programmable Brookfield Digital Viscometer DV-III+ Rheometer. Both viscometers rotate the spindle at a constant speed and register the torque. For the optimum strain of the spring, the spindle speed is manually adjusted on the manual viscometer; for the programmable viscometer, the Theta Viscometer Controller adjusts the speed automatically. Each viscometer was set up above a vertical tube Deltech furnace and was equipped with a Pt-Rh spindle—a cylindrical disc 14 mm in diameter and 2 mm thick, located 8 mm from the end of the rod. The temperature-recording thermocouple was located under the crucible. The setup is schematically shown in Figure 1.

The measurement was conducted according to the standard PNNL test procedure. Each sample was prepared by crushing the glass into pieces 5 to 25 mm in size. A 50±5-mL volume of glass, measured with liquid (ethanol) displacement, was dried and added to a 5-cm diameter platinum crucible; the sample was premelted in a DT-31 furnace and transferred to the viscometer furnace. After the position of the crucible on the alumina pedestal and the spindle in the crucible were carefully centered, the spindle was lowered into the melt so its lower end of the rod was 2 mm from the bottom of the crucible.

The maximum operating temperature for the furnace was 1550°C. The lowest temperature, T_{min}, at which the viscosity was measurable by the viscometers (~300 Pa·s for the manual viscometer and ~700 Pa·s for the programmable viscometer), was initially estimated using viscosity models for waste glasses [3,5]. After a sufficient database of commercial glass was accumulated, first-order models were developed for the commercial glass composition region and used for more accurate estimates. For glasses tested in this study, T_{min} could be as low as 1000°C and as high as 1400°C. For a majority of glasses, four measurements were taken at 100°C intervals, starting at 1200°C and ending at 1500°C. The temperature interval, the number of temperature set-points, or both, were decreased when $T_{min} > 1250$°C. The glass was heated to each set-point and allowed to stabilize, i.e., to reach thermal equilibrium, for 25 min.

When the melt stabilized in the manual viscometer, the temperature, torque, and speed were recorded three times over 5 minutes (the initial, intermediate, and final readings). Even with the most careful instrument setup, the torque value oscillated nearly sinusoidally with amplitude that increased as the viscosity increased; the average of maximum and minimum value of the oscillating torque was taken as the final reading.

The programmable viscometer was set to record the temperature, time, torque, and spindle speed at 30-s intervals throughout the run. The final torque value was obtained by averaging ~20 readings at each set temperature. Table II shows the torque readout statistics for three randomly selected viscosity measurements on the programmable Brookfield viscometer. The corresponding standard deviations for these viscosity values are 0.11, 4.19, and 0.78 Pa·s, representing 0.5, 5.6, and 0.5% of the viscosity value. LE Base glass data from both viscometers are compared in Figure 2.

TABLE I. COMPOSITION REGIONS OF GLASSES IN MASS %

	LE	TV	E	FW	F	C	All
SiO_2	63-84	50-83	41.6-71.5	48.6-78	67.5-77.3	62.5-81	41.6-84
B_2O_3	10-15		0-9	3-9			0-15
Al_2O_3	2-7	1.3-3.5	12-16	0-6	.1-2	1-3	0-16
MgO		0-1.5	.5-4.5	1-5	3-4	0-3	0-5
CaO	0-2	0-3.5	16-24	5-11	7-9	7-12	0-24
Na_2O	4-8	6-9	0-2	13-17	12-15	11-15	0-17
K_2O	0-3	6-9	0-.5	0-2	0-2	0-2	0-9
Fe_2O_3			0-.8	0-.6	0.1-1.5	0-.4	0-1.5
TiO_2		.1-.5	0-1		0-1	0-.5	0-1
F		0-.7	0-.6	0-.6			0-.7
SO_3				0-.2		0-.3	0-.3
Li_2O		0-.5				0-1	0-1
Cr_2O_3					0-.4	0-.3	0-.4
BaO	0-2	2-13				0-.1	0-13
CeO_2		0-.7					0-.7
ZrO_2		0-3					0-3
PbO		0-3					0-3
ZnO		0-1.5					0-1.5
As_2O_3		0-.3					0-.3
Sb_2O_3		.2-.6					0-.6
SrO		1-10					0-10
MnO_2					0-.3		0-.3
Se					0-.03		0-.03
Co_3O_4					0-.2		0-.2
FeO			0-.1	0-.1			0-.1
LE—low-expansion-borosilicate glass, TV panel glass, E glass, FW—fiberglass wool, F—float glass, C—container glass							

The viscometers were calibrated with National Institute for Standards and Technology (NIST) 710a soda-lime-silica glass, for which the VFT coefficients, Equation (1), are: T_0 = 513.95 K, A = −1.7290 (for η in dPa·s), and B = 4560.0 K. Calibration data were used to calculate the spindle factor, F, defined as

$$F = \eta \frac{\omega}{\tau} \tag{3}$$

where τ is the torque, and ω is the spindle speed.

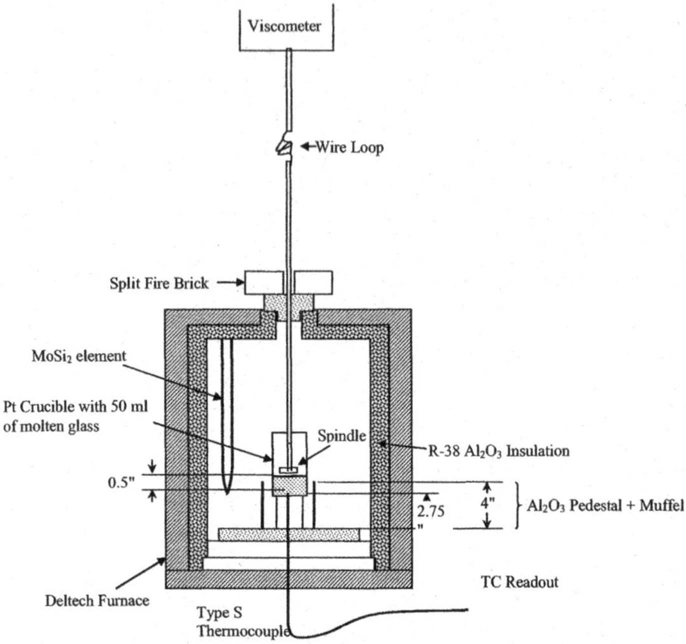

Figure 1. Viscometer setup

TABLE II. PROGRAMMABLE VISCOMETER MEASUREMENT STATISTICS

Glass ID	C 17	E 15	F 8
T, °C	1102	1351	1201
Mean Torque Reading	63.27	53.27	52.76
Standard Error	0.01	0.67	0.07
Median Torque Reading	63.28	54.44	52.77
Standard Deviation	0.051	2.983	0.276
Sample Variance	0.003	8.896	0.076
Kurtosis	-0.531	-1.969	-0.388
Skewness	0.075	-0.232	0.393
Range	0.180	7.215	0.970
Minimum Torque Reading	63.19	49.66	52.34
Maximum Torque Reading	63.37	56.87	53.31
Count	20	20	17

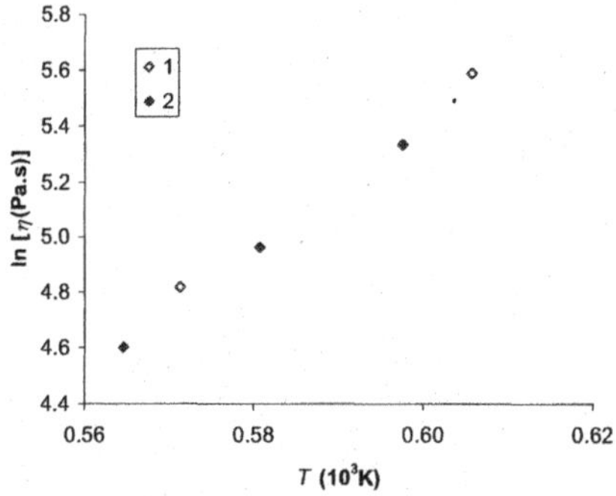

Figure 2. Viscosity data for LE Base glass measured with the manual (1) and programmable (2) viscometers

The temperature range for calibration on the manual viscometer varied from 1200°C to 1500°C. The measurement began at 1200°C. For each subsequent measurement, the temperature was increased by 100°C. The highest temperature was 1500°C. The measurement was repeated with temperature decreasing to 1200°C by 100°C intervals to assure that neither crystallization nor volatilization affected glass viscosity during measurement. As this precaution, necessary for waste glasses that are prone to crystallization, appeared superfluous for commercial glasses, the measurement during decreasing temperature was no longer followed for the programmable viscometer. The programmable viscometer operated at a wider viscosity range, from 1100°C to 1500°C; its calibration temperature range varied from 1200°C to 1500°C for the first calibration and from 1100°C to 1500°C for following calibrations. Viscometers were calibrated monthly at regular intervals determined by the number of measurements rather than time, and when a new crucibles or spindle were implemented. Manual viscometer furnace was repaired starting 8/7/02 and ending 8/12/02, and the programmable viscometer crucible and spindle were changed on 11/1/02.

When the viscometer was calibrated in response to an equipment adjustment, the new spindle factor was used for the glasses following the calibration. During the intervals when the spindle factor changed without replacing a part of the equipment, we reasoned that the crucible and spindle sustained progressive deformation with each additional test. Therefore, the spindle factor was determined by interpolation,

$$F = F_i + (F_{i+1} - F_i)\frac{n}{n_t} \tag{4}$$

where F_i is the spindle factor from the preceding calibration, F_{i+1} is the spindle factor from the subsequent calibration, n_t is the number of glasses tested between the calibrations, and n is the number of glasses tested since the preceding calibration.

The viscosity of glasses was calculated using Equation (3) rearranged as

$$\eta = F\frac{\tau}{\omega} \tag{5}$$

Table III shows the date of calibration, the temperatures at which measurements were taken, the calculated viscosities, and the corresponding spindle factors. The average standard deviation for the spindle factor for each calibration was 4.9%. This value can be considered as characteristic per the measurement accuracy. Disregarding the 7/12/02 calibration before the manual viscometer was rebuilt, the average standard deviation for the spindle factor was 3.3% for the manual viscometer and 2.3% for the programmable viscometer.

The accuracy of the viscosity measurement has been established in previous studies: samples of glasses were sent to Corning Engineering Laboratory Services (CELS) for viscosity overcheck testing; the CELS and PNNL data showed good agreement [5].

Measurement chronology is summarized in Tables IV (for the manual viscometer) and V (for the programmable viscometer).

RESULTS

Measured viscosity data are collected in Tables VI to XI. The temperature of measurements spanned an interval from 899°C to 1554°C and viscosity varied from 1 Pa·s to 747 Pa·s. However, most of the data points were taken within the temperature-viscosity region from 1100°C to 1550°C and 10 to 400 Pa·s (see the rectangle in Figure 3).

The Arrhenius coefficients, *A* and *B*, obtained by fitting Equation (2) to data, are listed in Table XII.

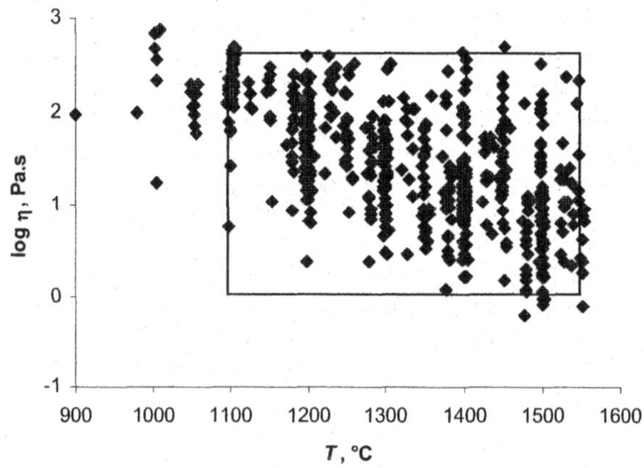

Figure 3. Map of viscosity-temperature data points

TABLE III. VISCOSITY CALIBRATION DATA[(a)]

#	Date	T,°C	η, Pa.s	F	F_a	#	Date	T,°C	η, Pa.s	F	F_a
1	7/12/02	1179	135.3	0.93		1	9/26/02	1198	108.4	0.86	
		1180	133.7	1.08				1200	105.9	0.85	
		1277	47.0	1.19				1296	39.1	0.86	
		1279	46.0	1.26				1299	38.0	0.86	
		1376	19.4	1.00				1396	16.5	0.86	
		1377	19.2	0.78				1398	16.3	0.88	
		1476	9.2	0.63	0.98			1496	8.0	0.89	0.87
1	8/6/02	1180	133.7	1.06		1	11/4/02	1200	105.9	0.83	
		1182	130.6	1.05				1295	39.5	0.86	
		1279	46.0	1.06				1395	16.7	0.87	
		1282	44.7	1.02				1496	8.0	0.87	
		1379	18.9	1.02				1398	16.3	0.86	
		1380	18.8	1.06				1300	37.7	0.83	
		1478	9.1	1.00	1.04			1198	108.4	0.85	0.85
1	8/12/02	1177	138.6	0.99		2	9/6/02	1202	103.5	9.92	
		1181	132.1	1.02				1298	38.2	9.90	
		1276	47.4	1.01				1399	16.1	10.13	
		1279	46.0	1.02				1499	7.8	10.49	10.11
		1376	19.4	0.96		2	10/9/02	1199	107.1	9.72	
		1378	19.1	1.01				1298	38.3	9.91	
		1476	9.2	1.00	1.00			1398	16.3	10.38	
1	9/24/02	1198	108.4	0.92				1498	7.9	10.47	10.12
		1201	104.7	0.90		2	11/4/02	1199	107.2	9.58	
		1297	38.8	0.90				1298	38.5	9.69	
		1301	37.3	0.88				1398	16.3	9.92	
		1397	16.4	0.90				1498	7.9	10.36	9.89
		1399	16.1	0.88							
		1497	8.0	0.82	0.89						

(a) #1 = Manual Viscometer, #2 = Programmable Viscometer; F_a is the spindle factor averaged over the temperature interval of calibration

TABLE IV. MEASUREMENT CHRONOLOGY: MANUAL VISCOMETER

Date	ID	Date	ID	Date	ID
7/11/02	LE Base, 3	8/26/02	FW 4, 5	9/23/02	TV 21, 22
7/22/02	LE 4	8/27/02	FW 6, 8	9/24/02	TV 23
7/23/02	LE 8, 9, 23	8/29/02	FW 9-11	9/27/02	C 14
7/26/02	LE 10, 24,21	8/30/02	FW 12, 14	9/30/02	C 16
7/30/02	LE 7	9/3/02	FW 15, 16	10/1/02	C 20
7/31/02	LE 11	9/4/02	FW 19, 20	10/3/02	TV Base
8/1/02	E 3	9/5/02	FW 22, 23	10/7/02	FW Base
8/7/02	E 7	9/6/02	FW 24	10/8/02	FW 7
8/13/02	E 22, 23	9/9/02	TV 1, 6	10/21/02	F 2
8/14/02	E 6, 24	9/10/02	TV 3, E 5	10/22/02	F 3
8/15/02	E 8, 9	9/11/02	TV 5-9	10/23/02	F 6
8/16/02	E 10, 11	9/13/02	TV 10, 11	10/24/02	F 7
8/19/02	E 12, 13	9/16/02	TV 12, 13	10/25/02	F 10
8/20/02	E 18, 19	9/17/02	LE 13, 19, TV 14	10/28/02	F 11
8/21/02	E base, 21	9/18/02	TV 15, 16	10/29/02	F 16
8/22/02	E 14, FW 1	9/19/02	TV 17, 18	10/30/02	F 21
8/23/02	FW 2, 3	9/20/02	TV 19, 20	10/31/02	F 24

TABLE V. MEASUREMENT CHRONOLOGY: PROGRAMMABLE VISCOMETER

Date	ID	Date	ID	Date	ID
9/11/02	LE 5	10/3/02	C 21	10/28/002	F 15, 17
9/12/02	LE 6	10/7/02	C Base, 22	10/29/02	F 18, 19
9/17/02	LE 20, E 2	10/8/02	C 23, 24	10/30/02	F 20, 22
9/18/02	E 15, 17	10/11/02	FW 13	10/31/02	F 23
9/19/02	TV 4, C 1-3	10/14/02	FW 17	11/8/02	FW 13
9/23/02	C 4-7	10/15/02	FW 21	11/14/02	TV 24, LE Base, 1
9/25/02	C 8, 9	10/21/02	F Base	11/15/02	LE 2
9/26/02	C 10, 11	10/22/02	F 1, 4	11/19/02	LE 12, 14, 15, 17
9/27/02	C 13, 12,18	10/23/02	F 5, 8	11/20/02	E 1, 4, 18, 22
9/30/02	C 15	10/24/02	F 9, 12	11/21/02	E 16, TV 2
10/1/02	C 17, 19	10/25/02	F 13, 14	12/23/02	LE 13, 19, FW 21

TABLE VI. VISCOSITY OF LOW-EXPANSION BOROSILICATE GLASSES

Glass	T, °C	η, Pa·s	Glass	T, °C	η, Pa·s	Glass	T, °C	η, Pa·s
LE Base	1401	206.82	LE 8	1282	53.10	LE 17	1401	281.01
LE Base	1449	143.19	LE 8	1381	23.46	LE 17	1448	186.36
LE Base	1498	99.76	LE 8	1481	12.86	LE 17	1498	124.44
LE 1	1402	356.15	LE 9	1180	62.47	LE 18	1398	91.42
LE 1	1449	233.77	LE 9	1278	22.93	LE 18	1446	62.87
LE 1	1499	153.34	LE 9	1377	10.78	LE 18	1497	43.81
LE 2	1399	168.13	LE 9	1476	6.26	LE 19	1258	20.34
LE 2	1448	109.60	LE 9	1526	4.84	LE 19	1352	8.07
LE 2	1499	73.27	LE 10	1234	233.62	LE 19	1452	3.48
LE 3	1234	97.34	LE 10	1334	108.25	LE 19	1552	1.82
LE 3	1334	40.48	LE 10	1433	53.77	LE 20	1302	281.51
LE 3	1433	19.67	LE 10	1534	25.14	LE 20	1376	148.45
LE 3	1533	10.81	LE 11	1260	325.92	LE 20	1451	79.48
LE 4	1531	235.97	LE 11	1358	148.70	LE 20	1526	45.76
LE 4	1546	213.14	LE 11	1458	67.97	LE 21	1184	107.75
LE 5	1226	199.51	LE 11	1547	35.85	LE 21	1283	43.57
LE 5	1325	79.78	LE 12	1397	424.57	LE 21	1382	20.14
LE 5	1426	36.30	LE 12	1545	122.57	LE 21	1482	10.72
LE 5	1527	18.82	LE 13	1252	8.12	LE 22	1450	500.20
LE 6	1229	257.95	LE 13	1350	3.23	LE 22	1498	323.29
LE 6	1326	104.81	LE 13	1451	1.49	LE 23	1187	116.31
LE 6	1426	46.56	LE 13	1551	0.78	LE 23	1286	42.70
LE 6	1526	23.24	LE 14	1400	54.29	LE 23	1385	24.23
LE 7	1237	294.37	LE 14	1448	37.61	LE 23	1482	11.98
LE 7	1337	105.56	LE 14	1498	26.51	LE 24	1183	72.84
LE 7	1436	41.76	LE 15	1400	72.91	LE 24	1282	29.00
LE 7	1536	17.72	LE 15	1448	49.82	LE 24	1381	14.43
LE 8	1184	149.87	LE 15	1497	34.02	LE 24	1480	8.63

TABLE VII. VISCOSITY OF E GLASSES

Glass	*T*, °C	η, Pa·s	Glass	*T*, °C	η, Pa·s	Glass	*T*, °C	η, Pa·s
E Base	1182	130.19	E 8	1278	12.56	E 16	1398	111.77
E Base	1281	28.97	E 8	1378	4.59	E 16	1447	61.91
E Base	1380	9.47	E 8	1478	1.99	E 16	1497	35.20
E Base	1479	4.15	E 9	1183	23.53	E 17	1252	251.63
E 1	1401	363.94	E 9	1281	6.81	E 17	1349	62.92
E 1	1448	199.35	E 9	1380	2.49	E 17	1450	19.71
E 1	1498	105.15	E 9	1479	1.18	E 17	1550	7.47
E 2	1306	332.14	E 10	1181	46.44	E 18	1230	90.85
E 2	1377	125.37	E 10	1279	12.29	E 18	1329	18.79
E 2	1451	49.07	E 10	1378	4.36	E 18	1428	5.99
E 2	1527	22.00	E 10	1478	1.74	E 18	1528	2.35
E 3	1237	53.71	E 11	1183	83.67	E 19	1231	169.53
E 3	1335	12.72	E 11	1281	20.60	E 19	1282	88.53
E 3	1435	4.81	E 11	1382	7.03	E 19	1381	26.87
E 3	1536	2.14	E 11	1480	2.83	E 19	1480	10.72
E 4	1400	87.60	E 12	1181	154.80	E 21	1181	33.77
E 4	1448	50.56	E 12	1279	38.87	E 21	1280	9.13
E 4	1498	29.66	E 12	1379	12.20	E 21	1379	2.70
E 5	1251	58.92	E 12	1479	4.87	E 21	1479	1.14
E 5	1351	15.80	E 13	1182	250.45	E 22	1229	216.05
E 5	1450	5.53	E 13	1282	52.46	E 22	1327	56.34
E 5	1550	2.36	E 13	1380	16.10	E 22	1427	20.93
E 6	1182	186.07	E 13	1480	6.10	E 22	1527	9.66
E 6	1280	40.88	E 14	1250	155.11	E 23	1181	28.65
E 6	1380	11.45	E 14	1349	34.73	E 23	1280	8.07
E 6	1479	3.64	E 14	1449	13.04	E 23	1379	3.07
E 7	1230	113.79	E 14	1540	5.93	E 23	1478	1.51
E 7	1328	78.71	E 15	1253	288.96	E 24	1179	8.56
E 7	1428	18.60	E 15	1351	74.84	E 24	1277	2.42
E 7	1528	10.79	E 15	1452	23.70	E 24	1377	1.19
E 8	1180	46.30	E 15	1553	8.99	E 24	1477	0.60

TABLE VIII. VISCOSITY OF FIBERGLASS WOOL GLASSES

Glass	*T*, °C	η, Pa·s	Glass	*T*, °C	η, Pa·s	Glass	*T*, °C	η, Pa·s
FW Base	1100	128.07	FW 8	1203	6.24	FW 16	1152	243.17
FW Base	1197	18.35	FW 8	1302	2.90	FW 16	1278	68.91
FW Base	1298	8.03	FW 8	1402	1.60	FW 16	1401	24.32
FW Base	1398	4.03	FW 8	1500	0.93	FW 16	1500	12.27
FW 1	1251	80.96	FW 9	1201	6.63	FW 17	1002	684.13
FW 1	1350	40.17	FW 9	1300	3.01	FW 17	1099	168.26
FW 1	1449	14.76	FW 9	1400	1.63	FW 17	1199	53.45
FW 1	1539	7.90	FW 9	1500	0.83	FW 17	1298	20.82
FW 2	1200	86.93	FW 10	1204	8.42	FW 19	1056	57.34
FW 2	1299	34.64	FW 10	1304	5.29	FW 19	1201	11.60
FW 2	1399	16.16	FW 10	1403	2.54	FW 19	1352	3.97
FW 2	1499	8.97	FW 10	1502	0.93	FW 19	1501	1.62
FW 3	1206	14.70	FW 11	1199	11.74	FW 20	1059	195.31
FW 3	1302	6.95	FW 11	1299	5.70	FW 20	1207	33.22
FW 3	1402	3.63	FW 11	1399	2.89	FW 20	1355	9.11
FW 3	1501	2.13	FW 11	1499	1.71	FW 20	1503	3.77
FW 4	1201	65.77	FW 12	1199	40.32	FW 21	1002	482.59
FW 4	1300	27.05	FW 12	1299	17.02	FW 21	1098	118.79
FW 4	1400	14.07	FW 12	1399	8.14	FW 21	1198	39.54
FW 4	1500	8.04	FW 12	1499	4.06	FW 21	1298	16.60
FW 5	1203	22.74	FW 13	1004	219.56	FW 22	1099	77.39
FW 5	1302	9.86	FW 13	1100	61.77	FW 22	1222	22.43
FW 5	1402	4.81	FW 13	1199	22.57	FW 22	1348	8.98
FW 5	1500	2.56	FW 13	1299	9.81	FW 22	1475	6.71
FW 6	1201	15.73	FW 14	1203	63.99	FW 23	980	96.75
FW 6	1300	6.80	FW 14	1302	23.54	FW 23	1155	10.94
FW 6	1400	3.35	FW 14	1402	10.58	FW 23	1327	2.81
FW 6	1500	1.80	FW 14	1501	5.74	FW 23	1500	1.11
FW 7	1100	27.03	FW 15	1201	25.51	FW 24	899	94.04
FW 7	1197	19.80	FW 15	1300	11.95	FW 24	1005	17.65
FW 7	1297	4.60	FW 15	1400	5.90	FW 24	1098	5.82
FW 7	1398	2.42	FW 15	1500	3.26	FW 24	1199	2.41

TABLE IX. VISCOSITY OF TV PANEL GLASSES

Glass	T, °C	η, Pa·s	Glass	T, °C	η, Pa·s	Glass	T, °C	η, Pa·s
TV Base	1198	90.39	TV 8	1350	7.20	TV 16	1497	7.70
TV Base	1296	32.88	TV 8	1500	2.24	TV 17	1054	116.14
TV Base	1397	14.08	TV 9	1055	94.50	TV 17	1199	21.13
TV Base	1497	6.60	TV 9	1202	18.65	TV 17	1349	5.68
TV 1	1249	85.11	TV 9	1351	5.59	TV 17	1497	2.29
TV 1	1348	33.49	TV 9	1500	2.07	TV 18	1099	154.39
TV 1	1447	14.84	TV 10	1051	110.94	TV 18	1247	28.43
TV 1	1548	8.36	TV 10	1197	24.88	TV 18	1397	7.59
TV 2	1401	49.85	TV 10	1347	8.14	TV 18	1548	2.54
TV 2	1448	33.84	TV 10	1498	3.22	TV 19	1202	87.44
TV 2	1498	23.41	TV 11	1127	111.14	TV 19	1300	34.38
TV 3	1248	153.35	TV 11	1251	32.14	TV 19	1399	14.99
TV 3	1347	65.83	TV 11	1375	11.19	TV 19	1498	7.54
TV 3	1447	28.97	TV 11	1498	4.83	TV 20	1197	129.89
TV 3	1546	14.67	TV 12	1050	162.15	TV 20	1297	46.61
TV 4	1227	389.90	TV 12	1197	31.64	TV 20	1398	20.51
TV 4	1323	139.52	TV 12	1346	7.53	TV 20	1497	9.51
TV 4	1424	53.96	TV 12	1496	2.42	TV 21	1128	106.90
TV 4	1524	24.01	TV 13	1152	89.08	TV 21	1251	27.10
TV 5	1248	159.76	TV 13	1276	22.10	TV 21	1375	9.27
TV 5	1348	51.69	TV 13	1400	7.31	TV 21	1499	3.72
TV 5	1447	20.57	TV 13	1524	2.91	TV 22	1126	157.68
TV 5	1547	11.44	TV 14	1053	71.62	TV 22	1249	40.71
TV 6	1054	148.42	TV 14	1198	14.53	TV 22	1374	13.96
TV 6	1203	21.36	TV 14	1348	4.29	TV 22	1499	5.86
TV 6	1351	5.39	TV 14	1499	1.61	TV 23	1251	86.18
TV 6	1499	1.81	TV 15	1152	168.11	TV 23	1349	34.47
TV 7	1056	156.01	TV 15	1253	50.46	TV 23	1449	16.11
TV 7	1203	20.83	TV 15	1402	12.34	TV 23	1550	8.66
TV 7	1350	5.02	TV 15	1551	4.10	TV 24	1399	103.77
TV 7	1500	1.62	TV 16	1199	103.51	TV 24	1448	69.31
TV 8	1050	195.26	TV 16	1298	37.38	TV 24	1498	46.96
TV 8	1199	29.72	TV 16	1398	15.75			

TABLE X. VISCOSITY OF CONTAINER GLASSES

Glass	T, °C	η, Pa·s	Glass	T, °C	η, Pa·s	Glass	T, °C	η, Pa·s
C Base	1202	107.07	C 8	1296	62.54	C 16	1397	29.31
C Base	1299	39.46	C 8	1400	25.63	C 16	1499	15.66
C Base	1399	16.66	C 8	1501	12.27	C 17	1102	136.69
C Base	1499	8.02	C 9	1105	112.36	C 17	1199	41.71
C 1	1303	127.22	C 9	1199	38.92	C 17	1298	15.66
C 1	1375	63.55	C 9	1298	15.76	C 17	1398	6.76
C 1	1449	34.07	C 9	1398	7.22	C 18	1106	183.58
C 1	1524	19.24	C 10	1102	453.37	C 18	1200	57.74
C 2	1152	294.51	C 10	1198	137.43	C 18	1300	21.42
C 2	1274	67.75	C 10	1299	49.10	C 18	1401	8.70
C 2	1400	20.30	C 10	1400	19.91	C 19	1107	469.31
C 2	1554	6.65	C 11	1102	134.29	C 19	1201	144.15
C 3	1173	45.07	C 11	1199	41.96	C 19	1300	50.21
C 3	1257	19.01	C 11	1299	15.37	C 19	1401	21.84
C 3	1354	7.92	C 11	1400	6.57	C 20	1198	251.98
C 3	1453	3.80	C 12	1201	166.39	C 20	1297	81.12
C 4	1203	93.35	C 12	1298	54.51	C 20	1397	30.84
C 4	1300	33.46	C 12	1398	20.90	C 20	1498	13.63
C 4	1401	14.03	C 12	1501	9.34	C 21	1103	286.36
C 4	1502	6.66	C 13	1106	251.77	C 21	1199	87.43
C 5	1153	79.76	C 13	1201	84.51	C 21	1298	32.40
C 5	1248	30.45	C 13	1300	32.98	C 21	1398	14.04
C 5	1348	12.99	C 13	1401	14.73	C 22	1203	241.93
C 5	1448	6.18	C 14	1296	64.25	C 22	1300	81.55
C 6	1102	121.54	C 14	1371	32.90	C 22	1400	31.78
C 6	1197	39.56	C 14	1446	17.89	C 22	1499	14.31
C 6	1297	14.85	C 14	1523	6.31	C 23	1102	246.94
C 6	1398	6.45	C 15	1103	158.22	C 23	1197	81.20
C 7	1102	325.72	C 15	1199	50.33	C 23	1298	30.78
C 7	1197	105.26	C 15	1298	19.31	C 23	1399	13.42
C 7	1298	36.63	C 15	1398	8.41	C 24	1005	353.44
C 7	1398	16.60	C 16	1198	210.45	C 24	1100	65.31
C 8	1202	163.12	C 16	1297	73.99	C 24	1199	21.24
						C 24	1300	8.19

TABLE XI. VISCOSITY OF FLOAT GLASSES

Glass	T, °C	η, Pa·s	Glass	T, °C	η, Pa·s	Glass	T, °C	η, Pa·s
F Base	1201	86.52	F 8	1101	176.34	F 16	1320	24.33
F Base	1299	31.99	F 8	1201	53.25	F 16	1422	10.99
F Base	1398	13.28	F 8	1300	19.87	F 17	1102	412.60
F Base	1499	6.41	F 9	1106	414.91	F 17	1199	109.14
F 1	1201	241.00	F 9	1200	123.13	F 17	1298	37.22
F 1	1298	80.42	F 9	1300	43.00	F 17	1398	14.67
F 1	1399	31.08	F 9	1400	17.80	F 18	1102	240.92
F 1	1499	13.35	F 10	1103	189.13	F 18	1198	73.92
F 2	1099	183.76	F 10	1197	54.63	F 18	1298	27.52
F 2	1196	59.48	F 10	1298	20.60	F 18	1398	11.75
F 2	1299	22.13	F 10	1398	8.94	F 19	1108	469.88
F 2	1398	9.78	F 11	1101	154.43	F 19	1201	141.47
F 3	1098	228.00	F 11	1197	46.99	F 19	1300	50.33
F 3	1195	74.47	F 11	1296	18.35	F 19	1401	20.46
F 3	1295	28.53	F 11	1397	8.28	F 20	1105	207.63
F 3	1396	23.52	F 12	1105	326.65	F 20	1201	63.50
F 4	1105	161.23	F 12	1200	96.26	F 20	1301	22.14
F 4	1201	50.67	F 12	1299	33.80	F 20	1402	9.21
F 4	1299	18.93	F 12	1399	14.01	F 21	1149	216.29
F 4	1400	8.06	F 13	1105	362.09	F 21	1246	71.67
F 5	1008	714.42	F 13	1201	95.56	F 21	1345	26.56
F 5	1102	171.39	F 13	1301	32.23	F 21	1447	11.84
F 5	1201	49.91	F 13	1401	11.76	F 22	1107	405.24
F 5	1300	19.65	F 14	1104	266.06	F 22	1201	116.14
F 6	1096	124.06	F 14	1200	78.69	F 22	1301	39.81
F 6	1199	41.84	F 14	1299	30.14	F 22	1401	15.57
F 6	1297	16.11	F 14	1399	12.82	F 23	1105	487.84
F 6	1398	7.24	F 15	1103	237.34	F 23	1200	141.60
F 7	1147	162.70	F 15	1200	71.10	F 23	1299	49.45
F 7	1247	50.67	F 15	1301	23.39	F 23	1400	19.48
F 7	1347	19.84	F 15	1401	9.64	F 24	1195	224.65
F 7	1448	8.66	F 16	1123	205.07	F 24	1296	65.06
F 8	1009	747.33	F 16	1221	66.74	F 24	1394	24.89
						F 24	1495	11.45

TABLE XII. ARRHENIUS COEFFICIENTS FOR VISCOSITY IN PA·S (MEASURED)

Glass	A	$10^{-3}B$ (K)	Glass	A	$10^{-3}B$ (K)	Glass	A	$10^{-3}B$ (K)
LE Base	-7.959	22.248	FW Base	-14.447	26.111	C Base	-10.834	22.843
LE 1	-9.494	25.743	FW 1	-10.522	22.844	C 1	-10.512	24.187
LE 2	-9.578	24.572	FW 2	-9.076	19.898	C 2	-11.624	24.587
LE 3	-8.742	20.038	FW 3	-8.962	17.203	C 3	-11.486	22.094
LE 4	-6.875	22.255	FW 4	-8.273	18.301	C 4	-11.150	23.114
LE 5	-8.861	21.191	FW 5	-9.903	19.221	C 5	-10.547	21.265
LE 6	-9.051	21.923	FW 6	-10.103	18.938	C 6	-11.799	22.792
LE 7	-11.297	25.663	FW 7	-11.011	19.947	C 7	-11.141	23.242
LE 8	-9.608	21.215	FW 8	-9.493	16.684	C 8	-10.276	22.645
LE 9	-9.113	19.125	FW 9	-10.297	19.968	C 9	-10.967	21.588
LE 10	-7.851	20.100	FW 10	-10.693	19.194	C 10	-11.433	24.103
LE 11	-8.205	21.485	FW 11	-8.996	16.866	C 11	-12.074	23.310
LE 12	-9.203	25.468	FW 12	-9.820	19.903	C 12	-11.992	25.170
LE 13	-12.227	21.799	FW 13	-11.225	21.158	C 13	-10.614	22.223
LE 14	-9.034	21.796	FW 14	-10.273	21.237	C 14	-13.650	28.095
LE 15	-9.579	23.203	FW 15	-9.001	18.049	C 15	-11.556	22.835
LE 16			FW 16	-9.779	21.751	C 16	-10.152	22.739
LE 17	-9.262	24.941	FW 17	-12.029	23.608	C 17	-12.063	23.306
LE 18	-8.561	21.844	FW 18			C 18	-12.051	23.775
LE 19	-12.076	23.056	FW 19	-10.171	18.795	C 19	-11.397	24.166
LE 20	-8.988	23.048	FW 20	-10.691	21.156	C 20	-11.733	25.360
LE 21	-8.966	19.853	FW 21	-11.777	22.808	C 21	-11.454	23.501
LE 22	-9.909	27.782	FW 22	-7.449	15.990	C 22	-11.457	24.980
LE 23	-8.503	19.282	FW 23	-10.836	19.148	C 23	-10.909	22.545
LE 24	-8.395	18.387	FW 24	-13.596	21.170	C 24	-14.188	25.465

TABLE XII. ARRHENIUS COEFFICIENTS FOR VISCOSITY IN PA·S (MEASURED)—CONT.

Glass	A	$10^{-3}B$ (K)	Glass	A	$10^{-3}B$ (K)	Glass	A	$10^{-3}B$ (K)
E Base	-14.447	29.428	TV Base	-10.966	22.730	F Base	-11.089	22.898
E 1	-16.675	37.794	TV 1	-9.853	21.708	F 1	-11.705	25.318
E 2	-16.399	35.039	TV 2	-9.929	23.161	F 2	-11.192	22.481
E 3	-15.519	29.282	TV 3	-9.346	21.891	F 3	-7.918	18.099
E 4	-15.026	32.616	TV 4	-10.924	23.325	F 4	-11.943	23.428
E 5	-15.623	29.951	TV 5	-11.210	24.668	F 5	-12.954	24.952
E 6	-17.943	33.682	TV 6	-12.607	23.277	F 6	-10.967	21.617
E 7	-10.439	23.080	TV 7	-13.267	24.238	F 7	-11.674	23.769
E 8	-14.687	26.826	TV 8	-12.348	23.259	F 8	-12.978	25.037
E 9	-14.701	25.921	TV 9	-10.687	20.174	F 9	-11.962	24.770
E 10	-15.513	28.071	TV 10	-9.331	18.532	F 10	-12.025	23.678
E 11	-15.555	29.007	TV 11	-10.313	21.015	F 11	-11.495	22.651
E 12	-15.392	29.646	TV 12	-11.645	22.153	F 12	-12.160	24.692
E 13	-16.411	31.812	TV 13	-12.067	23.548	F 13	-13.429	26.588
E 14	-15.316	30.850	TV 14	-10.842	19.977	F 14	-11.566	23.554
E 15	-15.494	32.237	TV 15	-11.887	24.184	F 15	-12.629	24.890
E 16	-15.990	34.597	TV 16	-10.836	22.749	F 16	-11.382	23.293
E 17	-16.020	32.794	TV 17	-11.076	20.913	F 17	-12.811	25.840
E 18	-17.636	29.428	TV 18	-11.639	22.843	F 18	-11.599	23.451
E 19	-14.448	36.774	TV 19	-10.245	21.691	F 19	-11.756	26.684
E 20			TV 20	-10.489	22.549	F 20	-12.305	24.280
E 21	-16.656	29.300	TV 21	-11.373	22.434	F 21	-11.482	23.947
E 22	-13.464	28.166	TV 22	-10.625	21.890	F 22	-12.599	25.642
E 23	-14.124	25.307	TV 23	-9.602	21.376	F 23	-12.076	25.135
E 24	-13.342	22.323	TV 24	-9.491	23.629	F 24	-12.220	25.814

ACKNOWLEDGMENT

This study was funded by the U.S. Department of Energy's Office of Science and Technologies through the G-Plus Program. Pacific Northwest National Laboratory is operated for the U.S. Department of Energy by Battelle under Contract DE-AC06-76RL01830. Clem See, Oanh Lam, and Kevin Minister are grateful for the student fellowship at Pacific Northwest National Laboratory. The interest and encouragement by Jim Shell, G-Plus facilitator, formerly Techneglass Company, is gladly appreciated. The authors thank Mike Schweiger and Don Smith for laboratory guidance. Thomas P. Seward and Terese Vascott from Alfred University are acknowledged for providing the glasses for testing.

REFERENCES

H. Scholze, *Glass*, Springer, New York 1990.

N. P. Bansal and R. H. Doremus, *Handbook of Glass Properties*, Academic Press, Orlando 1986.

P. Hrma, G.F. Piepel, P.E. Redgate, D.E. Smith, M.J. Schweiger, J.D. Vienna, and D-S. Kim, "Prediction of Processing Properties for Nuclear Waste Glasses," *Environmental Issues and Waste Management Technologies in the Ceramic and Nuclear Industries, Ceramic Transaction* 61, 505-513 (1995).

A. Fluegel, A. K. Varshneya, D. A. Earl, T. P. Seward, and D. Oksoy, "Improved Composition-Property Relations in Silicate Glasses, Part I: Viscosity," *Ceramic Transaction* [in print].

P. Hrma, G.F. Piepel, M.J. Schweiger, D.E. Smith, D-S. Kim, P.E. Redgate, J.D. Vienna, C.A. LoPresti, D.B. Simpson, D.K. Peeler, and M.H. Langowski, *Property/Composition Relationships for Hanford High-Level Waste Glasses Melting at 1150°C*, PNL-10359, Vol. 1 and 2, Pacific Northwest Laboratory, Richland, Washington (1994).

PARALLEL PLATE VISCOMETRY

Alexander Fluegel and Arun K. Varshneya
New York State College of Ceramics at Alfred University
Alfred, New York 14802

ABSTRACT

In this chapter the common techniques used to measure viscosity in the glass softening range are discussed and compared. These include the ball penetration, fiber elongation, and parallel plate methods. The parallel plate technique is applied for determining the viscosities of 146 commercial glass variations comprising low expansion borosilicate, textile fiber type E, borosilicate wool, TV panel, container, and float glasses in the range of $\log_{10}$(viscosity / Pa·s) = 6 – 9.

VISCOMETRY IN THE GLASS SOFTENING RANGE

For measuring the viscosity in the glass softening range, there are three techniques that are commonly used:

The parallel plate technique (viscosity range = log(viscosity / Pa·s) = 4 to 10)
The ball penetration technique (viscosity range = log(viscosity / Pa·s) = 8 to 12)
The fiber elongation technique (viscosity range = log(viscosity / Pa·s) = 6 to 15)

Parallel Plate Viscometry

The principles of parallel-plate viscometry are described by Dienes [1], Gent [2], Fontana [3], and Varshneya [4]. The main parts of the instrument are shown in Figure 1. A disk of glass, roughly 6-9 mm diameter and 4-6 mm high, is sandwiched between two parallel plates inside a well-insulated furnace as shown. Direct contact with the plates is avoided by employing alumina substrates (or platinum foil). The upper pedestal (marked "load rod") is loaded, and the rate of sagging is recorded as a function of time.

It is important to pay attention to the geometry of deformation during measurement. Figure 2 illustrates the extrema: either the sample shows a "perfect slip" on the substrates (i.e. the contact areas between sample and substrate increase, the sample remains a rectangular cylinder), or "no slip" where the contact areas to the substrates stay constant and the sample "bulges out". Varshneya [4] shows that superior results are obtained if no-slip condition is assumed using alumina substrates.

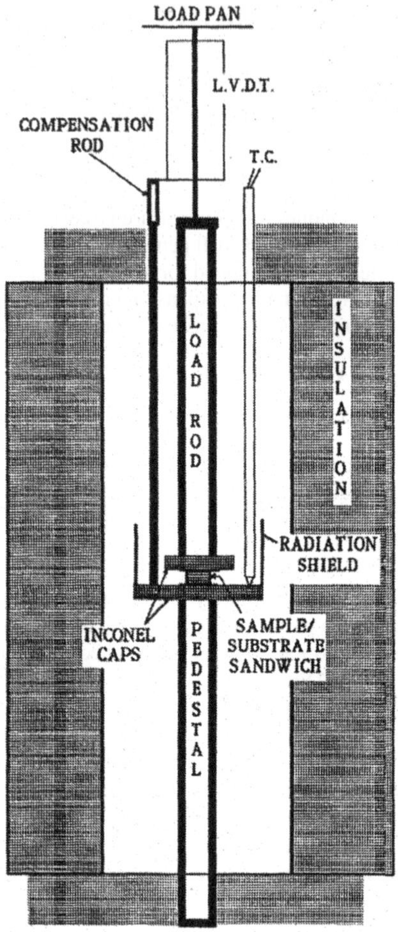

Fig. 1: Schematic of the parallel plate viscometer [5, p 192]

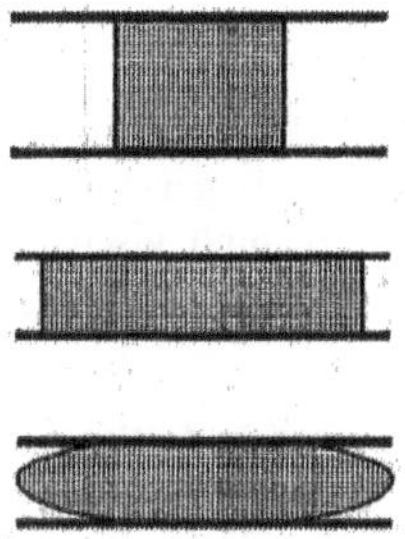

Fig. 2: Perfect slip and no-slip conditions during parallel plate measurement

Following assumptions are further made:

- The viscous sample is incompressible,
- The flow is Newtonian,
- The sample does not completely fill the area between the substrates during testing,
- The sample remains cylindrically symmetric during flow.

Under these assumptions, the glass viscosity may be calculated from the sag rate through equation (1):

$$\eta = \frac{2\,\pi\,M \cdot g \cdot h^5}{3 \cdot V \cdot dh/dt \cdot (2\,\pi\,h^3 + V)} \tag{1}$$

where η = viscosity in Poise or Pa•s; M = applied load; g = gravity acceleration; h = sample height; V = sample volume; dh/dt = deformation or sag rate.

Using the parallel plate technique, it is possible to measure viscosities in the glass softening range, $\log(\eta\,/\,\mathrm{Pa{\cdot}s}) = 4$ to 10. At the lower end of the range, low loads and large diameter sample may be needed. The heating rate should not be substantially slower than 1°C/min, else some glasses may crystallize during measurement which can lead to incorrect viscosity measurement.

Ball Penetration Viscometry

The ball penetration viscometry is described by Douglas et al. [6] and Brückner et al. [7].A hard, incompressible, chemically inert sphere about 3 mm diameter is allowed to penetrate a flat specimen of glass. Again, the sag rate is measured with time. Following expression is used to calculate the viscosity:

$$\eta = \frac{9 \cdot M \cdot g}{32 \cdot dL/dt \cdot (2 \cdot L \cdot R)^{0.5}} \tag{2}$$

with η = viscosity in Poise or Pa•s; M = applied load; g = gravity acceleration; L = penetration depth; dL/dt = deformation or sag rate; R = sphere radius. It should be noted that this equation is valid for small penetration only. The advantages of the technique lie in the simplicity of the method and the ease of the sample preparation. Ball/spheres as indenting body are preferred over cylindrical rods or cones because, in the latter, the friction that depends on the penetration depth and viscosity [7]. For chemical inertness, Brückner [7] recommended chromium nickel steel up to 650°C. Polished alumina or zirconia balls may be used at higher temperatures. The investigated material should have absolutely no irregular inclusions (e.g. bubbles), because the sampled material volume is generally very small. The ball penetration technique appears to be suitable for relatively high viscosities in the glass softening range ($\log(\eta/\text{Pa·s}) = 8$) up to the lower end of glass annealing ($\log(\eta/\text{Pa·s}) = 12$).

Fiber Elongation Viscometry

The fiber elongation viscometry is based upon the Littleton softening point determination and is described in ASTM C338-73. For most glasses the Littleton softening point is at a viscosity of $\log(\eta/\text{Pa·s}) = 6.6$, where high density lead, bismuth, barium, and telluride glass compositions may show a slightly higher viscosity at the Littleton softening point.

A 0.55 to 0.75 mm diameter glass fiber is drawn. One end is fused to make a ball. The fiber is cut to 23.5 cm length, and then suspended inside a specified furnace, which only covers the top 10 cm [5, p 193-194]. The furnace is heated at 5°C/min. The sagging of the lower end is sighted using a telescope and measured as a function of time. When the fiber extends under its own weight, the temperature when the rate of extension of the lower end is 1 mm/min corresponds to the softening point viscosity. To measure higher viscosities, one may need to load the fiber by placing weights in a pan attached to the lower end. Viscosities in this manner have been measured to the practically high limit of about $\log(\eta/\text{Pa·s}) = 15$.

For obtaining high accuracies, it should be noted that all viscometer calibration measurements using standard glasses should be performed under similar experimental conditions as the sample measurements, even though the theoretically derived equations allow wide variations of the sample dimensions, chemical composition, and heating rate.

Comparison of Viscometry Techniques in the Glass Softening Range

The parallel plate technique offers various advantages relative to the ball penetration and fiber elongation methods. The samples are easy to prepare and the sample size is small. It is not necessary to flame-work fibers that may possess extensive stresses and altered surface compositions which can cause changes in the glass structure and viscosity. In addition, the parallel plate technique does allow viscosity measurements over a wider range than the ball penetration viscometry, especially at lower viscosities. Therefore, the parallel plate viscometry was considered most effective and was chosen for this study.

EQUIPMENT SET-UP AND EXPERIMENTAL PROCEDURE

For this study, a Theta model "Rheotronic R" parallel-plate viscometer was used (Figure 3). For the cooling of the LVDT measuring head, an air-fan had to be used because the factory built-in water cooling system did not work efficiently at high temperatures. The inconel cup at the lower end of the load rod was removed, because having too many contact interfaces which were not perfectly parallel added to inaccuracy in the rigidity of translations. The electronic-grade alumina plates of the sample/substrate sandwich (thickness = 0.62 mm, see Figures 1 and 2) were stable sufficiently to hold the specimen in place. The inconel cup at the lower pedestal was not removed for providing a stable position of the compensation rod (Figure 1). The pedestals and compensation rod material was 100% silica glass.

The temperature reading and the LVDT displacement of the viscometer were calibrated first, using a thermocouple calibration multimeter, and samples of known height respectively. A NIST viscosity standard 710A (soda-lime glass) was used for calibration. The standard deviation of all repeated 710A measurements over the 620-820°C range was 0.4°C. The calibration was confirmed by NIST standard 717A (borosilicate glass) from 600-780°C. During calibration with the NIST standards, no error trend was observed over temperature and time, assuming a constant temperature distribution within the furnace during all measurements. In addition, no temperature trend was observed over the cross-section of the vertical furnace chamber.

Cylindrical samples (diameter 8.5-9 mm, height 4-6 mm) of annealed glass were prepared through core drilling and grinding off both ends until they were parallel to ± 10 μm. The accuracy of the sample dimensions was sufficient, as demonstrated by the repeated 710A calibration runs mentioned above. The samples were placed into the viscometer between the two alumina substrates mentioned above. The applied load was 500 g, and the heating rate was 1°C/min. A data logger recorded the temperature and the sample height every 20 seconds, where the expansion of the silica push rod was compensated automatically through the reference rod directly adjacent to it (Figure 1). Additionally, the expansion of the alumina substrates was corrected for mathematically (the compensation rod rested on the lower inconel plate and not on the alumina substrate).

Next, the viscosity was calculated through equation (1) within the range where the sag rate dh/dt was higher than circa 0.1 μm/s. The temperatures of constant viscosity (isokom) for log (η/Pa·s) = 6, 6.6, 7, 8, and 9 were calculated, after fitting the experimental data to the Vogel-Fulcher-Tammann equation. The standard deviation of the temperature errors during repeated viscosity measurements of different sections of the same glasses due to inhomogeneities in the glasses and due to other experimental irregularities was 2°C.

Fig. 3: The parallel plate viscometer

EXPERIMENTAL RESULTS

TABLE I: TEMPERATURES OF CONSTANT VISCOSITY (ISOKOM) IN °C OF THE TV PANEL GLASSES

#	Isokoms in °C for log(η/Pa·s)				
	6	6.6	7	8	9
TV 0	739	701	679	633	598
TV 1	741	701	678	629	591
TV 2	750	700	673	616	573
TV 3	732	684	658	604	564
TV 4	823	776	749	695	653
TV 5	826	787	765	717	679
TV 6	729	697	677	636	603
TV 7	719	687	668	626	592
TV 8	724	691	671	630	598
TV 9	653	622	603	563	530
TV 10	665	630	610	568	535
TV 11	705	668	646	602	566
TV 12	707	671	651	608	574
TV 13	744	709	689	644	608
TV 14	677	647	629	591	560
TV 15	744	704	680	631	592
TV 16	739	700	678	630	592
TV 17	681	649	629	588	555
TV 18	737	702	682	639	604
TV 19	689	651	629	582	545
TV 20	745	703	679	628	588
TV 21	702	664	642	597	561
TV 22	723	688	668	623	587
TV 23	730	688	664	615	577
TV 24	817	766	736	675	627

TABLE II: TEMPERATURES OF CONSTANT VISCOSITY (ISOKOM) IN °C OF THE TEXTILE FIBER E-GLASSES

#	Isokoms in °C for log(η/Pa·s)				
	6	6.6	7	8	9
E 0	877	846	828	789	756
E 1	1029*	1000*	981	939	902
E 2	962	928	908	866	832
E 3	887	861	846	812	784
E 4	959	921	900	854	818
E 5	930	901	884	848	818
E 6	896	870	855	821	793
E 7	850	820	802	766	737
E 8	855	830	815	782	755
E 9	803	778	764	732	705
E 10	853	827	812	778	749
E 11	851	822	805	768	738
E 12	886	857	840	803	772
E 13	903	873	855	818	789
E 14	945	914	896	858	827
E 15	944	907	886	843	810
E 16	995	959	937	892	855
E 17	931	896	876	833	798
E 18	938	910	894	859	832
E 19	875	844	826	787	757
E 20	Inhomogeneous				
E 21	848	823	808	777	750
E 22	907	873	854	811	776
E 23	837	811	795	762	735
E 24	804	780	765	733	707

* Extrapolated

TABLE III: TEMPERATURES OF CONSTANT VISCOSITY (ISOKOM) IN °C OF THE LOW EXPANSION BOROSILICATE GLASSES

#	Isokoms in °C for log(η/Pa·s)				
	6	6.6	7	8	9
LE 0	857	807	778	721	678
LE 1	882	832	803	746	702
LE 2	Phase separation				
LE 3	793	759	739	699	667
LE 4	888	833	802	740	693
LE 5	802	764	742	699	667
LE 6	807	770	748	706	673
LE 7	810	768	744	695	658
LE 8	787	753	733	692	660
LE 9	766	734	715	678	649
LE 10	790	749	726	681	647
LE 11	808	769	747	702	667
LE 12	886	830	800	740	696
LE 13	798	763	744	703	671
LE 14	825	786	764	719	686
LE 15	832	792	770	724	688
LE 16	Inhomogeneous				
LE 17	871	824	798	745	705
LE 18	821	780	757	710	676
LE 19	807	764	739	689	650
LE 20	Phase separation				
LE 21	774	741	723	685	657
LE 22	873	822	793	730	679
LE 23	768	734	714	675	644
LE 24	751	721	704	667	639

TABLE IV: TEMPERATURES OF CONSTANT VISCOSITY (ISOKOM) IN °C OF THE CONTAINER GLASSES

#	Isokoms in °C for log(η/Pa·s)				
	6	6.6	7	8	9
Co 0	768	731	710	668	637
Co 1	812	765	742	701	673
Co 2	799	763	742	699	667
Co 3	712	681	664	627	598
Co 4	746	712	693	652	620
Co 5	689	656	637	597	566
Co 6	718	686	668	630	599
Co 7	746	710	690	648	616
Co 8	754	715	693	647	612
Co 9	716	683	665	627	598
Co 10	759	721	699	655	621
Co 11	737	705	687	649	620
Co 12	813	777	756	714	681
Co 13	723	686	665	622	590
Co 14	755	710	687	642	612
Co 15	739	707	689	651	622
Co 16	793	752	729	682	645
Co 17	746	714	696	658	629
Co 18	739	705	685	646	616
Co 19	764	728	708	665	631
Co 20	825	789	768	725	691
Co 21	748	712	692	650	617
Co 22	796	755	732	685	650
Co 23	720	684	663	621	589
Co 24	692	662	646	611	583

TABLE V: TEMPERATURES OF CONSTANT VISCOSITY (ISOKOM) IN °C OF THE FLOAT GLASSES

#	Isokoms in °C for log(η/Pa·s)				
	6	6.6	7	8	9
FL 0	757	721	701	659	627
FL 1	788	747	724	679	645
FL 2	741	706	686	645	614
FL 3	727	690	670	630	599
FL 4	728	694	675	638	610
FL 5	722	689	670	632	603
FL 6	720	691	674	637	607
FL 7	765	730	710	668	636
FL 8	741	707	689	650	621
FL 9	771	734	713	670	637
FL 10	731*	703	686	649	618
FL 11	723	690	671	632	602
FL 12	764	728	708	666	635
FL 13	771	735	714	673	640
FL 14	746	709	688	647	616
FL 15	731	700	682	643	613
FL 16	756	720	699	657	625
FL 17	771	736	716	674	643
FL 18	745	709	689	648	617
FL 19	770	732	710	666	632
FL 20	747	712	692	651	620
FL 21	760	723	702	660	627
FL 22	779	743	722	680	647
FL 23	778	738	715	670	636
FL 24	774	736	714	670	635

* Extrapolated

TABLE VI: TEMPERATURES OF CONSTANT VISCOSITY (ISOKOM) IN °C OF THE WOOL GLASSES

#	Isokoms in °C for log(η/Pa·s)				
	6	6.6	7	8	9
Wo 0	705	678	662	630	604
Wo 1	762	724	704	661	629
Wo 2	726	693	674	634	603
Wo 3	683	658	642	610	583
Wo 4	729	696	677	638	608
Wo 5	707	679	664	631	605
Wo 6	709	682	667	635	608
Wo 7	696	671	657	627	602
Wo 8	706	682	667	637	611
Wo 9	686	663	650	623	600
Wo 10	700	676	662	632	607
Wo 11	676	649	635	604	581
Wo 12	749	718	701	665	637
Wo 13	684	655	638	603	575
Wo 14	762	729	711	673	643
Wo 15	703	673	655	618	590
Wo 16	751	714	693	650	617
Wo 17	732	701	683	646	618
Wo 18	740	709	691	655	627
Wo 19	715	688	672	639	613
Wo 20	725	696	680	645	617
Wo 21	734	706	690	656	629
Wo 22	735	706	689	655	628
Wo 23	684	661	648	621	599
Wo 24	649	628	615	588	566

REFERENCES

[1] G. J. Dienes, H. F. Klemm: "Theory and Application of the Parallel Plate Plastometer"; J. Appl. Phys. 17 (1946), 458-471

[2] A. N. Gent: "Theory of the Parallel Plate Viscometer"; Br. J. Appl. Phys. 11 (1960), 85-88

[3] E. H. Fontana: "A Versatile Parallel-Plate Viscometer for Glass Viscosity Measurements to 1000°C"; Ceram. Bull. 49 [6] (1970), 594-597

[4] A. K. Varshneya, N. H. Burlingame, W. H. Schultze: "Parallel Plate Viscometry to Study Deformation-Induced Viscosity Changes in Glass"; Glastechn. Ber. 63K (1990), 447-459

[5] A. K. Varshneya, "Fundamentals of Inorganic Glasses", Academic Press, Boston 1993

[6] R. W. Douglas, W. L. Amstrong, J. P. Edward, D. Hall: "A penetration viscometer"; Glass Tech., April 1965, vol. 6, no. 2, p 52-55

[7] R. Brückner, G. Demharter: "Systematic investigation of the use of penetration viscometers"; Glastechn. Ber., Jan. 1975, vol. 48, no. 1, p 12-18 (in German)

TRANSFORMATION RANGE VISCOSITY MEASUREMENTS

James E. Shelby
New York State College of Ceramics at Alfred University
Alfred, New York 14802

Transformation range viscosity was measured using the beam-bending method originally developed by Hagy [1] and later modified by the principal investigator [2]. The modified method allows use of a much smaller sample than that required by the original method.

The beam-bending method relies on the measurement of the rate of deflection due to viscous flow of the mid-point of a bar loaded in a 3-point configuration. Viscosity is calculated from the deflection rate and other information about the sample and apparatus, including the span width, the thickness and width of the bar, the density of the glass, and the load applied to the mid-point of the bar. The viscosity, η, is calculated using the expression

$$\eta = \frac{gL^3}{2.4 I_c V}\left(M + \frac{AL\rho}{1.6}\right)$$

where g is the acceleration of gravity, L is the span between the sample supports, I_c is the cross-sectional moment of inertia of the sample, V is the deflection rate of the mid-point of the beam at the temperature of measurement, M is the load applied to the mid-point of the beam, A is the cross-sectional area of the beam, and ρ is the glass density. Samples used in the present study were rectangular bars, 1 to 1.5 mm thick and 2 to 4 mm wide. The span was 17.8 mm. Since the mass of the beam is much less than the added load of ≈38 grams, it was ignored. Any error from ignoring the sample mass would result in a difference in viscosity smaller than the reproducibility of the measurement.

The beam-bending viscometer (a complete description of this apparatus is found in reference 2) was calibrated by use of NIST viscosity standards 710, 711, and 717. The viscosity of each glass was measured at ≈5 K intervals and plotted against temperature, as indicated by a thermocouple placed within 2 mm of the top surface of the sample. Temperatures corresponding to several viscosities were determined from a curve fit through the data. The experimental temperature corresponding to each of these viscosities was plotted against the "correct" temperature for that viscosity per the NIST data sheets. A combined plot of experimental versus NIST temperature for measurements using all 3 standards (Figure 1) was fit with a straight line which was then used to correct all experimental data. This correction is very accurate for glasses where the measurement temperatures are within the range of the calibration curve. Errors will be larger for the more viscous E-glass and borosilicate glasses, where the correction curve has to be extrapolated to temperatures significantly outside the range of available standards.

Viscosity versus temperature curves were obtained using the same method as that used for the NIST calibration samples. Typical samples are small bars, approximately 1 to 1.5 mm thick and 3 to 4 mm wide. The span was 17.8 mm for all samples. The hook and sample weight tray weigh 8.25 grams. An additional weight of 30 grams is placed on the tray for the higher viscosity measurements. The temperature is measured to within 1 K by a thermocouple placed directly above the midpoint of the sample, at a distance of 2 mm for the starting, unbent sample. This distance increases as the sample bends, reaching a value of approximately 4 mm by the end of the measurement. Any small effect of the change in position relative to the thermocouple is accounted for by the calibration procedure.

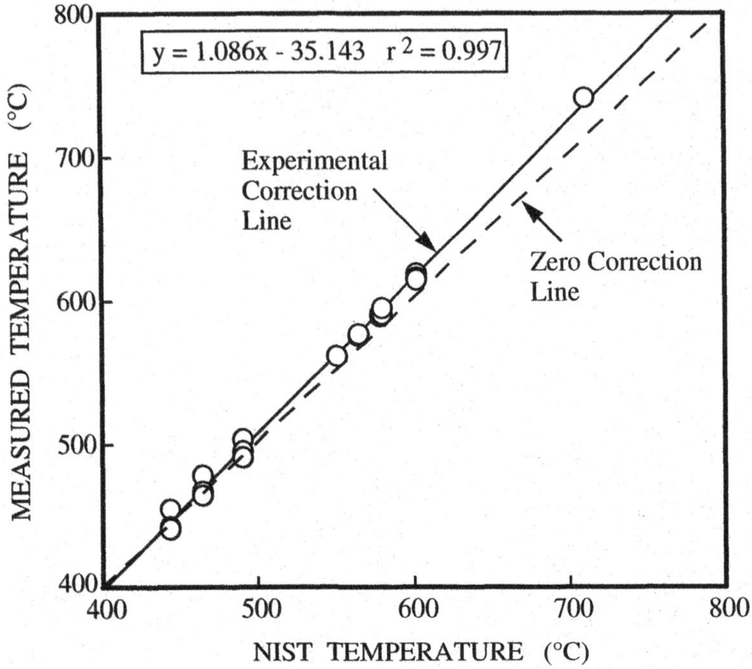

Figure 1. Calibration curve for viscosity measurements.

Each sample was heated to a temperature where flow began to occur at a detectable rate. The sample was held at that temperature until sufficient data were collected to determine the deflection rate. The temperature was increased by ≈5 K, held for several minutes to allow thermal equilibration, and the new deflection rate measured. Initially, a 30 gram weight was added to the sample tray (≈8 grams) hanging from the beam to increase the deflection rate to a measurable value. Once the viscosity had decreased to a point where the deflection rate was becoming too fast to measure, this weigh was removed, which resulted in a decrease in the deflection rate, as expected for a change in the force applied to the mid-point of the beam. The measurement continued at increasingly higher temperatures until the deflection rate once again became too great to measure, at which point the experiment was terminated. The point of removal of the 30 grams weight cannot be detected in the viscosity curves.

Viscosity data for several samples cut from a single piece of float glass are shown in Figure 2. The samples were measured on 4 different days spanning a period of 2 months. The data fall on the same line, which establishes the reproducibility of the measurement.

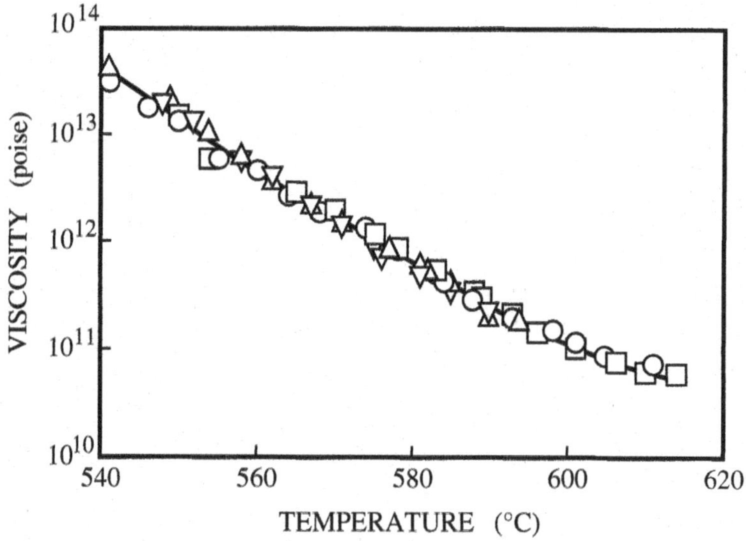

Figure 2. Viscosity curves for 4 series of measurements for a float glass.

Typical measurements consist of approximately 15 points measured at 5 to 10 K increments, depending on the slope of the viscosity curve for a particular glass. These data were fit by a smooth curve. The temperatures corresponding to fixed viscosities were determined from these curves. These isokoms are listed in Tables I-VI for all of the samples.

Several of the borosilicate glasses are obviously phase separated, as evidenced by visible light scattering. Several of the others, while not obviously phase separated, were determined to be phase separated by other measurements, as discussed in detail elsewhere [3]. The viscosities of those samples are affected by the morphology of the sample and are not representative of the homogeneous glass of the same composition. Any use of these data for the borosilicate glasses should be treated with caution.

Some of the E-glass samples had viscosities which required use of temperatures well in excess of the calibration range available for the NIST standards. The isokoms reported for those glasses should also be treated with caution, since the potential error in temperature is greater than for samples where the measurement temperatures are within the NIST calibration range.

REFERENCES

[1] H. E. Hagy, "Experimental Evaluation of Beam-Bending Method of Determining Glass Viscosities in the Range of 10^8 to 10^{15} Poises," *J. Am. Ceram. Soc.*, **46**[2] 93-97 (1963).

[2] J. E. Shelby, "Viscosity and Thermal Expansion of Alkali Germanate Glasses," *J. Am. Ceram. Soc.*, **57**[10] 436-9 (1974).

[3] J. E. Shelby, INTRODUCTION TO GLASS SCIENCE AND TECHNOLOGY, CH. 7, Royal Society of Chemistry, Cambridge, 1997.

TABLE I. VISCOSITY OF E-GLASS VARIATIONS

Sample Number	10^{13} Poise Isokom (°C)	10^{12} Poise Isokom (°C)	10^{11} Poise Isokom (°C)
Base Glass	663	690	717
1*	---	---	---
2	742	766	790
3	698	720	745
4	709	734	762
5	699	722	745
6	705	731	753
7	645	666	693
8**	---	---	---
9	632	653	675
10	665	685	706
11	654	677	701
12	677	702	728
13	699	724	748
14	725	750	777
15	705	730	755
16	740	769	794
17	707	724	750
18	737	761	786
19	658	681	706
20	656	680	705
21	661	683	708
22	681	708	736
23	652	673	695
24	630	650	671

*Viscosity above range of instrument

**No glass provided for measurement

TABLE II. VISCOSITY OF WOOL GLASS VARIATIONS

Sample Number	10^{13} Poise Isokom (°C)	10^{12} Poise Isokom (°C)	10^{11} Poise Isokom (°C)
Base Glass	543	562	582
1	556	574	601
2	540	560	585
3	540	558	578
3*	529	547	566
4	538	558	583
5	548	564	582
6	555	574	593
7	549	566	586
8	556	574	593
9	558	574	593
10	558	574	593
11	528	548	570
12	563	584	606
13	522	540	563
14	566	587	609
15	518	538	559
16	545	568	595
17	551	571	592
18	564	585	607
19	548	566	585
20	553	575	599
21	569	589	610
22	564	583	603
23	548	564	587
24	515	531	550

*Two samples with different properties were labeled #3

TABLE III. VISCOSITY OF BOROSILICATE TUBING GLASS VARIATIONS

Sample Number	10^{13} Poise Isokom (°C)	10^{12} Poise Isokom (°C)	10^{11} Poise Isokom (°C)
Base Glass	554	592	634
1	594	625	660
2	627	663	---
3	591	613	638
4	579	612	649
5	587	609	636
6	595	619	644
7	566	593	623
8	581	604	627
9	576	600	625
10	558	585	615
11	570	600	629
12	583	616	651
13	594	616	643
14	600	624	653
15	600	627	656
16	591	622	656
17	591	622	656
18	592	615	639
19	553	584	615
20	597	632	666
21	582	606	631
22	560	594	632
23	566	588	612
24	571	593	616

TABLE IV. VISCOSITY OF TV PANEL GLASS VARIATIONS

Sample Number	10^{13} Poise Isokom (°C)	10^{12} Poise Isokom (°C)	10^{11} Poise Isokom (°C)
Base Glass	521	544	570
1	507	530	554
2	483	512	543
3	487	513	540
4	558	585	612
5	591	616	644
6	532	556	580
7	520	542	567
8	529	551	576
9	474	495	515
10	474	496	519
11	496	519	542
12	498	522	547
13	538	558	579
14	496	518	540
15	507	530	554
16	515	537	560
17	489	510	533
18	532	556	581
19	473	494	518
20	506	528	550
21	491	509	529
22	522	540	559
23	502	527	554
24	520	550	583

TABLE V. VISCOSITY OF FLOAT GLASS VARIATIONS

Sample Number	10^{13} Poise Isokom (°C)	10^{12} Poise Isokom (°C)	10^{11} Poise Isokom (°C)
Base Glass	551	576	602
1	561	584	612
2	520	541	563
3	521	545	571
4	538	560	583
5	529	554	577
6	542	565	588
7	555	579	604
8	543	568	593
9	559	581	605
10	546	569	593
11	528	550	575
12	559	581	607
13	562	587	612
14	538	561	587
15	552	575	601
16	546	570	595
17	566	589	616
18	542	567	589
19	550	573	600
20	551	573	596
21	550	575	601
22	571	596	621
23	552	579	604
24	554	579	607

TABLE VI. VISCOSITY OF CONTAINER GLASS VARIATIONS

Sample Number	10^{13} Poise Isokom (°C)	10^{12} Poise Isokom (°C)	10^{11} Poise Isokom (°C)
Base Glass	564	588	611
1	573	602	637
2	589	608	630
3	531	551	574
4	540	564	594
5	496	517	544
6	530	552	575
7	538	563	587
8	532	556	583
9	538	558	580
10	538	563	589
11	552	572	596
12	604	626	649
13	524	547	572
14	524	550	579
15	557	577	597
16	560	583	608
17	559	579	599
18	543	565	589
19	559	582	607
20	611	631	661
21	548	569	592
22	569	589	622
23	523	541	586
24	517	538	560

Chapter 8

ELECTRICAL RESISTIVITY

Arun K. Varshneya, Terese Vascott, Ramesh Karuppanan, and Jeffrey M. Jones
New York State College of Ceramics at Alfred University
Alfred, New York 14802

ABSTRACT

Electrical resistance was measured at a frequency of 2 kHz in the range of 950 to 1450°C for 150 glass compositions of container, float, low-expansion borosilicate, TV panel, wool, and textile fiberglass types using a simple two-probe method. Resistivity was calculated from the measured resistance using a calibrated cell constant and is reported here as a function of temperature. Details of the specific compositions, designed as statistical variations around the six glass types, are presented in Chapter 2.

INTRODUCTION

Knowledge of the relationships between variations in resistivity in these glass systems and the statistically-designed changes in composition can be of great value in developing and improving glass process models. This report focuses on electrical resistivity in the glass melt (900-1450°C) for twenty-four compositional variations around each of six commercial base glasses: container glass, E glass, float glass, low-expansion borosilicate glass, TV panel glass, and fiber wool glass (Chapter 2).

In this study, two to four cycles of resistance measurements were made for each sample in 50-degree intervals from high temperature to low. The resistivity values reported below in Tables II-VII are averages of these measurements.

EXPERIMENTAL PROCEDURE

Equipment to measure the electrical resistance of a glass melt is shown in Figure 1. The furnace used was a bottom loading Lindberg/Blue M™ electrical resistance heated unit with a counter-weighted assembly where two platinum electrodes (0.5 mm dia., 1.9 cm center-to-center) along with a central type-S thermocouple could be lowered using a micrometer into a Pt-3.5% Rh crucible containing 70 ml of glass. The sample temperature was measured using the submerged type-S thermocouple.

Glasses were obtained in bulk-melted form as described in Chapter 2. To measure electrical resistivity, sufficient sample was used to make a 70-ml melt. The mass of glass needed was calculated from room-temperature density and thermal expansion data. Chunks of the glass were first heated in the Pt-3.5% Rh crucible to a high enough temperature (generally around 1200°C) to soften and flow within the crucible and held at temperature until no bubbles were visible. The temperature was then increased to 1450°C and the electrode assembly was lowered into the melt until contact with the surface was indicated by an electrical signal. The electrodes and thermocouple were then lowered further and adjusted to 1 cm depth using the micrometer. Measurements began when the temperature and resistance readings were stabilized.

Resistance was measured at 2 kHz frequency in descending 50-degree intervals to 900-950°C. The measurement frequency of 2kHz was selected for convenience as previous studies[1,2] have shown virtually no frequency dependence over the range 1 to 100 kHz for resistance in the glass melt. The sample was then reheated to 1450°C, the electrode depth was adjusted to 1 cm, and the cycle was repeated. When measurement was completed, the sample was reheated to melting temperature and the electrode assembly was raised out of the melt.

The Resistivity (ρ) was then calculated, from Eq. (2), using the cell constant (L/A) and the resistance (R) measured at each temperature.

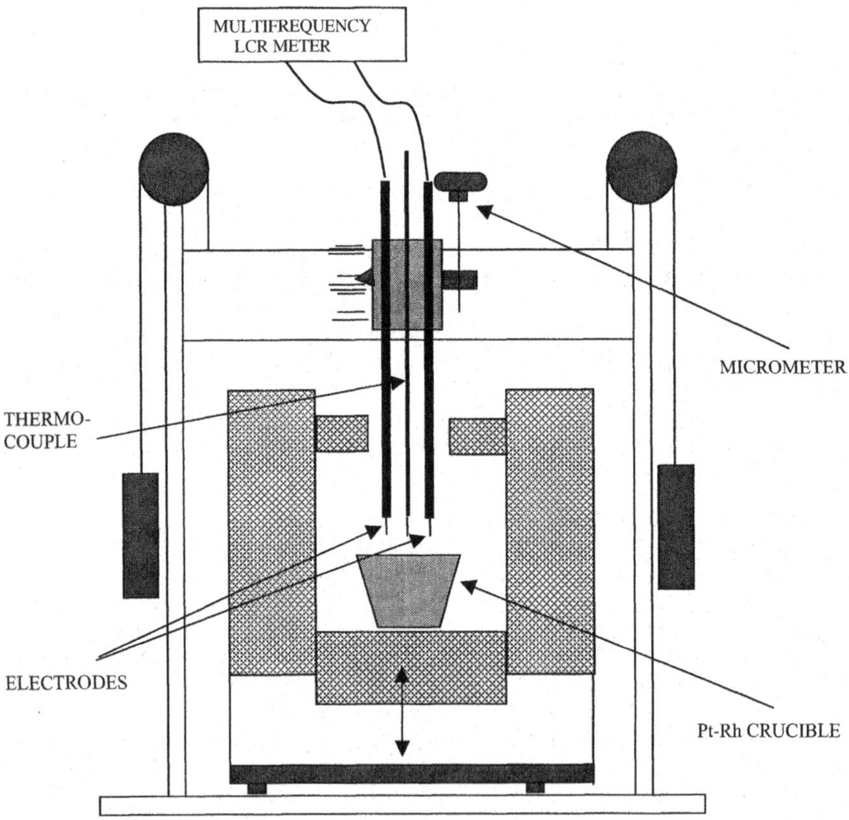

Figure 1. Electrical resistance measurement schematic.[3]

Calibration

The instrument was calibrated by determining the cell constant for a desired configuration from the resistance and reactance of KCl solutions (0.1 Demal: 7.4745g KCl + 1000g distilled water and 1.0 Demal: 76.583g KCl + 1000g distilled water)[4] measured in the platinum crucible to be used for the melts. Since the conductivity of the Demal solutions is frequency dependent, the resistance and the reactance were measured as a function of frequency from 100 Hz to 100 kHz (11 data points). At least three repeat measurements for each Demal solution were made using the same amount of liquid (70 ml) at the fixed electrode depth (1.0 cm) and separation (1.9 cm) each time.

Temperature was measured using a type-S thermocouple positioned between the electrodes and submerged to the same depth as the electrodes (1.0 cm). The thermocouple calibration was verified with an Omega CL-505A meter.

From these data, the reactance (X-axis) and the resistance (Y-axis) were plotted as shown in Figure 2. The cell constant can then be calculated using the following formula:

$$L/A = R(0)/\rho = Y_0/\rho, \text{ where } Y_0 \text{ is intercept at } x = 0 \quad (1)$$

The cell constant, L/A (cm^{-1}), depends on the geometry of the electrodes, melt, and crucible; R is the resistance measured (Ohms); and $1/\rho$ is the specific conductivity of Demal at T measured (°C). Specific conductivity values are shown below in Table I.

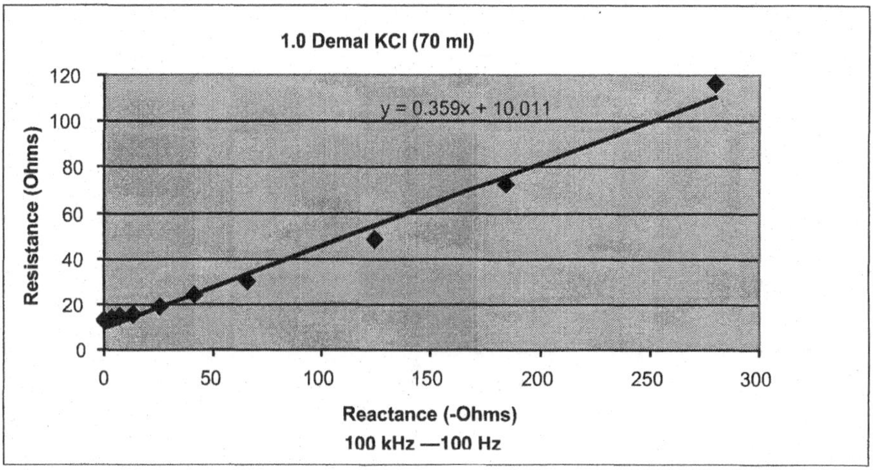

Figure 2. Resistance vs. Reactance from 0.1-100 kHz for 1.0 Demal solution KCl

TABLE I. SPECIFIC CONDUCTIVITY OF KCl SOLUTIONS

grams KCl per kg H_2O	specific conductivity = $1/\rho$ (Ohm-cm)$^{-1}$			
	0° C	18° C	25° C	25° C
76.5829 (1Demal)	6.514 x10^{-2}	9.781 x10^{-2}	11.131 x10^{-2}	
7.47458 (0.1Demal)	0.7134 x10^{-2}	1.1163 x10^{-2}	1.2852 x10^{-2}	1.2854 x10^{-2}*
0.745510 (0.01Demal)	0.07733 x10^{-2}	0.12201 x10^{-2}	0.14083 x10^{-2}	0.14086 x10^{-2}*
7.45510 (0.1Demal)				1.28217 x10^{-2}*
0.745510 (0.01Demal)				0.14079 x10^{-2}*

Reference: CRC Handbook, edition 79. [*Values are taken from Y. C. Wu, K. W. Pratt and W. F. Koch, J. Solution Chemistry **18**, 515 (1989).]

Verification of the cell constant over the entire temperature range was made by comparing the reported resistivity of a standard glass[5], in this case NIST SRM1414, to the values calculated from measuring the resistance of the melt over the temperature range of interest. The Resistivity (ρ) was calculated using the cell constant (L/A) determined from the Demal solution measurements and the resistance (R) measured at each temperature.

$$\rho \text{ (Ohm-cm)} = R/(L/A) \quad (2)$$

A typical calibration verification curve is shown below in Figure 3.

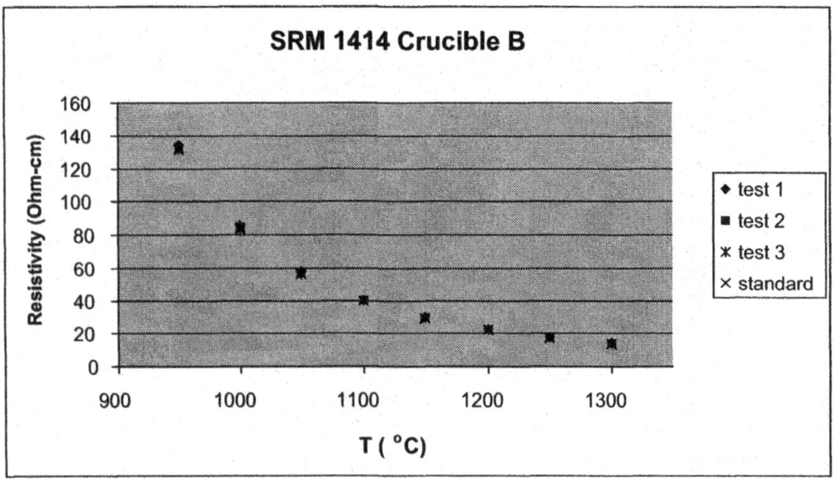

Figure 3. Resistivity vs. Temperature for cell constant calibration verification using a standard glass.

The calibration procedure was repeated at regular intervals throughout the sample measurement and also whenever the cell configuration was changed.

Sources of Error

Measurement errors can arise from changes in cell geometry, fluctuations in melt volume, volatization of the melt, and inaccuracy of melt temperature readings. Potential for error can be reduced by maintaining consistency in both equipment configuration and technique over the entire course of the measurements.

Changes in the shape of the crucible from repeated heating and cooling can affect the value of the cell constant and result in incorrect calculations of the resistivity. Repeating the calibration procedure with KCl solutions at regular intervals, and adjusting the cell constant accordingly, can minimize calculation errors. The standard glass, SRM1414[6], is used to verify the cell constant over the temperature range 950-1300°C. Additionally, the electrode/thermocouple configuration must be checked before each measurement to ensure that the distance between the electrodes remains constant and that the ends of the two electrodes and the thermocouple are aligned, thereby increasing the reliability of the resistance data.

Changes in sample volume can occur due to thermal expansion and changes in density of the glass melt and, in some compositions, volatization at high temperatures. In this procedure, the electrode assembly is positioned at 1.0 cm depth in the melt at the highest measurement temperature (1450°C). After the measurements have been taken, the sample is reheated from the lowest measurement temperature (900-950°C) to 1450°C, the electrode assembly is removed from the melt and then repositioned for repeat measurements.

The sample temperature is taken from a type-S thermocouple positioned between the electrodes (Figure 1) and at the same depth in the melt as the electrodes. This arrangement minimizes effects from temperature gradients in the crucible. The thermocouple readout is monitored by a thermometer and a multimeter.

RESULTS

Measurement results are shown in Tables II-VII at the end of this chapter. Data are presented as Resistivity (Ohm-cm) at Temperature (°C). The reported Resistivity values are averages of 2-4 measurement cycles. Figure 4 shows the log resistivity values for the six commercial base glasses plotted against inverse temperature (K). Because of the apparent lack of linear behavior, no efforts were made to derive an activation energy.

DISCUSSION

Effects of Composition

The predominant relationship is between a decrease in resistivity with increased alkali content. As illustrated in Table II, Na_2O is the primary contributor. The six base glasses fall into two groups, with E glass and low-expansion borosilicate as 'low alkali' and container, float, TV panel, and fiber wool as 'high alkali'. Figures 5a-c show resistivity for the six base glasses at 1450, 1200, and 950 °C. In the 'low alkali' glasses, resistivity values converge at high temperature with the resistivity of the E glass decreasing more than the borosilicate. The 'high alkali' glasses, however, maintain the same relative values over the entire temperature range as the resistivity decreases with increasing temperature. The TV panel base glass shows higher resistivity than the other three 'high alkali glasses', due to the lower sodium content.

TABLE II. ALKALI CONTENT OF COMMERCIAL BASE GLASSES

mol %	Li_2O	Na_2O	K_2O
E base		0.61	
LO base	trace	4.08	0.04
TV base	0.02	8.66	5.63
CO base		12.9	0.19
FL base		13.13	0.02
WO base		15.7	0.54

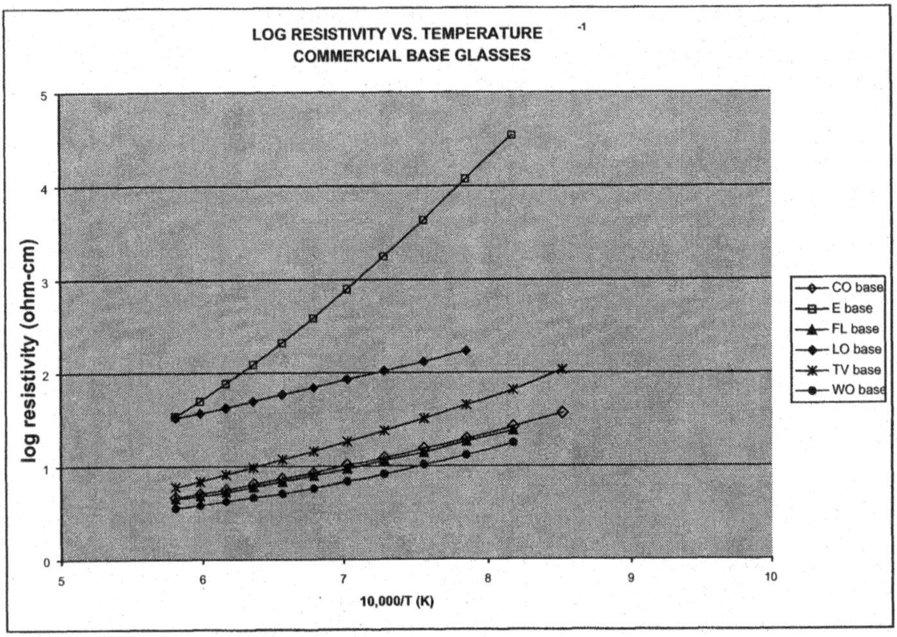

Figure 4. Log resistivity vs. inverse temperature for the commercial base glasses: container, E-glass, float, low-exp. borosilicate, TV panel, and fiber wool.

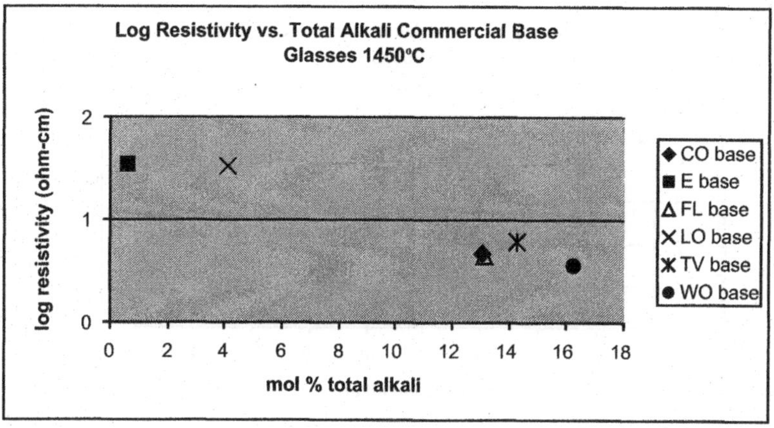

Figure 5a. Log resistivity at 1450 °C as a function of total alkali content in six commercial glasses.

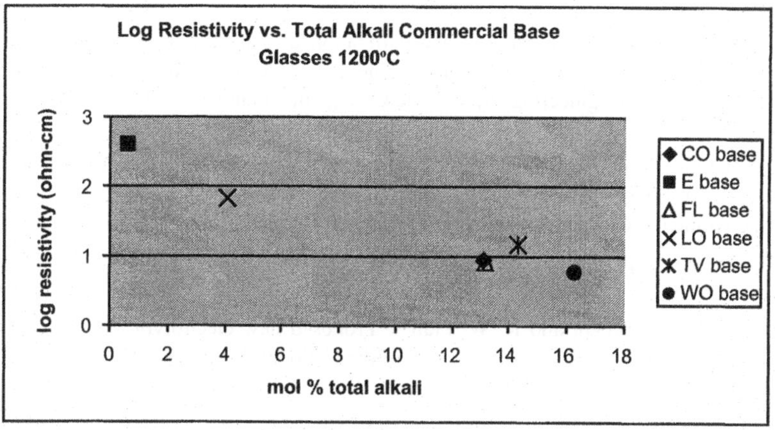

Figure 5b. Log resistivity at 1200 °C as a function of total alkali content in six commercial glasses.

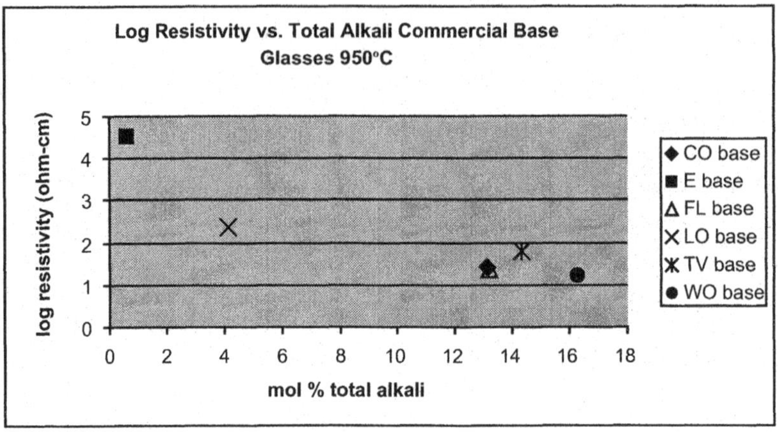

Figure 5c. Log resistivity at 950 °C as a function of total alkali content in six commercial glasses.

REFERENCES

1 D. McPherson, et al. "Preparation and calibration of standards and procedures useful for characterizing simulated nuclear waste glasses and melts". Monthly Report, April 1986. Institute of Glass Science and Engineering, N. Y. State College of Ceramics at Alfred University, Alfred, NY.

2 E. N. Boulos, J. W. Smith, and C. T. Moynihan. "Rapid and accurate measurements of electrical resistivity on glass melts". Glastechnische Berichte, Vol. 56K, Bk. 1, pp. 509-514. XIII. Internationaler Glaskongress, 4. bis 9. Juli 1983, Hamburg.

3 W. C. Eaton. Properties of simulated glass forming nuclear waste melts. M. S. Thesis, Alfred University, Alfred, NY, March 1992.

4 Y. C. Wu, W. F. Koch, and K. W. Pratt. "Proposed new electrolytic conductivity primary standards for KCl solutions". J. of Research of the National. Inst. of Standards and Technology, Vol. 96, No. 2, March-April 1991.

5 NIST Certificate, SRM 1414, Lead-Silica Glass, High-Temperature Resistivity, July 17, 1991. Note: High-temperature electrical resistivity/conductivity standards are no longer available from NIST; formerly SRM711 and SRM1414 were used

TABLE II. RESISTIVITY OF CONTAINER GLASS COMPOSITIONS

	Resistivity (Ohm-cm)										
T (°C)	1450	1400	1350	1300	1250	1200	1150	1100	1050	1000	950	900
CO base	4.726	5.194	5.746	6.427	7.425	8.634	10.273	12.327	15.380	19.649	26.446	37.656
CO 1	6.680	7.384	8.283	9.419	10.872	12.734	15.227	18.524	23.061	29.500	39.164	
CO 2	7.007	7.979	9.183	10.736	12.802	15.595	19.464	24.959	32.988	44.923	64.074	
CO 3	3.329	3.629	3.987	4.428	4.995	5.722	6.694	8.025	9.855	12.498	16.499	
CO 4	4.219	4.587	5.045	5.599	6.308	7.203	8.392	9.973	12.184	15.282	19.814	
CO 5	3.147	3.392	3.706	4.096	4.605	5.254	6.135	7.302	8.896	11.190	14.528	
CO 6	4.573	5.132	5.840	6.757	7.975	9.587	11.798	14.936	19.441	26.335	36.994	
CO 7	3.588	3.883	4.233	4.673	5.232	5.940	6.894	8.183	9.936	12.411	16.063	
CO 8	6.935	7.738	8.742	10.014	11.621	13.715	16.544	20.422	25.849	33.624	45.459	
CO 9	3.143	3.406	3.738	4.160	4.718	5.431	6.408	7.725	9.555	12.252	16.194	
CO 10	4.785	5.229	5.791	6.444	7.399	8.524	10.078	12.176	15.101	19.246	25.306	
CO 11	3.747	4.114	4.573	5.150	5.890	6.853	8.193	10.005	12.661	16.499	22.371	
CO 12	8.338	9.696	11.421	13.479	16.058	19.641	24.537	31.131	40.736	54.891	75.477	
CO 13	3.933	4.269	4.682	5.204	5.863	6.712	7.866	9.432	11.594	14.741	19.273	
CO 14	5.341	5.958	6.667	7.543	8.642	10.073	11.975	14.555	18.025	23.093	30.431	
CO 15	3.569	3.901	4.332	4.873	5.540	6.431	7.639	9.255	11.580	14.900	20.064	
CO 16	6.222	6.966	7.866	8.965	10.418	12.302	14.859	18.383	23.397	30.441	41.131	
CO 17	3.606	3.955	4.396	4.964	5.681	6.639	7.929	9.673	12.062	15.577	20.881	
CO 18	4.655	5.177	5.867	6.753	7.911	9.455	11.540	14.491	18.724	25.104	34.814	
CO 19	3.955	4.264	4.641	5.095	5.686	6.421	7.411	8.733	10.545	13.120	16.857	
CO 20	7.393	8.370	9.541	11.090	13.093	15.686	19.351	24.273	31.467	41.944	57.943	
CO 21	5.027	5.813	6.698	7.788	9.246	11.104	13.815	17.425	22.816	30.786	42.938	
CO 22	5.577	6.208	7.007	8.006	9.305	11.008	13.288	16.362	20.672	26.844	36.149	
CO 23	3.597	3.887	4.282	4.796	5.427	6.231	7.298	8.733	10.668	13.433	17.539	
CO 24	3.020	3.297	3.642	4.074	4.619	5.336	6.303	7.620	9.505	12.298	16.621	

TABLE III. RESISTIVITY OF E GLASS COMPOSITIONS

	Resistivity (Ohm-cm)										
T (°C)	1450	1400	1350	1300	1250	1200	1150	1100	1050	1000	950
E base	34.723	50.416	76.576	123.248	211.171	390.624	785.912	1745.97	4320.035	11786.455	34135.091
E 1	255.226	411.150	702.107	1281.476	2526.707	5261.577	11984.03				
E 2	106.698	164.727	269.179	466.592	878.249	1806.343	4000.609	10010.91			
E 3	16.076	22.318	32.295	51.321	86.729	155.975	310.692	679.717	1685.534	4715.409	14842.77
E 4	55.503	76.887	109.937	161.950	241.509	378.302	605.346	983.962	1629.560	2767.296	4847.170
E 5	30.591	43.720	64.682	103.783	174.057	309.119	591.195	1214.465	2711.321	6361.635	15705.35
E 6	24.172	35.694	55.244	93.340	171.377	344.472	823.390	2123.271	6722.704	26040.25	120221.4
E 7	33.852	43.962	62.264	90.255	130.638	194.025	308.176	492.767	818.868	1452.516	2525.157
E 8	21.721	29.791	43.100	66.708	109.793	197.995	392.515	875.415	2325.815	7243.983	28781.88
E 9	12.784	16.301	21.762	30.134	47.290	75.871	126.917	253.996	534.422	1309.380	3673.364
E 10	21.632	30.009	43.772	68.003	112.608	202.585	397.777	895.976	2317.838	7192.212	27460.1
E 11	25.277	34.227	48.272	72.016	112.843	184.496	331.849	615.328	1256.655	2749.916	6342.264
E 12	38.038	55.962	86.829	144.364	254.712	495.200	1084.415	2648.631	7600.225	23655.73	110599.4
E 13	27.070	37.628	54.801	85.049	137.238	235.918	437.896	861.329	1829.600	4121.598	9749.346
E 14	49.902	74.651	112.185	189.711	350.070	706.322	1616.589	4137.000	12667.05		
E 15	44.101	62.090	89.591	134.441	209.228	336.785	561.591	958.515	1696.496	3022.073	5367.412
E 16	125.297	195.028	322.106	569.550	1076.124	2235.584	5023.876	12524.92			
E 17	39.265	55.771	81.520	125.405	198.871	330.756	587.786	1071.468	2033.575	4061.286	8223.659
E 18	27.715	41.161	64.503	109.909	204.179	414.895	970.085	2622.653	8582.769		
E 19	38.456	51.339	70.723	100.751	146.932	221.967	346.001	554.776	914.092	1532.962	2632.977
E 20	36.911	52.259	78.303	122.973	207.894	380.115	758.226	1700.122	4343.107	11941.52	33110.4
E 21	17.963	24.802	35.823	54.910	90.591	160.782	318.724	717.355	1881.783	6073.260	24660.8
E 22	75.986	114.972	180.932	302.500	536.525	1026.500	2160.000	4986.700	13122.00		
E 23	17.102	22.752	31.430	45.627	70.341	115.594	206.077	403.441	870.711	2053.932	5518.385
E 24	9.537	12.202	16.272	22.836	33.931	53.952	94.383	180.776	381.928	969.250	2889.133

TABLE IV. RESISTIVITY OF FLOAT GLASS COMPOSITIONS

	Resistivity (Ohm-cm)										
T (°C)	1450	1400	1350	1300	1250	1200	1150	1100	1050	1000	950
FL base	4.460	4.890	5.410	6.070	6.910	8.010	9.480	11.480	14.260	18.280	24.220
FL 1	5.722	6.312	7.020	7.884	8.967	10.343	12.221	14.744	18.314	23.346	30.841
FL 2	4.310	4.780	5.270	5.920	6.630	7.560	8.770	10.280	12.410	15.340	19.660
FL 3	4.022	4.356	4.785	5.329	6.015	6.897	8.076	9.630	11.778	14.822	19.360
FL 4	3.862	4.196	4.607	5.137	5.809	6.691	7.866	9.456	11.787	14.941	19.799
FL 5	4.109	4.456	4.872	5.393	6.056	6.938	8.149	9.845	12.134	15.457	20.503
FL 6	3.910	4.150	4.580	5.070	5.700	6.560	7.710	9.240	11.510	14.820	19.690
FL 7	5.610	6.310	7.240	8.290	9.700	11.510	13.790	17.000	21.180	28.010	37.490
FL 8	3.889	4.292	4.730	5.288	5.987	6.874	8.071	9.657	11.892	15.101	19.877
FL 9	5.380	5.820	6.430	7.210	8.380	9.800	11.700	14.330	17.830	23.630	32.030
FL 10	3.853	4.223	4.685	5.233	5.905	6.737	7.870	9.410	11.449	14.470	19.072
FL 11	3.752	4.118	4.584	5.110	5.782	6.627	7.724	9.191	11.216	14.168	18.505
FL 12	5.384	5.960	6.686	7.587	8.734	10.251	12.326	15.192	19.314	25.343	34.461
FL 13	5.580	6.110	7.030	8.130	9.440	11.160	13.260	16.300	20.300	26.420	35.510
FL 14	4.004	4.328	4.712	4.250	5.727	6.463	7.427	8.757	10.580	13.204	17.011
FL 15	4.490	4.950	5.500	6.020	6.830	7.830	9.100	10.860	13.190	16.640	21.970
FL 16	5.201	5.896	6.824	7.939	9.310	11.001	13.254	16.266	20.475	26.668	36.280
FL 17	5.608	6.239	7.043	8.039	9.346	11.033	13.350	16.458	20.772	27.157	36.974
FL 18	3.999	4.314	4.689	5.142	5.727	6.495	7.537	8.949	10.873	13.652	17.710
FL 19	5.242	5.836	6.595	7.509	8.665	10.151	12.093	14.735	18.336	23.697	31.627
FL 20	3.890	4.150	4.540	4.970	5.720	6.410	7.370	8.900	10.580	13.230	17.240
FL 21	5.580	6.210	6.870	7.770	8.700	10.070	11.850	14.400	18.030	23.430	31.580
FL 22	5.710	6.280	7.060	8.030	9.260	10.860	13.080	16.030	20.000	25.980	34.850
FL 23	5.558	6.188	6.993	7.930	9.168	10.736	12.761	15.585	19.420	24.886	33.186
FL 24	5.960	6.581	7.367	8.359	9.607	11.225	13.368	16.243	20.334	26.065	34.173

TABLE V. RESISTIVITY OF LOW-EXPANSION BOROSILICATE GLASS COMPOSITIONS

	Resistivity (Ohm-cm)									
T (°C)	1450	1400	1350	1300	1250	1200	1150	1100	1050	1000	950
LO base	33.228	37.359	42.727	49.573	58.762	69.496	84.571	104.887	132.734	173.219	
LO 1	32.855	36.841	41.790	49.013	56.951	67.388	80.973	99.209	124.340	160.017	207.195
LO 2	44.477	50.158	57.656	67.142	79.558	95.400	116.367	145.438	186.268	245.367	331.991
LO 3	11.197	12.684	14.535	17.553	20.584	24.909	30.854	39.178	51.333	69.400	95.756
LO 4	27.605	30.927	34.502	40.426	47.076	55.504	66.838	81.418	102.031	130.657	172.636
LO 5	10.091	11.112	12.438	14.206	16.427	19.246	23.374	29.134	36.764	47.502	63.871
LO 6	10.045	11.080	12.419	14.127	16.358	18.987	22.590	27.357	34.085	43.968	58.384
LO 7	26.743	31.231	36.957	44.354	54.293	67.662	86.434	113.273	153.592	214.459	
LO 8	12.890	14.520	16.603	19.241	22.648	27.158	33.231	41.643	53.729	71.553	
LO 9	9.860	11.137	12.721	14.763	17.500	18.992	26.331	34.049	44.537	61.180	87.420
LO 10	10.160	11.056	12.174	13.638	15.476	17.837	21.008	25.206	30.992	39.389	51.507
LO 11	35.767	41.085	48.166	57.215	68.960	84.667	106.619	137.378	181.289	248.300	
LO 12	24.014	26.640	29.917	34.167	39.558	46.398	55.361	67.709	84.616	108.257	144.002
LO 13	10.155	11.353	12.979	14.823	17.457	20.973	25.759	32.560	42.503	56.783	80.081
LO 14	12.547	13.916	15.653	17.759	20.411	23.761	28.252	34.237	41.581	54.383	71.409
LO 15	32.386	37.892	44.995	54.427	67.008	84.163	108.338	142.871	193.952	274.223	
LO 16	not measured										
LO 17	26.922	30.574	35.282	41.100	48.699	58.440	71.815	89.837	115.445	152.054	
LO 18	9.254	10.191	11.391	12.874	14.766	17.158	20.329	24.577	30.364	38.600	
LO 19	29.152	33.796	39.866	47.678	57.872	71.649	90.695	117.490	156.447	215.448	
LO 20	27.633	31.486	36.450	42.447	50.421	60.868	75.104	94.925	123.174	165.292	
LO 21	9.789	11.019	12.617	14.598	17.244	20.727	25.531	32.191	41.818	56.402	
LO 22	24.761	27.603	31.153	35.555	41.182	48.330	57.708	70.100	87.010	111.010	
LO 23	14.096	15.967	18.301	21.268	25.153	30.464	37.694	47.861	62.761	84.488	
LO 24	8.440	9.459	10.789	12.469	14.646	17.579	21.636	27.306	35.416	47.837	

TABLE VI. RESISTIVITY OF TV PANEL GLASS COMPOSITIONS

	Resistivity (Ohm-cm)											
T (°C)	1450	1400	1350	1300	1250	1200	1150	1100	1050	1000	950	900
TV base	6.174	7.058	8.224	9.758	11.789	14.545	18.289	24.254	32.470	45.262	65.311	106.763
TV 1	6.481	7.382	8.471	9.901	11.729	14.117	17.480	22.044	28.974	39.207	54.781	80.656
TV 2	8.387	9.525	11.006	12.519	15.286	18.343	23.003	28.921	37.903	51.170	71.528	102.626
TV 3	6.808	7.519	8.541	9.755	11.352	13.406	16.362	20.377	25.157	33.616	45.000	62.642
TV 4	10.217	11.808	13.195	15.745	18.947	23.399	29.632	38.928	50.975	72.358	101.777	159.057
TV 5	11.787	14.067	17.126	21.256	26.910	34.903	46.455	63.948	90.787	134.084	208.448	338.350
TV 6	6.673	7.940	9.540	11.846	15.314	19.915	27.233	38.412	56.509	88.019	143.711	253.145
TV 7	5.942	7.020	8.550	10.650	13.605	17.882	24.126	34.638	50.731	79.142	130.878	227.184
TV 8	4.616	5.214	6.053	7.110	8.660	10.667	13.645	17.956	25.396	35.148	52.003	82.437
TV 9	3.582	3.972	4.479	5.124	6.067	7.227	8.923	11.283	14.917	20.370	29.259	44.709
TV 10	4.014	4.432	5.013	5.673	6.595	7.830	9.560	11.974	15.533	21.062	29.361	44.119
TV 11	4.695	5.226	5.924	6.904	8.108	9.715	12.018	15.070	19.527	26.601	37.263	55.513
TV 12	4.224	4.748	5.403	6.287	7.332	8.969	11.176	14.265	18.943	25.978	37.753	57.634
TV 13	6.102	7.074	8.349	10.134	12.437	15.844	20.623	28.250	39.679	57.904	89.663	147.871
TV 14	3.909	4.362	5.006	5.802	6.868	8.302	10.406	13.258	17.947	23.299	37.475	58.267
TV 15	6.163	7.074	8.250	9.784	12.023	14.910	19.133	25.751	34.758	49.691	73.462	116.954
TV 16	5.092	5.676	6.565	7.649	9.089	10.938	13.515	17.209	22.287	29.549	42.335	61.419
TV 17	4.107	4.572	5.145	6.024	7.223	8.554	10.828	13.612	18.753	24.448	36.056	54.733
TV 18	6.973	7.968	9.367	11.303	14.052	17.897	22.958	31.193	44.556	64.668	99.606	165.578
TV 19	7.212	8.313	9.788	11.559	14.097	17.262	21.671	28.260	37.925	52.388	75.426	116.274
TV 20	5.871	6.580	7.530	8.739	10.283	12.361	15.067	18.911	24.395	32.753	45.066	65.391
TV 21	4.289	4.754	5.395	6.195	7.265	8.700	10.693	13.505	17.547	23.888	33.297	49.299
TV 22	6.055	6.972	8.140	9.780	11.833	14.831	19.117	25.173	34.409	49.670	73.369	118.092
TV 23	9.544	10.921	12.830	15.330	18.431	22.626	28.805	37.720	50.349	69.975	101.447	155.346
TV 24	14.711	16.818	19.575	23.063	27.682	33.761	41.343	53.006	69.321	93.472	129.720	188.145

TABLE VII. RESISTIVITY OF FIBER WOOL GLASS COMPOSITIONS

	Resistivity (Ohm-cm)										
T (°C)	1450	1400	1350	1300	1250	1200	1150	1100	1050	1000	950
WO base	3.623	3.875	4.195	4.610	5.170	5.915	6.903	8.315	10.285	13.215	17.588
WO 1	5.279	5.759	6.353	7.121	8.103	9.351	11.037	13.254	16.293	20.622	27.249
WO 2	5.201	5.676	6.312	7.116	8.154	9.493	11.321	13.757	17.244	22.148	29.877
WO 3	3.275	3.513	3.760	4.180	4.628	5.190	6.003	7.083	8.653	10.980	14.473
WO 4	3.853	4.122	4.429	4.826	5.320	5.964	6.801	7.916	9.465	11.581	14.561
WO 5	3.328	3.548	3.838	4.230	4.690	5.330	6.168	7.328	8.830	11.388	15.055
WO 6	3.748	4.088	4.535	5.100	5.848	6.758	8.163	10.058	12.750	16.968	23.435
WO 7	3.431	3.613	3.859	4.162	4.536	5.049	5.757	6.726	8.104	10.115	13.208
WO 8	4.274	4.708	5.261	5.841	6.808	7.950	9.611	12.047	15.615	21.043	29.970
WO 9	3.410	3.656	4.013	4.447	5.037	5.814	6.878	8.304	10.311	13.423	17.930
WO 10	4.150	4.485	4.895	5.483	6.055	7.215	8.588	10.533	13.348	17.745	24.290
WO 11	3.131	3.341	3.606	3.949	4.397	4.977	5.718	6.709	8.126	10.091	13.140
WO 12	5.169	5.667	6.261	7.043	8.012	9.250	10.969	13.277	16.463	21.289	28.318
WO 13	3.005	3.175	3.345	3.578	3.910	4.345	4.903	5.590	6.595	8.033	10.030
WO 14	5.891	6.495	7.276	8.309	9.639	11.380	13.761	17.061	21.874	28.501	39.049
WO 15	3.515	3.803	4.141	4.543	5.064	5.727	6.577	7.774	9.401	11.700	15.055
WO 16	5.169	5.640	6.261	7.002	7.934	9.168	10.777	12.888	16.005	20.160	26.115
WO 17	2.820	3.010	3.293	3.670	4.103	4.668	5.443	6.510	7.790	9.823	12.610
WO 18	3.720	4.153	4.703	5.420	6.353	7.568	9.208	11.548	14.658	19.363	26.728
WO 19	4.113	4.415	4.813	5.283	5.992	6.787	8.012	9.561	11.801	14.963	19.941
WO 20	5.356	5.855	6.476	7.281	8.327	9.717	11.577	14.205	17.843	23.254	31.792
WO 21	4.035	4.408	4.963	5.640	6.508	7.623	9.135	11.270	14.238	18.743	25.543
WO 22	4.781	5.183	5.718	6.426	7.317	8.496	10.105	12.226	15.014	19.452	25.804
WO 23	4.484	4.899	5.393	6.033	6.837	7.911	9.324	11.307	14.255	18.432	25.037
WO 24	2.994	3.236	3.524	3.885	4.337	4.936	5.740	6.901	8.510	10.900	14.392

Chapter 9

STATISTICAL ANALYSIS OF VISCOSITY, ELECTRICAL RESISTIVITY, AND FURTHER GLASS MELT PROPERTIES

Alexander Fluegel, David A. Earl, Arun K. Varshneya, and Dolun Öksoy
New York State College of Ceramics at Alfred University
Alfred, New York 14802

ABSTRACT

The application of multiple regression techniques for glass property modeling is described. Multiple regression was applied for the development of models for the viscosity at $\log(\eta/\text{Pa·s}) = 1.5$ to 12.0 and for the high temperature electrical resistivity at T = 1000°C to 1400°C. The models are based on 150 composition-property data of commercial glasses, including low-expansion borosilicate, textile fiber type E, fiber wool, TV panel, container, and float glasses. The model errors were found to depend on the glass type and on the property measured. The best predictions were possible for soda lime container and float glass viscosities (error 1-8°C), and less accurate results were obtained for borosilicate and TV panel glass viscosities (error 2-30°C). High temperature electrical resistivities could be modeled with errors of $\log(\rho/\Omega\cdot\text{cm}) = 0.005$-0.015 for container and float glasses, and $\log(\rho/\Omega\cdot\text{cm}) = 0.01$-0.13 for borosilicate glasses. The models were evaluated in comparison to experimental data from the scientific literature. The influences of all glass components are discussed, and relations between the high temperature viscosity and electrical resistivity are examined. In Appendix C of this chapter initial modeling results for the water solubility and surface tension of glass melts are presented.

INTRODUCTION

Methodical studies of glass properties were initiated in the 19th century by E. Abbe, O. Schott, and A. Winkelmann. They considerably improved the use of glasses for special applications, e.g., for optics, thermometers, and as thermal shock resistant material. Winkelmann and Schott published the first model that allowed the prediction of glass properties based on the chemical composition, using the additivity principle [1, 2], i.e., multiple regression using linear functions. This principle is based on the assumption that the relation between the glass composition and a specific property is linear to all component concentrations, and all component influences can be summed:

$$\text{Property} = \beta_0 + \sum_{j=1}^{n} \beta_j \cdot C_j \qquad \textit{Eq. (1)}$$

β_0 in Equation (1) is the model intercept, n is the total number of significant glass components excluding the main component (usually silica), the j values are the individual numbers of the significant glass components, βj are the component-specific coefficients, and Cj are the concentrations of the glass components (also called model factors).

The additivity principle does allow very precise and accurate predictions within limited concentration ranges [1, 3, 4, 5, 6, 7, 8, 9], not reached by structural [10, 11, 12, 13, 14], thermodynamic [15, 16, 17, 18, 19, 20, 21, 22, 23, 24, 25, 26, 27, 28, 29, 30, 31, 32, 33, 34, 35, 36, 37], or molecular dynamics [38, 39, 40] modeling approaches. Because of the simplicity of the technique, the ease of interpretation, and good prediction results within specified limits, Equation (1) is most widely used for glass property modeling. The additivity principle cannot be applied for modeling glass properties over wide concentration ranges however, due to component interactions.

In this study, a Plackett-Burman two-level experimental design was intended to result in multiple regression models on the basis of the additivity principle in Equation (1). The experimental design gives a perfect statistical independence of the regression factors (component concentrations), but component interactions can not be investigated in general. It is possible to model component interactions through multiple regression using non-linear functions, as described in the references listed in the following paragraph of this chapter.

REGRESSION ANALYSIS IN GLASS SCIENCE [41, 42]

Single regression using linear functions

$$\text{Property} = \beta_0 + \beta_1 \cdot C \qquad \textit{Eq. (2)}$$

The terms β_0 and β_1 in Equation (2) are defined as regression coefficients, where C is the regression factor (concentration). Equation (2) can be used for glass property modeling in binary systems within narrow concentration limits.

Single regression using polynomial functions

$$\text{Property} = \beta_0 + \beta_1 \cdot C + \beta_2 \cdot C^2 + \beta_3 \cdot C^3 \ldots \qquad \textit{Eq. (3)}$$

Equation (3) can be used for modeling in binary glass systems within wide concentration ranges, as long as significant crystallization, phase separation, or sharp property extrema [43, 49] do not occur.

Multiple regression using linear functions

$$\text{Property} = \beta_0 + \beta_1 \cdot C1 + \beta_2 \cdot C2 + \beta_3 \cdot C3 \ldots \qquad \textit{Eq. (4)}$$

Equation (4) is identical to Equation (1). It may be applied for glass property modeling in multi-component systems within narrow concentration limits. Component interactions are not considered.

It needs to be mentioned that the intercept in the Equations (1) to (4) may be eliminated, if the main glass component silica (SiO_2) is included in the equation. For example in the binary system SiO_2-Na_2O Equation (2) may be modified as follows:

$$\text{Property} = \beta_0 + \beta_1 \cdot C(Na_2O) = \beta_2 \cdot C(SiO_2) + \beta_3 \cdot C(Na_2O) \qquad \textit{Eq. (5)}$$

The approach using an intercept is called "slack variable" technique, and the approach without results in a "canonical" or "Scheffé" type model. Both approaches are nearly identical mathematically; however, the canonical form of the model is expected to produce slightly more accurate predictions if the main component concentration varies significantly. In this study the slack variable technique was used because the concentration ranges were relatively narrow, and the modeling results could be interpreted easily: A specific coefficient shows the property variation due to the exchange of 1% of the considered component for silica. The multiple regression coefficients obtained through the slack variable technique are identical to the values commonly depicted in "spider graphs" [5, 6].

Multiple regression using polynomial functions

$$\text{Property} = \beta_0 + \beta_1 \cdot C1 + \beta_2 \cdot C2 + b_3 \cdot C3 \ldots + \beta_4 \cdot C1^2 + \beta_5 \cdot C2^2 + \beta_6 \cdot C3^2 \ldots + \beta_7 \cdot C1 \cdot C2 + \beta_8 \cdot C2 \cdot C3 \ldots + \beta_9 \cdot C1^3 \ldots + \beta_{10} \cdot C1^2 \cdot C2 + \beta_{11} \cdot C1^2 \cdot C3 \ldots \qquad \textit{Eq. (6)}$$

Equation (6) is given analogous to the Equations (2) to (5) for comparison. The exact form of Equation (6) for the slack variable regression model using polynomial functions of the second order is Equation (7). β0-β3 are the coefficients, with β0 being the intercept, β1 the single-component coefficients, β2 the coefficients of squared influences, and β3 the coefficients of two-component interactions. The variable n in Equation (7) is the total number of the significant glass components excluding silica; j and k are individual numbers of the significant glass components, and C are the component concentrations (excluding silica):

$$\text{Property} = \beta 0 + \sum_{j=1}^{n} \left\{ \beta 1_j \cdot C_j + \beta 2_j \cdot C_j^2 + \sum_{k=j+1}^{n} \beta 3_k \cdot C_j \cdot C_k \right\} \qquad \textit{Eq. (7)}$$

Equation (7) or its canonical variations may be used for glass property modeling in multi-component systems over wide concentration ranges [44, 45, 46, 47, 48]. Sudden property changes, e.g., due to crystallization and/or phase separation are difficult to describe through multiple regression using polynomial functions [49], therefore advanced non-linear functions should be applied in these cases.

Multiple regression using advanced non-linear functions

It is possible to introduce advanced functions into Equation (7), or the structure of the equation may be changed completely. Widely applied is neural networks regression [50]:

Logistic activation functions such as $e^x / (1 + e^x)$ or tanh(x) are used as "neurons" that make graduated yes/no "decisions" according to the input signal. Several neurons are connected following specified rules, partially comparable to neural networks in biology. The connections between the neurons are weighted, i.e., the signal is amplified or reduced. The "weights" are the fitting coefficients that are optimized in neural networks regression. Dreyfus et al. [49] demonstrate the application of neural networks regression for glass property modeling. They show that neural networks regression is advantageous compared to multiple regression using polynomials functions, if the glass property does not change gradually with the chemical composition, but if sharp property extrema occur. Polynomial functions do not fit sharp extrema well, that may appear in glasses due to crystallization or phase separation.

For building a network, component interactions have to be assumed and the total number of adjustable weights in neural networks regression needs to be always larger than two times the number of significant glass components. Therefore, neural networks regression requires an experimental design where interactions can be investigated, similar to multiple regression polynomial using polynomial functions. Neural networks regression cannot be compared directly to regression using linear functions.

Limits of regression analysis for glass property prediction

In principle regression analysis can be applied in glass science as long as a systematic relation exists between the experimental conditions and the resulting properties. However, even though regression analysis can be used in almost all cases, it could be used incorrectly. Most important is the knowledge about glass property extrema. Besides crystallization and phase separation effects, extrema may occur in glasses with a network former content higher than 90 mol%. For example, the Littleton softening point of 100% pure silica glass may be estimated as 1666±50°C from 54 datapoints in SciGlass [59]. If as little as 0.06 mol% sodium oxide is introduced, according to V. K. Leko [43] the Littleton softening point decreases dramatically to 1280°C. Besides high-silica glasses property extrema may also be expected for glasses with high concentrations of B_2O_3 [51], P_2O_5, and GeO_2, or if extreme compositions have to be quenched fast to prevent crystallization (e.g. 50 mol% Na_2O + 50 mol% SiO_2).

Selection of the appropriate regression technique

In general, regression analysis should be used according to the available data. Within narrow concentration limits the linear additivity approach is most appropriate, where wider concentration limits require polynomial functions. If property extrema appear (e.g. through crystallization, phase separation, or in glasses with a network former content higher than 90 mol%) advanced regression techniques should be applied. The transition between the regression approaches is gradual. Most important for deciding which technique should be used is the study of the linear correlation matrix (see below) of the available concentration data; as well as the analysis of binary, ternary, and other related systems for evaluating the crystallization and phase separation tendency, and property extrema. Additionally, all the modeling techniques not based on regression analysis that are summarized in the introduction should be evaluated, according to the scientific experience.

Correlation analysis

The linear correlation matrix is made up of the simple, or two-way, correlation coefficients. They are denoted by the letter "r" and have a range of $-1 < r < +1$ (Pearson's "r"). The correlation coefficient for two factors (independent variables) is a measurement of the linear relationship between the two factors. If r is close to one (1) then a plot of the two factors against one another would look like a straight line with positive slope. If r is close to negative one (-1) then the plot of the two factors against one another would look like a straight line with a negative slope. If r is close to zero then a plot of the two factors would be a "shotgun" pattern with no discernible trends [52].

For selecting the appropriate modeling approach, correlations between changes in the component's concentrations and/or their interactions (concentration cross products) have to be considered if the data was not collected using a statistical design which is orthogonal for all of the factors of interest. If the absolute value of r is larger than approximately 0.5-0.6, the influences of the two considered factors are "partially correlated" (i.e. linked but not completely aliased) and may be difficult to separate. If the absolute value of r is larger than approximately 0.8-0.9, the influences are correlated so strongly that they may not be separated at all in most cases, and the factors should be combined or one factor should be excluded. A correlation is considered statistically significant if r exceeds the value determined by Equation (8), with "n" being the number of experiments. Squared or cubic component influences are often strongly correlated to the corresponding linear influences. To follow the system of hierarchy (see below), they should only be considered after analyzing simpler terms (linear effects and 2-factor interactions), and if the component concentration is widely varied.

$$\text{Significant correlation if: } |r| > 2 / [\,(n + 2)^{1/2}\,] \qquad \textit{Eq. (8)}$$

Evaluation of the factor significance

The t-value is an indicator of the significance of an independent variable (factor, component concentration). In other words, it is a measure of how much information a factor adds to the model. In general, a t-value with absolute value greater than or equal to two is considered to be significant, with a statistical confidence level of approximately 95% [52]. Most minor components are insignificant, i.e., their influence is less than the standard error ("noise"). The t-value is calculated from the quotient of the considered coefficient and the standard error of the coefficient. In general, the absolute t-value tends to increase with 1) increasing number of experiments, 2) decreasing standard error of the regression model (see below), 3) decreasing correlation between the levels of the different components, and 4) increasing variation of component levels (wider component concentration limits).

Factor hierarchy

The hierarchic factor principle implies that higher order terms (e.g. $M \times N \times O$ or M^2) should only be introduced if the significance of all related lower order terms was evaluated first (e.g. M, N, O, $M \times N$, $M \times O$, $N \times O$). Low order terms are always preferred over corresponding high order terms in cases where they are partially correlated and significant. Statistical modeling techniques follow Ockham's razor, a scientific principle that when one is confronted with several possible interpretations of a phenomenon, the simplest interpretation is considered the correct one.

Evaluation of the leverage of a single data point

The Cook value is a measure of how much impact each data point has on the effects (coefficients) quantified by the multiple regression analysis. The Cook value compares the effects (coefficients) with and without each experiment. As a rule-of-thumb, if the Cook value is larger than 1 (one) for a particular experiment, it has a high degree of influence or leverage [52]. High leverage of a single datapoint may occur due to a unique component combination.

Model residuals

The model residuals "Δ" are the difference between the observed value and the fitted value for each experiment. The residuals are assumed to be normally distributed with a mean of zero and a constant variance. For a good model, the residuals should not be influenced by the observed values, nor on any model factor.

Model standard error, prediction error, confidence interval

The model standard error (S_{yx}) for regression analysis is given in Equation (9). It is the standard deviation of the residuals (Δ) in terms of the regression degrees of freedom, which depends on the number of experiments (n) and the number of factors in the model (p). Approximately 68% of all residuals fall within the limits of $\pm S_{yx}$.

$$S_{yx} = \left(\frac{\Sigma \Delta^2}{n - p - 1}\right)^{1/2} \qquad \textit{Eq. (9)}$$

The standard prediction error for the *mean* or "true" response (PE) for a specified glass composition of interest can be determined using Equation (10). X is the matrix including all significant factors and examined compositions, X^T is its transpose matrix, x_o is the composition vector of interest with x_o^T being its transpose, and the operation "$^{-1}$" indicates matrix inversion [53]. The prediction error PE is relevant for practical model application, and it can be estimated using property calculators based on this study [54]:

$$PE = S_{yx} \cdot (x_o^T \cdot (X^T \cdot X)^{-1} \cdot x_o)^{1/2} \qquad \textit{Eq. (10)}$$

The prediction confidence interval for the mean response is obtained through multiplication of the prediction error PE with the t distribution value $t_{\alpha,DF}$ (α = desired confidence, DF = degree of freedom = n − p − 1). For a 95% confidence and DF > 15, $t_{\alpha,DF}$ can be approximated as 2 (two).

Naturally, the error for predicting a single future experiment (PS) is higher than the error for the predicting the mean response (PE). PS may be estimated through Equation (11). The comparison of PE and PS demonstrates the fact that single experiments are less valuable than complete measurement series for evaluating model accuracy.

$$PS = (S_{yx}^2 + PE^2)^{1/2} \qquad \textit{Eq. (11)}$$

The model standard error S_{yx} must be always larger than the standard deviation of repeated measurements, otherwise the model is "over-fitted" (i.e. unrealistically good fitted). The prediction error of the mean (PE) is lower than the standard deviation of repeated individual experiments.

Outlier analysis

Regression analysis assumes the residuals are normally distributed. Thus, a datapoint may be regarded as an outlier if the residual is larger than three times the model standard error (= standardized residual larger than three), if the largest residual is higher than 1.5 times the next largest residual, or if the externally studentized residual [53] is higher than three.

Procedures for obtaining optimal regression fits

Regression analysis applies a "least squares" technique to derive coefficients for each factor which minimize the sum of squares of the residuals. Robust regression [55, 56] considers that the least square fitting technique may not result in accurate results if many datapoints with high leverage are also close to being outliers, i.e., if the data are not normally distributed. Since the squared error value increases rapidly with the distance from the optimal fit, some outlying datapoints may influence the model incorrectly. They may "pull" the fit in their direction. In robust regression the squared error value is replaced by less influential functions. It is possible to avoid the mentioned unwanted effects through a careful analysis of the model residuals and the datapoint leverage values (Cook values).

In this study, robust regression fits were evaluated. However, robust regression was found to result in no significant changes of the models, therefore, the least square technique was applied. The Cook values in all models were smaller than 1 (one), i.e., all datapoints contributed fairly evenly to derivation of the model coefficients. One exception in Table XX is identified.

Two-level Plackett-Burman experimental design

Plackett-Burman designs are normally used for screening a large number of factors to identify those that may be important. In these designs, interactions are partially correlated to main factor effects. Table I shows an example of a two-level Plackett-Burman design for investigating the influences of the components M and N on a glass property. Both components are represented in two concentration levels. M and N are completely statistically independent, i.e., it is not possible to construct a linear relation between M and N. Therefore the influences of both components can be analyzed without mutual interference.

TABLE I: TWO-LEVEL PLACKETT-BURMAN EXPERIMENTAL DESIGN

	Component M	Component N	Property	$M \times N$
Base components +	Level 1	Level 1	10	*1*
Base components +	Level 2	Level 1	20	*2*
Base components +	Level 1	Level 2	5	*2*
Base components +	Level 2	Level 2	14	*4*

The example in Table I makes it also clear, that it is impossible to examine non-linear influences of M or N (e.g. M^2 or N^2); at least one additional concentration level of M and N would have to be introduced. The interaction between M and N (i.e. concentration product $M \times N$) can not be quantified exactly based on the design in Table I due to the partial interference (correlation) of the factor $M \times N$ with the factor M, and of the factor $M \times N$ with the factor N.

In the present study, two-level Plackett-Burman experimental designs were applied for up to 16 glass components, where each design contained 24 experiments (plus one baseline composition, see Tables III to VIII in Chapter 2). Since more than one concentration was changed simultaneously, it was not possible to interpret the data without multiple regression. In case of data inconsistency, strong non-linear component interactions, sharp property extrema, phase separation, or crystallization the multiple regression analysis would lead to a large model standard error; it even may appear that all component influences are insignificant. On the other hand, successful modeling in this study indicates data consistency and suggests that all component interactions, crystallization, and phase separation effects could be reduced to linear component influences within the concentration limits investigated. Since crystallization and phase separation lead to significant property changes and sharp extrema, successful modeling would make it very likely that they do not occur.

Comparative modeling

The term "comparative modeling" is used in this study in case different datasets were combined for the multiple regression analysis. For example, if the compositions in the system Base-M-N in Table I are so complex that only data from one investigator exists, the statistical analysis may be limited. In this case, if the systems of only Base-M or Base-N are widely studied and the interaction between M and N is well known from a system Base2-M-N study, it is possible to combine all data from the systems Base-M-N, Base-M, Base-N, and Base2-M-N. If the data for the system Base-M-N do not show an offset or data trend in the combined model they are assumed to be correct. Examples for comparative modeling are given in this chapter and in reference [48].

In the scientific literature, most properties are reported for glasses containing SiO_2, B_2O_3, Al_2O_3, CaO, MgO, Na_2O, and K_2O. Comparative modeling can be performed without difficulty for the related systems. More complex glasses require a detailed study of the correlation matrix.

Model validation

Statistical model validation can be obtained through:

1) Splitting of the source data into one set for modeling and a second set for comparison of predicted and experimental data,

2) Comparison of the model predictions to experimental data from another investigator,

3) Comparative modeling of two datasets from different investigators where coefficients and residual trends are compared with and without the second dataset,

4) Development of two independent models based on datasets from different investigators in similar composition regions, and comparison of the model coefficients, and

5) Development of two independent models including all possible component interactions based on datasets from different investigators in different composition regions (compositions in mol%), and comparison of the model coefficients.

If the standard errors are comparable to the errors found during model evaluation, it can be assumed that the model is correct, i.e., it is "validated". Method 1) can be used for an internal validation of the model precision, and the methods 2) to 5) allow conclusions concerning the total accuracy through comparison to other investigators.

SOURCE DATA HANDLING

First, the investigated glass compositions were evaluated. It appears that some of the low-expansion borosilicate and fiber wool glass compositions may exhibit phase separation tendency within specific temperature ranges [57]. In addition, thelow-expansion borosilicate glasses have comparatively low concentrations of basic network modifying oxides, e.g., alkali and alkali earth oxides, which can lead to special behavior, e.g. a long viscosity curve and high electrical resistivity. Some textile fiber type E glasses could show a strong crystallization tendency, according to the limits of glass formation reported in the literature [58, 59]. Regarding the TV panel glasses, 16 components were varied systematically, some at very low concentration levels. This fact may lead to an increase of the model standard error, in case the minor components cause insignificant property changes and due to the fact that all errors of the chemical glass analysis add up. The soda lime container and float glasses have very similar compositions, and it may be possible to combine them for modeling.

The majority of the investigated compositions are expected to show good glass formation and sufficiently narrow concentration variations for applying multiple regression using linear functions according to Equation (1). The correlation matrices confirmed excellent linear factor independence, but interactions were strongly aliased.

Second, the data consistency was evaluated separately according to the different investigators and techniques involved. It was noted that all the reported values cold be fitted well through simple functions, but regarding the high temperature viscosity data scattering appeared. In addition it can be seen from the experimental results in the Chapters 7 and 8 that the data were reported systematically where either the temperature (resistivity measurements) or the viscosity are given in constant intervals. However, concerning the high temperature viscosity data both, the temperature *and* viscosity intervals are not constant, which makes modeling less reliable. In addition it appears that the high temperature viscosity data are of various quality: mostly four values are reported, sometimes three, but for the Low-expansion borosilicate glasses only two datapoints are given in some cases ranging from $\log_{10}(\text{viscosity/Pa·s}) = 2.7$ (LE 22) to -0.1 (LE 13). The Low-expansion borosilicate glasses are expected to cause problems during modeling of the viscosity at high temperatures, where the modeling of all remaining data may be more reliable.

Next, the consistency of the viscosity data was examined over all three temperature ranges / applied techniques / responsible investigators, reported in Chapter 7. It was realized that most data compared well and could be fitted over the full measured temperature range through the Vogel-Fulcher-Tammann equation (VFT, see Equation (12) and Appendix B). However, some discrepancies appeared for the textile fiber type E glasses during comparison of the low-temperature beam bending results, and the medium temperature parallel plate values. If both series were extended to a medium viscosity level of $\log_{10}$(viscosity/Pa·s) = 9.5, the parallel plate estimate was ~20°C higher than the one obtained from beam bending. A closer analysis did show that this difference depended systematically on the temperature over all investigated glasses, where it approached zero for the container, float, and TV panel glasses. It appears that this discrepancy may be caused by a slightly incorrect calibration. The beam bending viscometer was at its upper temperature limit during measurement of the E glasses, hence the data might be less reliable. Likewise, the parallel plate technique is less reliable at low deformation rates at $\log_{10}$(viscosity/Pa·s) = 9-9.5. The beam bending viscometer calibration data were re-analyzed and compared to annealing point (T_g) data (see Appendix I at the end of this book, Tables I to VI), plus additional high temperature measurements of the NIST standard glass SRM 714 were carried out for the parallel plate viscometer, without resolving the discrepancy. It needs to be noted at this point that the NIST high temperature viscosity standard SRM 714 is less reliable for calibration than the common standards SRM 710A, SRM 717A or SRM 711, because only a few viscosity fixpoints are mentioned that do not fit well through the VFT equation. In addition, most viscosity data reported in the literature have errors of approximately 20°C, if the same glass composition measured by several investigators is compared [59, 60]. However, it appears from comparative modeling [48] that the beam bending data compare well to findings of other investigators at the viscosity level $\log(\eta/\text{Pa·s}) = 12$, similar to the parallel plate results at $\log(\eta/\text{Pa·s}) = 6.6$. It might be that at $\log(\eta/\text{Pa·s}) = 12$ and 6.6 the data in Chapter 7 are more accurate than at $\log(\eta/\text{Pa·s}) = 9.5$. Research is continuing. For the high temperature viscosity values discrepancies could not be found due to data scattering. On the whole, all high temperature viscosities from Chapter 7 compare well to findings of other investigators [48].

Next, an evaluation was performed to determine if the data should be modeled on the basis of weight% (wt%), mol%, or molar fraction to SiO_2, and whether the designed or chemically analyzed concentrations should be used. Since the differences between initial modeling errors were insignificant in all cases, the original analyzed glass compositions in weight fraction as given in the Tables III to VIII in Chapter 2 were selected as modeling basis (not normalized to a total sum of 1) and converted to wt% through multiplication by 100. Even though modeling on the basis of wt% is convenient for practical industrial applications in glass melting, it should be noted that all model results depend on the glass density. The use of wt% is acceptable within the narrow concentration limits studied here. However, calculations based on mol% concentrations are more appropriate over wider composition ranges.

It was considered whether the viscosity data at low, medium, and high viscosities should be combined for each glass composition through the Vogel-Fulcher-Tammann equation (VFT, see Equation (12)) *before* or *after* modeling. For example, for each TV panel glass, the VFT constants A, B, and T_o were determined from the experimental data of the complete viscosity curve, and regression models were developed for predicting them. Alternatively, other models were developed for estimating the temperatures of constant viscosities (isokom) at $\log(\eta/\text{Pa·s}) =$ 1.5, 6.6, and 12.0, without fitting the complete viscosity curve. It was found that the modeling of the isokom temperatures did result in significantly lower prediction errors, compared to modeling the VFT constants of the complete viscosity curve. Therefore, the viscosity models presented here predict isokom temperatures in °C at log(η/Pa·s) = 1.5, 6.6, and 12.0 (T(1.5), T(6.6), T(12.0)), which define the VFT constants through the Equations (12) and (15) to (17).

For the high temperature electrical resistivity data in Chapter 8, models were developed for predicting the decadic logarithm of the resistivities ρ at 1000°C, 1200°C, and 1400°C (ρ(1000), ρ(1200), ρ(1400)). Among all the equations used for describing the temperature dependence of the electrical resistivity of glass melts [61], the VFT equation was found to be accurate over the complete temperature range studied. Accordingly, ρ(1000), ρ(1200), ρ(1400) allow the calculation of the VFT constants through the Equations (13) and (18) to (20).

$$\log_{10}(\text{viscosity} / \text{Pa·s}) = A + B / (T - T_o) \quad \textit{Eq. (12)}$$
$$\log_{10}(\text{resistivity} / \Omega\text{·cm}) = A + B / (T - T_o) \quad \textit{Eq. (13)}$$

Prior to modeling, the original viscosity values were fitted to the VFT equation separately for each of the three temperature ranges / investigators / techniques. The fits described the experimental data well. Following the VFT fits, the temperatures of constant viscosities (isokom) were calculated for all glasses for log(η/Pa·s) = 1.5, 6.6, and 12.0. The rotating spindle measurements done by P. Hrma et al. had to be analyzed with exceptional care, firstly because the original data varied significantly in the viscosity levels examined for log(η/Pa·s) < 3, and secondly because the number of datapoints was only 2-4 for each glass. If 3 or 4 datapoints were reported to cover the range around log(η/Pa·s) = 1.5 ± 0.25, a VFT fit was used; otherwise a linear fit with 2 datapoints was employed. In about 20% of all cases, however, the reported viscosity levels were far off from log(η/Pa·s) = 1.5. Then, a datapoint from parallel plate measurements at log(η/Pa·s) = 6.0, obtained by Fluegel and Varshneya, was included in the VFT fit. As a result, the rotating spindle measurements done at PNNL may appear less accurate than other data. Table II lists all isokom temperatures at log(η/Pa·s) = 1.5 used for modeling:

TABLE II: ISOKOM TEMPERATURES IN °C AT LOG(η/PA·S) = 1.5; VALUES IN PARENTHESES ARE EXTRAPOLATED; 00 - BASE GLASSES, TV - TV PANEL GLASSES, E - TEXTILE FIBER E GLASSES, LO - LOW-EXPANSION BOROSILICATE GLASSES, CO - CONTAINER GLASSES, FL - FLOAT GLASSES, WO - FIBER WOOL GLASSES

#	TV	E	LO	CO	FL	WO
00	1301	1274	(1690)	1323	1299	1163
01	1352	(1613)	(1725)	1458	1396	(1359)
02	1457	1492	(1624)	1350	1260	1309
03	1438	1269	1366	1206	(1316)	(1125)
04	1489	1492	(1935)	1306	1246	1282
05	1398	1295	1446	1244	1248	(1167)
06	1169	1300	1480	1218	1226	(1130)
07	1169	(1354)	1468	1318	1295	1087
08	1195	1206	1342	1374	1250	(1054)
09	1150	1164	1243	1220	1333	(1062)
10	1172	1207	1503	1346	1250	(1096)
11	1252	1249	1564	1225	1236	(1100)
12	1195	1295	(1749)	1353	1306	1226
13	1242	1321	(1137)	1305	1300	1163
14	1122	1359	1472	1376	1292	1270
15	1298	1425	1507	1245	1273	1176
16	1316	1507	/	1393	1294	1369
17	1159	1406	(1703)	1225	1313	1252
18	1237	1293	1548	1258	1282	/
19	1309	1368	(1221)	1354	1350	1104
20	1343	/	1578	1395	1266	1210
21	1237	1186	1322	1301	1328	1221
22	1276	1382	(1799)	1401	1323	1174
23	1359	1175	1334	1294	1346	1058
24	(1552)	(1088)	1271	1159	1368	963

The original data used for modeling are summarized in the Chapters 7 and 8. Regarding the high temperature viscosity the values in Table II were considered. For model validation and comparison additional data are listed in the Tables III to VI:

TABLE III: GLASS COMPOSITIONS AND VISCOSITY DATA INVESTIGATED BY LAKATOS ET AL. [4]

Glass composition in wt%, as analyzed						Temperatures in °C at constant $\log_{10}$(viscosity / Pa·s)		
SiO_2	Al_2O_3	MgO	CaO	Na_2O	K_2O	1.5	6.6	12.0
77.41	0.20	0.00	10.17	12.09	0.13	1351.4	742.7	566.4
66.57	8.25	1.68	10.49	10.77	2.24	1386.5	778.0	597.2
66.70	6.01	0.74	11.96	10.58	4.01	1324.9	754.8	585.8
67.08	4.00	2.16	9.61	10.88	6.27	1314.9	733.6	556.7
66.55	2.24	1.25	11.18	10.51	8.27	1261.0	715.7	549.1
68.84	4.08	1.85	13.33	11.88	0.02	1308.2	758.7	592.6
71.76	2.21	0.78	10.14	12.50	2.60	1322.8	730.3	555.7
69.05	0.36	2.17	11.87	12.11	4.44	1243.6	709.6	549.4
65.78	7.13	1.19	8.52	11.31	6.06	1350.1	739.7	558.2
60.56	4.99	6.72	9.36	10.52	7.83	1291.9	719.4	551.6
67.50	7.32	0.64	11.77	12.75	0.03	1352.2	766.5	598.9
68.49	5.39	2.31	8.82	12.68	2.31	1343.8	744.6	563.8
67.64	3.38	1.20	10.68	12.78	4.32	1280.6	718.4	554.5
66.65	2.26	0.00	12.10	12.49	6.49	1240.0	706.9	550.0
67.22	0.00	1.69	9.65	12.71	8.74	1227.9	684.8	521.7
70.17	2.24	2.27	11.13	14.16	0.05	1280.5	726.8	559.2
69.43	0.17	1.09	12.96	13.87	2.46	1236.4	712.0	555.1
65.53	7.18	0.00	9.43	13.46	4.40	1323.1	727.9	561.3
63.60	5.11	1.72	10.56	12.77	6.24	1264.2	714.1	550.9
65.49	3.59	0.71	8.58	13.44	8.20	1262.6	694.3	527.7
68.17	5.45	1.25	9.78	15.33	0.02	1317.9	735.5	567.2
67.49	4.16	0.00	11.31	14.63	2.41	1271.1	717.3	555.9
68.05	1.96	1.75	8.88	14.95	4.42	1248.4	692.1	525.6
67.37	0.38	0.54	10.59	14.69	6.43	1205.9	678.5	523.0
59.54	6.73	1.98	10.98	13.16	7.60	1234.3	704.9	545.1
72.04	2.18	1.99	10.08	11.21	2.50	1342.7	745.2	563.4
69.38	2.35	1.27	8.21	11.98	6.81	1305.7	710.6	534.1
73.02	2.03	0.66	11.22	12.99	0.08	1326.6	743.2	571.0
70.51	2.15	0.00	8.86	13.85	4.62	1290.7	704.7	534.0
63.60	2.30	2.19	10.32	13.67	7.92	1209.5	684.6	526.5
59.54	0.00	0.00	8.21	10.51	0.02	**Minimum in wt%**		
67.71	3.53	1.39	10.42	12.69	4.26	**Average in wt%**		
77.41	8.25	6.72	13.33	15.33	8.74	**Maximum in wt%**		
3.48	2.38	1.26	1.32	1.37	2.90	**St. dev. in wt%**		

TABLE IV: HIGH LEVEL WASTE (HLW) GLASS COMPOSITIONS INVESTIGATED BY HRMA ET AL. [6]

#	Glass composition in wt%, as batched									
	SiO_2	B_2O_3	Na_2O	Li_2O	Fe_2O_3	Al_2O_3	ZrO_2	Bi_2O_3	P_2O_5	Rem*
HLW 01	53.33	0.05	12.34	2.86	6.80	4.87	4.38	1.37	0.79	13.21
HLW 02	60.00	0.04	10.57	2.45	5.83	4.17	3.75	1.18	0.68	11.33
HLW 03	45.00	0.05	14.54	3.37	8.01	5.74	5.16	1.62	0.94	15.57
HLW 04	40.00	0.06	15.86	3.67	8.74	6.26	5.63	1.76	1.02	17.00
HLW 05	37.00	0.06	16.65	3.86	9.18	6.57	5.91	1.85	1.07	17.85
HLW 06	48.89	8.37	11.31	2.62	6.23	4.46	4.01	1.26	0.73	12.12
HLW 07	51.22	4.00	11.85	2.74	6.53	4.68	4.20	1.32	0.76	12.70
HLW 08	47.10	0.04	22.58	2.52	6.00	4.30	3.87	1.21	0.70	11.68
HLW 09	57.79	0.05	5.00	3.09	7.37	5.27	4.74	1.49	0.86	14.34
HLW 10	50.97	0.04	11.79	7.16	6.50	4.65	4.18	1.31	0.76	12.64
HLW 11	52.15	0.05	12.06	5.00	6.65	4.76	4.28	1.34	0.78	12.93
HLW 12	54.90	0.05	12.70	0.00	7.00	5.01	4.51	1.41	0.82	13.60
HLW 13	48.64	0.04	11.25	2.60	15.00	4.44	3.99	1.25	0.72	12.07
HLW 14	50.93	0.04	11.78	2.73	11.00	4.65	4.18	1.31	0.76	12.62
HLW 15	57.22	0.05	13.24	3.06	0.00	5.22	4.70	1.47	0.85	14.19
HLW 16	42.04	0.04	9.72	2.25	5.36	25.00	3.45	1.08	0.63	10.43
HLW 17	45.97	0.04	10.63	2.46	5.86	18.00	3.77	1.18	0.68	11.41
HLW 18	48.77	0.04	11.28	2.61	6.22	13.00	4.00	1.26	0.73	12.09
HLW 19	56.06	0.05	12.97	3.00	7.15	0.00	4.60	1.44	0.84	13.89
HLW 20	46.85	0.04	10.84	2.51	5.97	4.28	16.00	1.21	0.70	11.60
HLW 21	49.08	0.04	11.35	2.63	6.26	4.48	12.00	1.26	0.73	12.17
HLW 22	51.31	0.04	11.87	2.75	6.54	4.68	8.00	1.32	0.76	12.73
HLW 23	55.77	0.05	12.90	2.99	7.11	5.09	0.00	1.44	0.83	13.82
HLW 24	49.66	0.04	11.49	2.66	6.33	4.53	4.08	1.95	1.13	18.13
HLW 25	57.42	0.05	13.28	3.07	7.32	5.24	4.71	0.73	0.42	7.76
HLW 27	48.64	0.04	11.25	2.60	15.00	4.44	3.99	1.25	0.72	12.07
HLW 28	51.70	0.05	11.96	2.77	6.59	4.72	4.24	4.37	0.77	12.83
HLW 29	50.62	0.05	11.71	2.71	6.45	4.62	4.16	6.37	0.75	12.56
HLW 30	47.92	0.05	11.09	2.57	6.11	4.37	3.93	11.37	0.71	11.88
HLW 31	45.22	0.04	10.46	2.42	5.76	4.13	3.71	16.37	0.67	11.22
HLW 37	52.14	0.05	12.06	2.79	6.65	4.76	4.28	1.34	3.00	12.93
HLW 38	51.07	0.05	11.81	2.74	6.51	4.66	4.19	1.31	5.00	12.66
HLW 39	49.99	0.05	11.56	2.68	6.37	4.56	4.10	1.29	7.00	12.40
HLW 40	48.92	0.05	11.32	2.62	6.24	4.47	4.02	1.26	9.00	12.10
Min	37.00	0.04	5.00	0.00	0.00	0.00	0.00	0.73	0.42	7.76
Avg	50.13	0.41	12.15	2.90	7.08	5.88	4.84	2.32	1.39	12.90
Max	60.00	8.37	22.58	7.16	15.00	25.00	16.00	16.37	9.00	18.13
St.dev.	4.98	1.56	2.62	1.03	2.56	4.41	2.62	3.16	1.88	1.99

*Rem: Waste simulant remainder in wt%: MgO: 0.34, CaO: 9.53, CdO: 0.42, CeO_2: 12.78, Cr_2O_3: 4.01, F: 2.57, K_2O: 0.98, La_2O_3: 2.00, MnO_2: 8.51, MoO_3: 0.34, Nd_2O_3: 45.54, NiO: 10.63, SO_3: 1.55, WO_3: 0.79

TABLE V: HIGH TEMPERATURE ELECTRICAL RESISTIVITY DATA OF HLW GLASSES INVESTIGATED BY HRMA ET AL. [6] AT AC FREQUENCY 10 KHZ

#	$\log_{10}(\rho/\Omega\cdot cm)$ at T in °C		
	1000°C	**1200°C**	**1400°C**
HLW 01	/	0.537	0.264
HLW 02	/	0.740	0.469
HLW 03	0.722	0.349	0.090
HLW 04	0.587	0.249	/
HLW 05	0.488	0.163	/
HLW 06	0.969	0.572	0.312
HLW 07	/	0.560	0.304
HLW 08	0.399	0.107	/
HLW 09	/	0.912	0.647
HLW 10	0.530	0.176	/
HLW 11	0.721	0.345	0.117
HLW 12	0.952	0.789	0.547
HLW 13	/	0.553	0.298
HLW 14	/	0.536	0.272
HLW 15	/	0.524	0.279
HLW 16	/	0.796	0.549
HLW 17	/	0.695	0.436
HLW 18	/	0.629	0.357
HLW 19	/	0.375	-0.009
HLW 20	/	0.604	0.303
HLW 21	/	0.559	0.221
HLW 22	/	0.700	0.418
HLW 23	/	0.468	0.232
HLW 24	/	0.577	0.316
HLW 25	/	0.410	0.081
HLW 27	/	0.553	0.292
HLW 28	/	0.566	0.280
HLW 29	/	0.562	0.312
HLW 30	0.965	0.561	0.300
HLW 31	0.965	0.559	/
HLW 37	/	0.523	0.275
HLW 38	/	0.518	0.272
HLW 39	/	0.621	0.367
HLW 40	/	0.546	0.311

Since, compared to the viscosity, only few high temperature electrical resistivity data exist in the scientific literature, those simple glass forming systems were used as guide for comparison which were studied by many investigators: SiO_2-Na_2O, SiO_2-K_2O, and SiO_2-CaO-Na_2O. All properties were obtained from ref. [62] and SciGlass [59]. Detailed references are given in Appendix (D) at the end of this chapter. It was assumed that the majority of the data in the literature are comparable, i.e., the investigators did report the electrical resistivity considering ionic and electrode polarization effects appropriately, as well as capacitive and inductive losses [61, 62]. The data of the analyzed systems were combined and grouped according to reference resistivities at 1000°C, 1200°C, and 1400°C. Three independent polynomial slack variable models (excluding silica, with intercept, see Equation (7)) were developed at 1000°C, 1200°C, and 1400°C from 110-160 data points at each of the three temperatures. The models were based on mol% (because of the relatively wide concentration range) including all linear factors Na_2O, K_2O, and CaO, and the following higher order terms: $(Na_2O)^2$, $(Na_2O)^3$, $(K_2O)^2$, $(K_2O)^3$, and the $Na_2O \times CaO$ interaction. The model standard errors were log $(\rho / \Omega \cdot cm) = 0.052$ at 1000°C, 0.094 at 1200°C, and 0.105 at 1400°C. The regression R^2 were 0.986 at 1000°C, 0.946 at 1200°C, and 0.928 at 1400°C. The three independent intermediate models at 1000°C, 1200°C, and 1400°C were combined through applying regression using quadratic functions, with the squared term set to a constant average value. The regression between the models resulted in an additional error of $\log(\rho / \Omega \cdot cm) = 0.013$. Tables VI to VIII show the final average values and 95% confidence intervals for the mean resistivity with the concentrations in mol% and wt%:

TABLE VI: AVERAGE VALUES FOR THE HIGH TEMPERATURE ELECTRICAL RESISTIVITY IN THE SYSTEM SIO_2-NA_2O (REFERENCES IN APPENDIX (D) OF THIS CHAPTER), ERROR = 95% CONFIDENCE INTERVAL

C(Na_2O)		log(ρ/Ω·cm) at T in °C 1000°C		1200°C		1400°C	
mol%	wt%	Value	Error	Value	Error	Value	Error
16	16.42	0.974	0.038	0.710	0.068	0.512	0.064
17	17.44	0.895	0.035	0.635	0.062	0.441	0.061
18	18.46	0.821	0.033	0.565	0.057	0.374	0.058
19	19.48	0.751	0.031	0.498	0.054	0.311	0.056
20	20.50	0.685	0.030	0.436	0.052	0.252	0.055
21	21.52	0.623	0.029	0.377	0.051	0.197	0.054
22	22.54	0.565	0.028	0.323	0.051	0.145	0.054
23	23.55	0.510	0.028	0.271	0.051	0.097	0.053
24	24.57	0.459	0.028	0.223	0.051	0.052	0.053
25	25.59	0.412	0.028	0.178	0.051	0.010	0.053
26	26.60	0.367	0.028	0.136	0.052	-0.029	0.053
27	27.62	0.326	0.028	0.097	0.052	-0.066	0.053
28	28.63	0.287	0.029	0.061	0.053	-0.100	0.054
29	29.64	0.250	0.030	0.027	0.054	-0.131	0.054
30	30.66	0.217	0.031	-0.004	0.055	-0.161	0.055
30	31.67	0.185	0.032	-0.034	0.056	-0.188	0.056
32	32.68	0.156	0.034	-0.061	0.058	-0.213	0.058
33	33.69	0.128	0.035	-0.087	0.060	-0.237	0.059
34	34.70	0.102	0.036	-0.111	0.061	-0.258	0.061
35	35.71	0.078	0.038	-0.133	0.063	-0.279	0.063
36	36.72	0.055	0.039	-0.154	0.065	-0.298	0.064
37	37.73	0.033	0.040	-0.174	0.066	-0.316	0.066
38	38.73	0.012	0.041	-0.193	0.068	-0.333	0.068
39	39.74	-0.007	0.041	-0.211	0.069	-0.349	0.070
40	40.75	-0.027	0.041	-0.228	0.069	-0.365	0.071

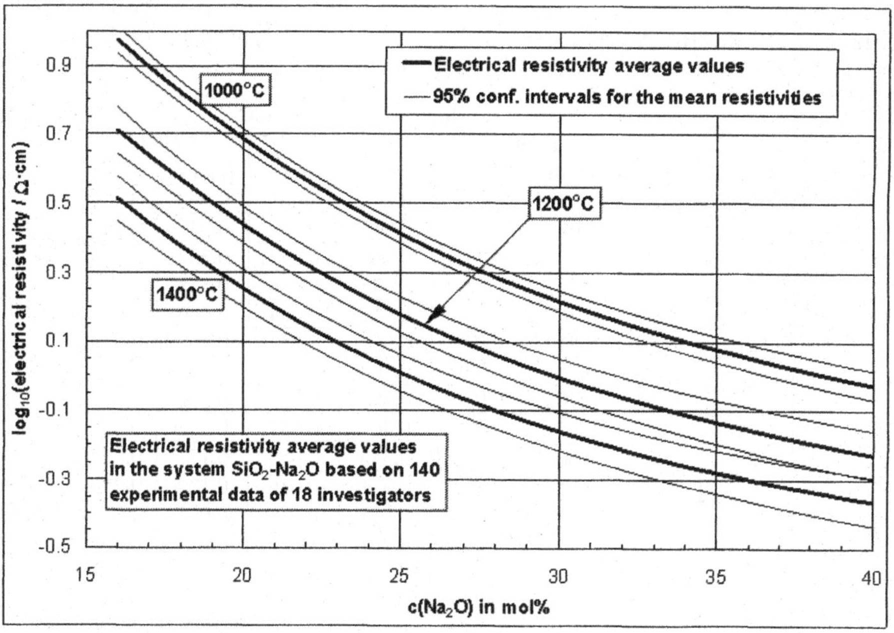

Figure 1: High temperature electrical resistivity average values from Table VI

TABLE VII: AVERAGE VALUES FOR THE HIGH TEMPERATURE ELECTRICAL RESISTIVITY IN THE SYSTEM SIO_2-K_2O (REFERENCES IN APPENDIX (D) OF THIS CHAPTER), ERROR = 95% CONFIDENCE INTERVAL

$C(K_2O)$		log(ρ/Ω·cm) at T in °C					
		1000°C		1200°C		1400°C	
mol%	wt%	Value	Error	Value	Error	Value	Error
16	16.42	1.139	0.035	0.883	0.061	0.692	0.060
17	17.44	1.066	0.033	0.811	0.059	0.621	0.058
18	18.46	0.997	0.032	0.742	0.057	0.552	0.056
19	19.48	0.931	0.031	0.677	0.056	0.488	0.055
20	20.50	0.869	0.030	0.615	0.054	0.426	0.056
21	21.52	0.810	0.029	0.557	0.053	0.368	0.056
22	22.54	0.754	0.029	0.502	0.053	0.314	0.056
23	23.55	0.702	0.029	0.450	0.053	0.263	0.057
24	24.57	0.651	0.029	0.401	0.053	0.215	0.057
25	25.59	0.604	0.030	0.355	0.055	0.171	0.058
26	26.60	0.558	0.031	0.311	0.056	0.129	0.058
27	27.62	0.515	0.032	0.271	0.059	0.091	0.059
28	28.63	0.473	0.034	0.232	0.062	0.056	0.061
29	29.64	0.433	0.036	0.197	0.064	0.025	0.063
30	30.66	0.395	0.039	0.163	0.068	-0.004	0.066
30	31.67	0.358	0.041	0.132	0.071	-0.029	0.070
32	32.68	0.322	0.043	0.103	0.075	-0.051	0.076

TABLE VIII: AVERAGE VALUES FOR THE HIGH TEMPERATURE ELECTRICAL RESISTIVITY IN THE SYSTEM SIO_2-CAO-NA_2O (REFERENCES IN APPENDIX (D) OF THIS CHAPTER), ERROR = 95% CONFIDENCE INTERVAL

C(CaO)		$C(Na_2O)$		log(ρ/Ω·cm) at T in °C					
				1000°C		1200°C		1400°C	
mol%	wt%	mol%	wt%	Value	Error	Value	Error	Value	Error
5	4.65	21	21.59	0.682	0.017	0.409	0.029	0.200	0.035
5	4.65	23	23.63	0.580	0.018	0.317	0.030	0.118	0.037
5	4.65	25	25.67	0.493	0.021	0.238	0.034	0.048	0.039
7	6.53	15	15.47	1.093	0.018	0.773	0.031	0.518	0.036
7	6.53	17	17.52	0.946	0.014	0.639	0.023	0.396	0.031
7	6.52	19	19.57	0.817	0.016	0.522	0.025	0.291	0.032
7	6.52	21	21.62	0.705	0.018	0.421	0.029	0.202	0.035
7	6.52	23	23.66	0.608	0.021	0.335	0.033	0.126	0.039
7	6.51	25	25.71	0.525	0.024	0.262	0.038	0.064	0.045
9	8.41	15	15.49	1.104	0.016	0.768	0.026	0.498	0.032
9	8.41	17	17.55	0.961	0.017	0.640	0.028	0.384	0.032
9	8.40	19	19.60	0.836	0.020	0.529	0.033	0.286	0.036
9	8.39	21	21.65	0.729	0.024	0.433	0.039	0.203	0.042
9	8.39	23	23.70	0.636	0.028	0.353	0.046	0.135	0.050
9	8.38	25	25.74	0.558	0.034	0.286	0.053	0.079	0.058
11	10.29	15	15.51	1.114	0.020	0.764	0.034	0.478	0.036
11	10.29	17	17.57	0.976	0.022	0.641	0.037	0.371	0.037
11	10.28	19	19.63	0.856	0.026	0.535	0.044	0.280	0.043
11	10.27	21	21.68	0.752	0.031	0.446	0.051	0.205	0.051
11	10.27	23	23.73	0.664	0.037	0.371	0.060	0.143	0.062
13	12.18	15	15.53	1.124	0.026	0.759	0.045	0.459	0.042
13	12.17	17	17.59	0.990	0.028	0.642	0.048	0.358	0.044
13	12.17	19	19.65	0.875	0.032	0.542	0.055	0.274	0.051
13	12.16	21	21.71	0.776	0.038	0.458	0.064	0.206	0.062
13	12.15	23	23.76	0.692	0.046	0.389	0.075	0.152	0.075
15	14.07	15	15.55	1.134	0.033	0.754	0.058	0.439	0.050
15	14.07	17	17.62	1.005	0.034	0.643	0.061	0.345	0.053
15	14.06	19	19.68	0.894	0.039	0.549	0.068	0.269	0.061
15	14.05	21	21.74	0.799	0.046	0.471	0.077	0.207	0.074

MODELING PROCEDURE

For modeling of the viscosity and high temperature electrical resistivity data in the Chapters 7 and 8 a slack-variable multiple regression model using linear functions and an intercept while excluding silica was selected (Equation (1)). All glass types studied in this work were modeled independently, except the soda lime container and float glasses were lumped together because of the nearly identical composition ranges. The independent variables were the concentrations in wt% (model factors); the dependent variables were the isokom temperatures in °C for log (η/Pa·s) = 1.5, 6.6, and 12.0; as well as the decadic logarithm of the resistivities at 1000°C,

1200°C, and 1400°C. No squared and interaction terms were considered due to correlations. If applicable, dataset offsets were analyzed during modeling through Equation (14), where the offset was represented by a categorical model factor with the value of one for the specific dataset and zero for all other datasets:

$$\text{Property, observed} = \text{Property, calculated} + \text{Offset} \qquad \textit{Eq. (14)}$$

The data preparation and part of the data analysis were performed using Microsoft Excel. Multiple regression analysis was done with the programs MCA (Nextbridge Softwares, version 1999), as well as SPSS (Standard version for Windows, release 11.5.0, 6 Sept 2002, LEAD Technologies Inc).

Factors were excluded stepwise from the model if there was less than a ~95% confidence level in their significance (t-value < 2). Next, the data were analyzed for any outliers. If the residual was larger than three times the model standard error, i.e. if the standardized residual was larger than three, the value was considered an outlier and deleted from the model. Finally, the complete modeling procedure was iterated, until no further outliers were present.

RESULTS

The Tables IX to XIV list all significant coefficients of the viscosity models based on wt%, as well as other statistical indicators. The Tables XV to XX show the results for the high temperature electrical resistivity models. Any factors not listed were found to be statistically insignificant over the composition range tested. One may wonder why in many cases the listed errors are higher than some of the model coefficients (which express the property change caused by a 1 wt% composition change). This occurs because the coefficients are based on several experiments and the component concentration ranges are often wider than 1 wt%.

TABLE IX: VISCOSITY MODELS FOR THE TV PANEL GLASSES

TV panel glasses	**Coefficients based on wt% for $\log_{10}$(viscosity / Pa·s)**		
Factors	**1.5**	**6.6**	**12.0**
Al_2O_3	0	6.34	4.99
CaO	-14.94	5.69	8.40
SrO	-13.30	-1.36	1.34
BaO	-11.65	-2.72	0
Li_2O	-53.28	-43.14	-32.24
Na_2O	-27.77	-14.00	-8.68
K_2O	-15.65	-6.16	-3.33
TiO_2	-53.48	-28.60	0
ZrO_2	0	5.43	4.37
PbO	-11.38	-5.76	-3.10
F	-50.05	-37.45	-30.51
Intercept	1864.95	880.20	584.845
Prediction error in °C*, depending on composition	10-25	4-12	3-8
Model standard error in °C	34.6	14.9	10.2
Std. dev. of residuals in °C	27.4	11.0	8.1
R^2	0.94	0.92	0.91
Average isokom in °C	1287	691	510
Model standard error in %	2.69	2.15	2.00
Glasses not in model	none		
Outlier	none		
Approximate concentration limits in wt% for TV panel glass compositions after glass melting: SiO_2: 51-81, Al_2O_3: 1.2-3.7, MgO: 0-1.5, CaO: 0-3.5, SrO: 1-10, BaO: 2-13, Li_2O 0-0.5:, Na_2O: 5.5-9.5, K_2O: 5.8-9, TiO_2: 0.1-0.6, CeO_2: 0-0.8, ZrO_2: 0-3, PbO: 0-3.2, ZnO: 0-2, As_2O_3: 0-0.35, Sb_2O_3: 0.2-0.7, F: 0-0.85, Fe total as Fe_2O_3: 0-0.04			

*Exact prediction errors can be determined using property calculators [54].
The excluded insignificant factors are not mentioned.

TABLE X: VISCOSITY MODELS FOR THE TEXTILE FIBER E GLASSES

Textile fiber E glasses	**Coefficients based on wt% for $\log_{10}$(viscosity / Pa·s)**		
Factors	**1.5**	**6.6**	**12.0**
B_2O_3	-18.52	-9.79	-6.97
MgO	-13.43	-4.14	0
CaO	-19.19	-4.42	-1.72
Na_2O	-17.44	-14.88	-12.49
Intercept	1837.40	1024.51	764.84
Prediction error in °C*, depending on composition	11-30	3-8	2-4
Model standard error in °C	23.8	12.7	8.0
Std. dev. of residuals in °C	21.4	11.4	7.4
R^2	0.96	0.95	0.95
Average isokom in °C	1314	865	686
Model standard error in %	1.81	1.47	1.17
Glasses not in model	E 01, E 08, E 20		
Outlier	none		
Approximate concentration limits in wt% for textile fiber E glass compositions after glass melting: SiO_2: 41-66, B_2O_3: 0-9.3, Al_2O_3: 9-16, MgO: 0.5-6, CaO: 13.5-25.5, Na_2O: 0-2.1, K_2O: 0-0.6, Fe total as Fe_2O_3: 0-0.8, TiO_2: 0-1, F: 0-0.6			
Strong crystallization was observed in the glass E 20.			

*Exact prediction errors can be determined using property calculators [54].
The excluded insignificant factors are not mentioned.

TABLE XI: VISCOSITY MODELS FOR THE LOW-EXPANSION BOROSILICATE GLASSES

Low-expansion borosilicate glasses	**Coefficients based on wt% for $\log_{10}$(viscosity / Pa·s)**
Factors	**6.6**
B_2O_3	-6.31
Al_2O_3	3.04
Na_2O	-13.37
K_2O	-9.22
Intercept	942.18
Prediction error in °C*, depending on composition	3-6
Model standard error in °C	11.7
Std. dev. of residuals in °C	10.5
R^2	0.91
Average isokom in °C	777
Model standard error in %	1.51
Glasses not in model	LE 02, LE 16, LE 20
Outlier	none
The modeling of the data for the viscosity levels log(η/Pa·s) = 1.5 and 12.0 is not possible.	
Approximate concentration limits in wt% for low-expansion borosilicate glass compositions after glass melting: SiO_2: 62-84, B_2O_3: 9-15.5, Al_2O_3: 2-7.3, CaO: 0-2, Na_2O: 3.7-8.7, K_2O: 0-3.2, BaO: 0-2	
Strong phase separation was observed in the glasses LO 02, LO 16, LO 20. The glasses LO 12 and LO 21 phase separate during prolonged heat treatment.	

*Exact prediction errors can be determined using property calculators [54].
The excluded insignificant factors are not mentioned.

TABLE XII: VISCOSITY MODELS FOR THE FIBER WOOL GLASSES

Fiber wool glasses	**Coefficients based on wt% for $\log_{10}$(viscosity Pa·s)**	
Factors	**1.5**	**6.6**
B_2O_3	-21.54	-2.78
MgO	-5.61	-2.95
CaO	-15.88	-2.60
Na_2O	-25.18	-8.68
K_2O	-22.75	-8.59
Intercept	1847.55	871.33
Prediction error in °C*, depending on composition	7-16	2-6
Model standard error in °C	29.1	11.1
Std. dev. of residuals in °C	25.7	9.9
R^2	0.94	0.85
Average isokom in °C	1172	685
Model standard error in %	2.48	1.62
Glasses not in model	WO 18	none
Outlier	none	
The modeling of the data for the viscosity level $\log(\eta/\text{Pa·s}) = 12.0$ is not possible.		
Approximate concentration limits in wt% for wool fiber glass compositions after glass melting: SiO_2: 49-78, B_2O_3: 3-9.3, Al_2O_3: 0-6.5, MgO: 1-5, CaO: 5-13.5, Na_2O: 13-18, K_2O: 0-2.2, Fe total as Fe_2O_3: 0-0.7, F: 0-0.8, SO_3: 0-0.3		

*Exact prediction errors can be determined using property calculators [54].
The excluded insignificant factors are not mentioned.

TABLE XIII: VISCOSITY MODELS FOR THE SODA LIME CONTAINER AND FLOAT GLASSES

Soda lime container and float glasses	**Coefficients based on wt% for $\log_{10}$(viscosity / Pa·s)**		
Factors	**1.5**	**6.6**	**12.0**
Al_2O_3	3.72	4.49	4.75
MgO	-11.13	-2.34	-2.15
CaO	-15.74	0.00	3.59
Li_2O	-65.18	-51.78	-43.61
Na_2O	-23.44	-11.26	-6.22
K_2O	-14.13	-8.25	-5.55
Fe_2O_3	-14.89	0	0
TiO_2	-21.06	0	3.57
Intercept	1811.00	882.82	610.47
Model standard error in °C	13.8	5.2	4.5
Std. dev. of residuals in °C	12.6	4.9	4.1
R^2	0.96	0.97	0.96
Average isokom in °C	1301	717	549
Model standard error in %	1.06	0.73	0.82
Glasses not in model	none		
Outlier	none	FL 15	FL 02
Approximate concentration limits in wt% for container and float glass compositions after glass melting: SiO_2: 63-81, Al_2O_3: 0.1-3.3, MgO: 0-4, CaO: 6.5-12.3, Li_2O: 0-1.1, Na_2O: 10-15.3, K_2O: 0-2.2, Fe total as Fe_2O_3: 0-1.5, Cr_2O_3: 0-0.4, TiO_2: 0-1.1, Mn total as MnO_2: 0-0.27, Co total as Co_3O_4: 0-0.2, SO_3: 0-0.3, Se: 0-0.02			

The excluded insignificant factors are not mentioned.

TABLE XIV: COMBINED VISCOSITY MODELS FOR THE SODA LIME CONTAINER AND FLOAT GLASSES INCLUDING THE EXPERIMENTAL DATA OF LAKATOS ET AL. [4]

Soda lime container and float glasses incl. Lakatos	**Coefficients based on wt% for $\log_{10}$(viscosity / Pa·s)**		
Factors	**1.5**	**6.6**	**12.0**
Al_2O_3	5.46	4.13	4.03
MgO	-9.91	-1.94	-1.67
CaO	-16.52	0	3.70
Li_2O	-64.65	-53.35	-43.02
Na_2O	-23.21	-11.29	-6.42
K_2O	-12.96	-6.51	-4.73
Fe_2O_3	-13.64	-2.37	0
Cr_2O_3	0	0	9.09
TiO_2	-19.37	0	3.43
Intercept	1807.23	882.15	610.18
Offset in °C, Lakatos	0	0	-3.85
Prediction error in °C*, depending on composition	3-8	1-3	3-5
Model standard error in °C	10.7	4.2	3.9
Std. dev. of residuals in °C	10.1	4.0	3.7
R^2	0.97	0.98	0.97
Average isokom in °C	1297	720	551
Model standard error in %	0.82	0.58	0.71
Glasses not in model	none		
Outlier	FL 15	FL 15, CO 03	FL 02
Approximate concentration limits in wt% for soda lime glass compositions in the range of container and float glasses after glass melting: SiO_2: 60-81, Al_2O_3: 0-8.2, MgO: 0-6.7, CaO: 6.5-13.3, Li_2O: 0-1.1, Na_2O: 10-15.3, K_2O: 0-8.7, Fe total as Fe_2O_3: 0-1.5, Cr_2O_3: 0-0.4, TiO_2: 0-1.1, Mn total as MnO_2: 0-0.27, Co total as Co_3O_4: 0-0.2, SO_3: 0-0.3, Se: 0-0.02			

*Exact prediction errors can be determined using property calculators [54].
The excluded insignificant factors are not mentioned.

Figure 2 demonstrates the good agreement between the experimental data and model predictions at log(η/Pa·s) = 6.6 for this study and the findings of Lakatos et al. [4]:

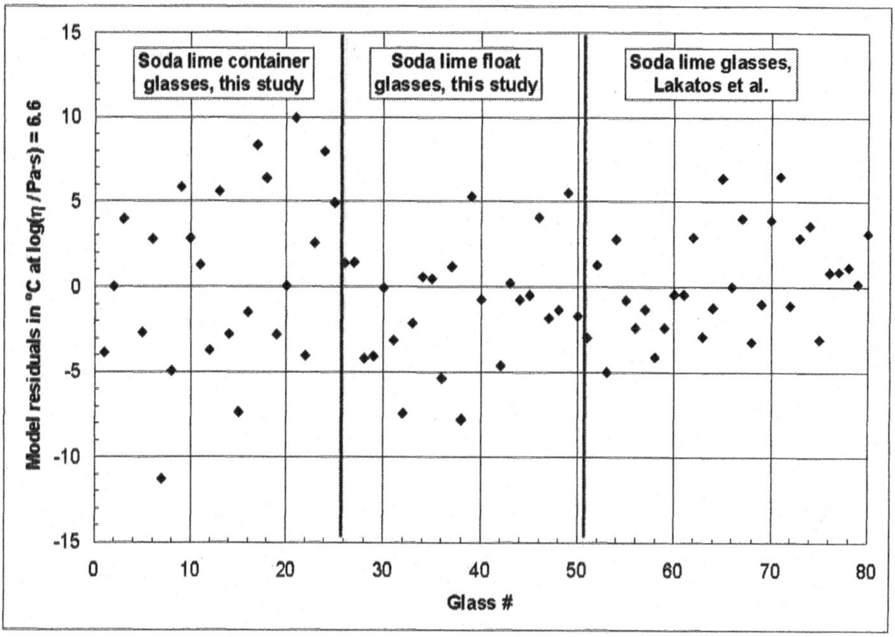

Figure 2: Model residuals (= observed value – calculated value) at at log(η / Pa· s) = 6.6, soda lime glass viscosity model from Table XIV

TABLE XV: HIGH TEMPERATURE ELECTRICAL RESISTIVITY MODELS FOR THE TV PANEL GLASSES

TV panel glasses	Coefficients based on wt% for $\log_{10}(\rho/\Omega\cdot cm)$ at T in °C		
Factors	**1000°C**	**1200°C**	**1400°C**
CaO	0.0201	0	0
SrO	0	0	-0.0068
BaO	0	-0.0071	-0.0094
Li_2O	-0.1352	-0.1179	-0.1121
Na_2O	-0.1314	-0.1029	-0.0808
K_2O	-0.0533	-0.0456	-0.0381
PbO	-0.0182	-0.0198	-0.0212
ZnO	-0.0394	-0.0344	-0.0296
Intercept	3.0874	2.4106	1.9288
Prediction error*, depending on composition	0.015-0.035	0.011-0.026	0.010-0.026
Model standard error	0.0586	0.0455	0.0413
Std. dev. of residuals	0.0507	0.0394	0.0348
R^2	0.95	0.96	0.95
Average resistivity in $\log_{10}(\rho/\Omega\cdot cm)$	1.635	1.152	0.842
Model standard error in %	3.58	3.95	4.90
Glasses not in model	none		
Outlier	none		
Approximate concentration limits in wt% for TV panel glass compositions after glass melting: SiO_2: 51-81, Al_2O_3: 1.2-3.7, MgO: 0-1.5, CaO: 0-3.5, SrO: 1-10, BaO: 2-13, Li_2O 0-0.5:, Na_2O: 5.5-9.5, K_2O: 5.8-9, TiO_2: 0.1-0.6, CeO_2: 0-0.8, ZrO_2: 0-3, PbO: 0-3.2, ZnO: 0-2, As_2O_3: 0-0.35, Sb_2O_3: 0.2-0.7, F: 0-0.85, Fe total as Fe_2O_3: 0-0.04			

*Exact prediction errors can be determined using property calculators [54].
The excluded insignificant factors are not mentioned.

TABLE XVI: HIGH TEMPERATURE ELECTRICAL RESISTIVITY MODELS FOR THE TEXTILE FIBER E GLASSES

Textile fiber E glasses	**Coefficients based on wt% for $\log_{10}(\rho/\Omega\cdot cm)$ at T in °C**		
Factors	**1000°C**	**1200°C**	**1400°C**
B_2O_3	-0.0690	-0.0550	-0.0406
MgO	0	-0.0448	-0.0494
CaO	-0.0367	-0.0553	-0.0537
Na_2O	-0.5415	-0.2510	-0.1430
Intercept	5.5506	4.2651	3.2226
Prediction error*, depending on composition	0.053-0.130	0.032-0.095	0.022-0.065
Model standard error	0.2579	0.1511	0.1033
Std. dev. of residuals	0.2405	0.1374	0.0939
R^2	0.88	0.91	0.93
Average resistivity in $\log_{10}(\rho/\Omega\cdot cm)$	3.941	2.535	1.688
Model standard error in %	6.54	5.96	6.12
Glasses not in model	E 20		
Outlier	none		
Approximate concentration limits in wt% for textile fiber E glass compositions after glass melting: SiO_2: 41-66, B_2O_3: 0-9.3, Al_2O_3: 9-16, MgO: 0.5-6, CaO: 13.5-25.5, Na_2O: 0-2.1, K_2O: 0-0.6, Fe total as Fe_2O_3: 0-0.8, TiO_2: 0-1, F: 0-0.6			

*Exact prediction errors can be determined using property calculators [54].
The excluded insignificant factors are not mentioned.

TABLE XVII: HIGH TEMPERATURE ELECTRICAL RESISTIVITY MODELS FOR THE LOW-EXPANSION BOROSILICATE GLASSES

Low-expansion borosilicate glasses	**Coefficients based on wt% for $\log_{10}(\rho/\Omega\cdot cm)$ at T in °C**		
Factors	**1000°C**	**1200°C**	**1400°C**
Al_2O_3	-0.0216	-0.0115	0
Na_2O	-0.1262	-0.1222	-0.1153
K_2O	0.0280	0	0
BaO	0.0544	0.0416	0.0277
Intercept	2.7600	2.3301	1.9792
Prediction error*, depending on composition	0.018-0.040	0.015-0.031	0.013-0.024
Model standard error	0.0807	0.0698	0.0645
Std. dev. of residuals	0.0734	0.0651	0.0617
R^2	0.93	0.94	0.94
Average resistivity in $\log_{10}(\rho/\Omega\cdot cm)$	1.989	1.575	1.303
Model standard error in %	4.06	4.43	4.95
Glasses not in model	LE 16		
Outlier	none		
Approximate concentration limits in wt% for low-expansion borosilicate glass compositions after glass melting: SiO_2: 62-84, B_2O_3: 9-15.5, Al_2O_3: 2-7.3, CaO: 0-2, Na_2O: 3.7-8.7, K_2O: 0-3.2, BaO: 0-2			

*Exact prediction errors can be determined using property calculators [54].
The excluded insignificant factors are not mentioned.

TABLE XVIII: HIGH TEMPERATURE ELECTRICAL RESISTIVITY MODELS FOR THE FIBER WOOL GLASSES

Fiber wool glasses	**Coefficients based on wt% for $\log_{10}(\rho/\Omega\cdot cm)$ at T in °C**		
Factors	**1000°C**	**1200°C**	**1400°C**
Na_2O	-0.0658	-0.0512	-0.0394
K_2O	-0.0294	-0.0241	-0.0189
Intercept	2.2156	1.6417	1.2544
Prediction error*, depending on composition	0.011-0.022	0.010-0.020	0.010-0.020
Model standard error	0.0570	0.0500	0.0512
Std. dev. of residuals	0.0546	0.0479	0.0490
R^2	0.86	0.83	0.74
Average resistivity in $\log_{10}(\rho/\Omega\cdot cm)$	1.182	0.836	0.634
Model standard error in %	4.82	5.98	8.08
Glasses not in model	none		
Outlier	none		
Approximate concentration limits in wt% for wool fiber glass compositions after glass melting: SiO_2: 49-78, B_2O_3: 3-9.3, Al_2O_3: 0-6.5, MgO: 1-5, CaO: 5-13.5, Na_2O: 13-18, K_2O: 0-2.2, Fe total as Fe_2O_3: 0-0.7, F: 0-0.8, SO_3: 0-0.3			

*Exact prediction errors can be determined using property calculators [54]. The excluded insignificant factors are not mentioned.

TABLE XIX: HIGH TEMPERATURE ELECTRICAL RESISTIVITY MODELS FOR THE SODA LIME CONTAINER AND FLOAT GLASSES

Soda lime container and float glasses	Coefficients based on wt% for $\log_{10}(\rho/\Omega\cdot cm)$ at T in °C		
Factors	1000°C	1200°C	1400°C
Al_2O_3	-0.0130	-0.0081	-0.0058
CaO	0.0192	0.0060	0
Li_2O	-0.1374	-0.1226	-0.1144
Na_2O	-0.0801	-0.0669	-0.0533
K_2O	0	-0.0076	-0.0112
Intercept	2.2171	1.7921	1.4429
Prediction error*, depending on composition	0.004-0.010	0.005-0.012	0.004-0.008
Model standard error	0.0227	0.0262	0.0221
Std. dev. of residuals	0.0217	0.0248	0.0212
R^2	0.98	0.96	0.96
Average resistivity in $\log_{10}(\rho/\Omega\cdot cm)$	1.288	0.922	0.700
Model standard error in %	1.76	2.84	3.16
Glasses not in model	none		
Outlier	CO 08, CO 12 (in addition at 1000°C: CO 15, FL 20)		
Approximate concentration limits in wt% for container and float glass compositions after glass melting: SiO_2: 63-81, Al_2O_3: 0.1-3.3, MgO: 0-4, CaO: 6.5-12.3, Li_2O: 0-1.1, Na_2O: 10-15.3, K_2O: 0-2.2, Fe total as Fe_2O_3: 0-1.5, Cr_2O_3: 0-0.4, TiO_2: 0-1.1, Mn total as MnO_2: 0-0.27, Co total as Co_3O_4: 0-0.2, SO_3: 0-0.3, Se: 0-0.02			

*Exact prediction errors can be determined using property calculators [54].
The excluded insignificant factors are not mentioned.

TABLE XX: HIGH TEMPERATURE ELECTRICAL RESISTIVITY MODELS FOR HIGH LEVEL WASTE GLASSES [6] AT AC FREQUENCY 10 KHZ

High level waste glasses	**Coefficients based on wt% for $\log_{10}(\rho/\Omega\cdot cm)$ at T in °C**
Factors	**1200°C**
Al_2O_3	0.0040
Li_2O	-0.0931
Na_2O	-0.0501
Fe_2O_3	-0.0074
Bi_2O_3	-0.0057
Intercept	1.4442
Model standard error	0.0372
Std. dev. of residuals	0.0342
R^2	0.96
Average resistivity in $\log_{10}(\rho/\Omega\cdot cm)$	0.527
Standard error in %	7.05
Glasses not in model	none
Outlier	HLW 22
HLW 08 has a significant leverage (Cook value > 1). It was not excluded from the model due to a unique composition (high Na_2O concentration).	
Approximate concentration limits in wt% for high level waste glass compositions after glass melting: SiO_2: 37-60, B_2O_3: 0-8.4, Al_2O_3: 0-25, MgO: 0.03-0.06, CaO: 0.7-1.7, Li_2O: 0-7, Na_2O: 5-22.5, K_2O: 0.08-0.18, Fe total as Fe_2O_3: 0-15, ZrO_2: 0-16, Bi_2O_3: 0.7-16, P_2O_5: 0.4-9, CdO: 0.03-0.08, CeO_2: 1-2.3, Cr_2O_3: 0.3-0.7, F: 0.2-0.5, La_2O_3: 0.2-0.4, Mn total as MnO_2: 0.7-1.5, Mo total as MoO_3: 0.03-0.06, Nd_2O_3: 3.5-8.3, NiO: 0.8-1.9, SO_3: 0.1-0.3, WO_3: 0.06-0.14	

The excluded insignificant factors are not mentioned.

The Figures 3 and 4 show the agreement between experimental data of the six base glasses, and the model predictions for these selected compositions. For the viscosity predictions of the container and float base glasses the combined model in Table XIV was applied based on all 50 soda lime container and float glasses investigated in this study, plus the experimental data of Lakatos et al. listed in Table III.

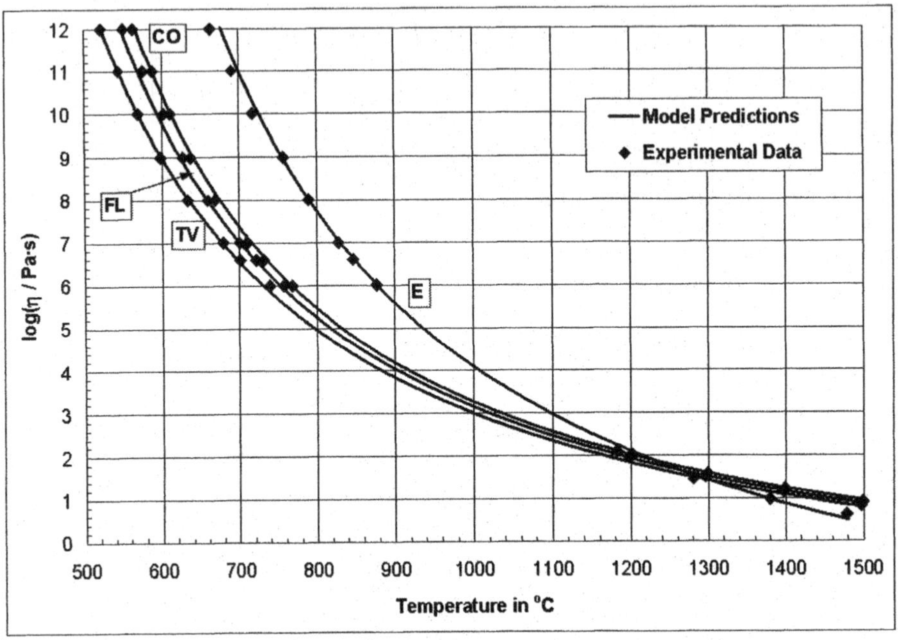

Figure 3: Comparison of viscosity model predictions with experimental data for selected base glasses (TV - TV panel glass, E - textile fiber E glass, CO - container glass, FL - float glass)

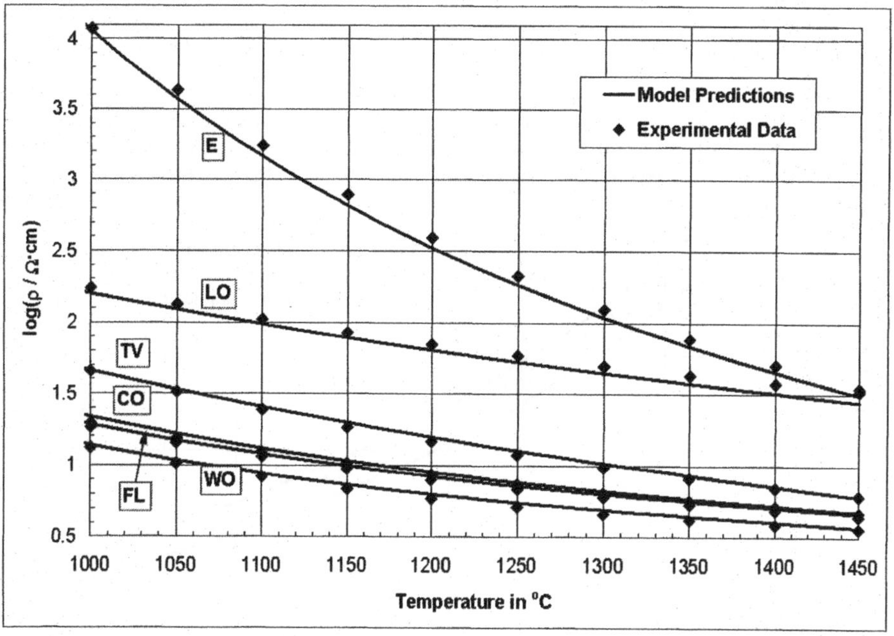

Figure 4: Comparison of electrical resistivity model predictions with experimental data for the base glasses (TV - TV panel glass, E - textile fiber E glass, LO - low-expansion borosilicate glass, CO - container glass, FL - float glass, WO - fiber wool glass)

In Figure 5 the various error limits derived from the Equations (9) to (11) are demonstrated for a part of the viscosity curve of the container base glass. The prediction error of the mean (PE) reflects the similarity to the estimated "true" viscosity, and it should be used in most cases.

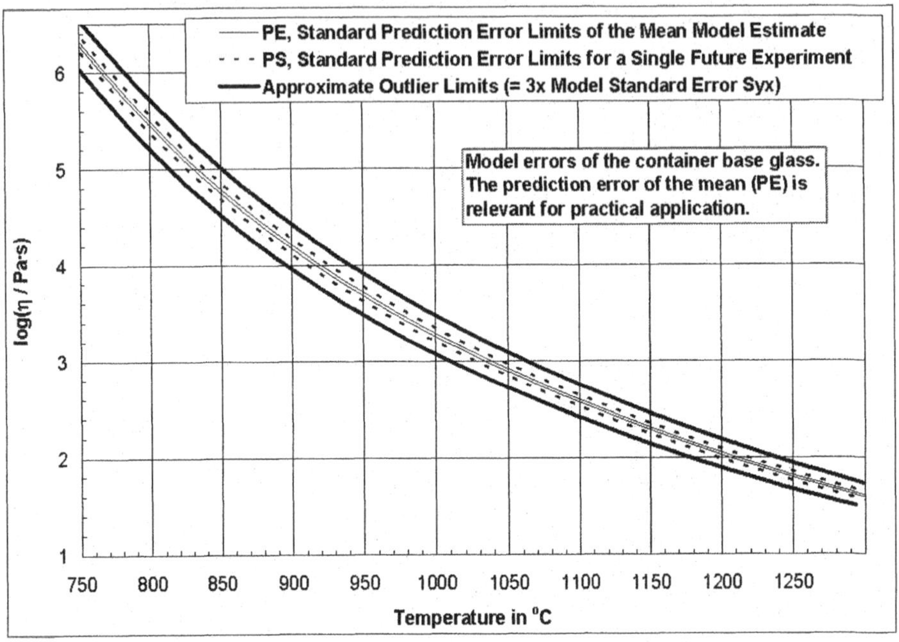

Figure 5: Error limits of the container base glass

From the temperatures of constant viscosity (isokom modeling results, Tables IX to XIV) at $\log(\eta/\mathrm{Pa{\cdot}s}) = 1.5$, 6.6, and 12.0 (T(1.5), T(6.6), T(12.0) in °C) the complete viscosity curve can be determined through calculating the VFT constants in Equation (12) with various mathematics programs, or through ref. [54], or by applying the Equations (15) to (17). A calculation example is given below.

$$T_o = \frac{18 \cdot T(6.6) \cdot T(12.0) - 35 \cdot T(1.5) \cdot T(12.0) + 17 \cdot T(6.6) \cdot T(1.5)}{35 \cdot T(6.6) - 18 \cdot T(1.5) - 17 \cdot T(12.0)} \qquad \textit{Eq. (15)}$$

The constant T_o from Equation (15) can be used for the calculation of B in Equation (16):

$$B = \frac{5.1 \cdot \{(T(6.6) - T_o) \cdot (T_o - T(1.5))\}}{T(6.6) - T(1.5)} \qquad \textit{Eq. (16)}$$

Finally, the constant A is obtained through Equation (17):

$$A = 6.6 - B / \{T(6.6) - T_o\} \qquad \textit{Eq. (17)}$$

Similarly, the electrical resistivity curve between 1000°C and 1400°C can be derived from the resistivity reference points modeled through Tables XV to XIX, or ref. [54]. The resistivity values in $\log(\rho/\Omega\cdot cm)$ at T = 1000°C, 1200°C, and 1400°C (ρ(1000), ρ(1200), ρ(1400)) need to be inserted into the Equations (18) to (20) for the calculation of the VFT constants, which in turn give the resistivity curve through Equation (13):

$$T_o = \frac{200 \cdot \{7 \cdot \rho(1400) + 5 \cdot \rho(1000) - 12 \cdot \rho(1200)\}}{\rho(1400) + \rho(1000) - 2 \cdot \rho(1200)} \qquad \textit{Eq. (18)}$$

$$B = \{\rho(1000) - \rho(1200)\} \cdot (T_o - 1000) \cdot (T_o - 1200) / 200 \qquad \textit{Eq. (19)}$$

$$A = \frac{B - 1200 \cdot \rho(1200) + \rho(1200) \cdot T_o}{T_o - 1200} \qquad \textit{Eq. (20)}$$

Calculation examples

1) The viscosity curve of a soda lime glass with the following composition in wt% has to be determined: SiO_2 66.57, Al_2O_3 8.25, MgO 1.68, CaO 10.49, Na_2O 10.77, K_2O 2.24. The coefficients in Table XIV, multiplied with the concentration in wt% result in the isokom temperatures after addition of the appropriate model intercept:

$\log(\eta/Pa{\cdot}s) = 1.5$:

$5.46 \times 8.25 - 9.91 \times 1.68 - 16.52 \times 10.49 - 23.21 \times 10.77 - 12.96 \times 2.24 + 1807.23 =$ **1383°C**

$\log(\eta/Pa{\cdot}s) = 6.6$:

$4.13 \times 8.25 - 1.94 \times 1.68 + 0 \times 10.49 - 11.29 \times 10.77 - 6.51 \times 2.24 + 882.15 =$ **777°C**

$\log(\eta/Pa{\cdot}s) = 12.0$:

$4.03 \times 8.25 - 1.67 \times 1.68 + 3.70 \times 10.49 - 6.42 \times 10.77 - 4.73 \times 2.24 + 610.18 =$ **600°C**

The calculated isokom temperatures compare well to experimental data in Table III. Next, it would be possible to fit the isokom temperatures to the VFT equation through common mathematics programs, or ref. [54]. In case such a program is not available, the Equations (15) to (17) can be used as alternative. With T(1.5) = 1383°C, T(6.6) = 777°C, and T(12.0) = 600°C, Equation (15) results in $T_o = 301.73$. This value has to be inserted in Equation (16) and it follows B = 4324.88. The constant A is obtained from Equation (17): A = –2.500. Finally, the complete VFT equation according to Equation (12) is: $\log(\eta/Pa{\cdot}s) = -2.5 + 4324.9 / (T \text{ in } °C - 301.7)$.

If required, the viscosity curve may be extended beyond $\log(\eta/Pa{\cdot}s) = 1.5\text{-}12.0$, but the errors stated in Tables IX to XIV may increase. For conversion to Poise viscosity units use the equation $\log(\eta/Poise) = \log(\eta/dPa{\cdot}s) = \log(\eta/Pa{\cdot}s) + 1$.

2) The high temperature electrical resistivity curve can be determined from the chemical glass composition in wt% in the way explained below. The composition may be given as follows: SiO_2 64.9, Al_2O_3 3.1, MgO 3.2, CaO 11.6, Na_2O 14.7, K_2O 2.1, Fe_2O_3 0.1, Cr_2O_3 0.3. From the coefficients in Table XIX we obtain:

1000°C: $\log(\rho/\Omega\cdot cm) =$ **1.222** = ρ(1000)

1200°C: $\log(\rho/\Omega\cdot cm) =$ **0.837** = ρ(1200)

1400°C: $\log(\rho/\Omega\cdot cm) =$ **0.618** = ρ(1400)

From the Equations (18) to (20) the VFT curve can be derived:

$\log(\rho/\Omega\cdot cm) = -0.179 + 739.2 / (T \text{ in } °C - 472.3)$.

The electrical resistivity curve may not be extended far beyond 1000-1400°C without loosing accuracy. Conversions are possible: $\log(\rho/\Omega\cdot cm) = \log(\rho/\Omega\cdot m) - 2 = 2 - \log(\text{conductivity} / S/m)$.

DISCUSSION

Modeling approach

The regression coefficients in the Tables IX to XX and the Figures 2 to 6 suggest that the simple linear modeling approach is well suited within most of the narrow concentration ranges studied. A few low-expansion borosilicate and fiber wool glasses may have undergone phase separation and some textile fiber E glasses could have partially crystallized during specific measurements, preferably at lower temperatures (T ~ 1000°C). Those assumptions derived from the modeling results could be verified through experimental observations mentioned in the Tables X and XI.

Viscosity models could not be established over the complete measurable viscosity range for the low-expansion borosilicate and fiber wool glasses. This might be explained in terms of possible phase separation in glass, and in terms of measurement errors in the case of the high temperature viscosity data for the low-expansion borosilicate glasses. For estimating viscosities of low-expansion borosilicate and fiber wool glasses, a non-linear comparative modeling approach [48] is recommended.

Model accuracy

The accuracy of the models relative to findings of other investigators depends on systematic errors in the experimental data. A large majority of the predicted viscosity values at $\log(\eta/Pa\cdot s) = 1.5$, 6.6, and 12.0 appear to be accurate within 1-30°C, depending on the glass composition and temperature (see Tables IX to XIV). This fact is demonstrated through the comparative models in Table XIV and reference [48] which show that the findings in this study compare well to various datasets from other investigators, including Lakatos et al. [4], Owens-Illinois Glass Company [63], Öksoy et al. [5], NIST viscosity standards, and others. The systematic temperature offset of 3.85°C in Table XIV between data in this study and findings of Lakatos et al. [4] is insignificant for practical application. The soda lime glass viscosity models in Table XIV are considered most accurate. One example is shown in Figure 6 below. For further improvement of the accuracy of the model predictions, comparative studies that consider data of various investigators in similar glass composition ranges should be performed.

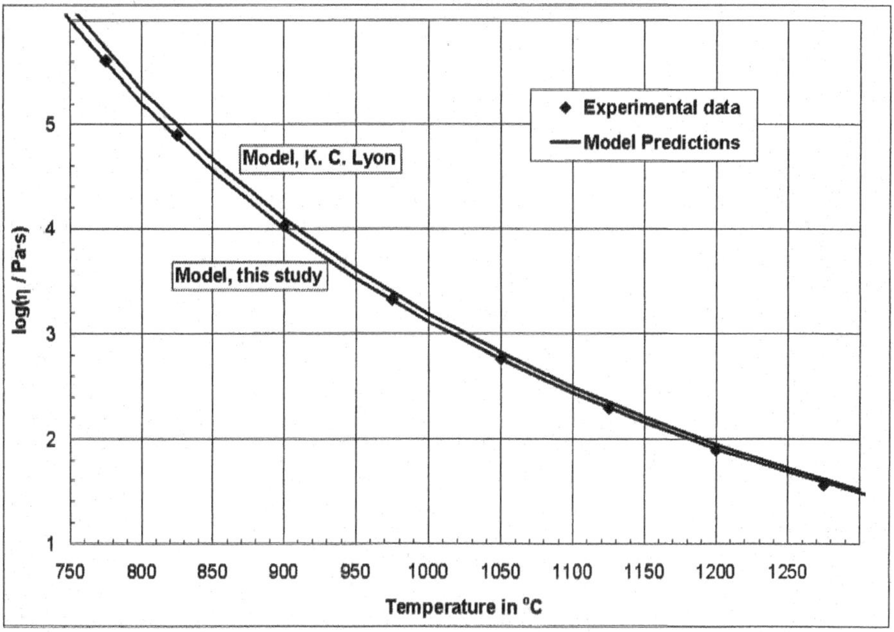

Figure 6: Comparison of experimental data by Ford Motor Company (Öksoy et al [5], wt%: SiO_2 72.60, Na_2O 13.77, CaO 8.65, MgO 3.73, Fe_2O_3 0.85, Al_2O_3 0.15, SO_3 0.19, K_2O 0.04, TiO_2 0.02) with model predictions by Lyon [44] and model predictions from this study (Table XIV)

With respect to the high temperature electrical resistivity data, it is difficult to make a definite statement about their accuracy. Unfortunately, only few experimental data exist in the scientific literature for comparison and some reported values seem to be inaccurate because of systematic errors, e.g. electrode polarization [59, 61, 62]. The findings by Hrma et al. [6] listed in the Tables IV and V, and the model in Table XX are only partially comparable to this study because they involve different glass compositions and densities. The coefficients for Na_2O and Li_2O in Table XX are similar to the coefficients found in this study. The coefficient for Al_2O_3 is less influencial than the ones for Na_2O and Li_2O, but differences between the value in Table XX and the values in this study may occur due to Al_2O_3 interactions and glass density variations. For additional model validation, the high temperature electrical resistivity data for the container and float glasses were analyzed through comparative modeling (see introduction) with data from similar compositions in the Tables VI and VIII, and unpublished data from Mazurin and Prokhorenko [62] in the system SiO_2-Al_2O_3-CaO-MgO-Na_2O-K_2O. It appears that at 1000°C and 1200°C all data compare well, but at 1400°C the predicted container and float glass resistivity values in this study may be too high for $\log(\rho/\Omega{\cdot}cm) = 0.1$. However, the reliability of this statement is not very high, because the model standard errors during comparative modeling may be estimated to $\log(\rho/\Omega{\cdot}cm) = 0.05$. In addition, the NIST high temperature electrical resistivity standard SRM 1414 is a low melting lead silicate glass, certified only up to 1300°C.

No conclusion can be made at present regarding the accuracy of the high temperature electrical resistivity models for the textile fiber E, low-expansion borosilicate, TV panel, and fiber wool glasses because no directly comparable data exist in the scientific literature and there is no data similarly sufficiently for comparative multiple regression using only linear functions. However, it may be possible to apply comparative multiple regression using non-linear functions in future studies [64].In addition to the references mentioned in this chapter, further electrical resistivity data, e.g., by Mazurina, Kostanyan, Baucke, Tickle, Kim, Kirakosyan, and Verzhkhovskaya are available in the SciGlass database [59].

Model Precision (Standard Errors, Standard Deviation Of Residuals, Confidence Intervals)

The measurement precision (consistency) largely influences the internal modeling error, i.e., how well the models predict values within the composition range of the source data. The measurement precision includes viscosity and electrical resistivity measurements, as well as the chemical analysis of the glass compositions. In addition, phase separation, crystallization, and inhomogeneity decrease the model precision.

For most glasses in this study, phase separation and crystallization do not occur. Therefore, the model precision is the sum of 1) chemical analysis errors, 2) viscosity or resistivity measurement errors, and 3) glass inhomogeneity due to insufficient glass melting.

Glasses with many variable components, e.g., the TV panel glasses, can be modeled less precisely than others because of the high number of random chemical analysis errors and the increased probability of glass inhomogeneity.

From the data in Chapters 7 and 8, the random measurement errors for the viscosity and high temperature electrical resistivity are estimated to be 1-15°C.

Considering normal measurement errors, the models in the Tables IX to XX are reasonably precise. Overall, the viscosity models have a higher precision than the high temperature electrical resistivity models because the viscosity is less sensitive to slight variations in chemical composition. The model standard errors can be compared to other multiple regression models [4, 5, 6, 7, 44, 48], and the standard deviation of the residuals can be used for comparison to models that are *not* based on multiple regression, e.g. [2, 13, 19]. The individual prediction errors (PE, Equation (10)) are of interest for practical applications [54].

Effects of Glass Components on the Viscosity and High Temperature Electrical Resistivity

The models in this study were developed in the way that the coefficients in the Tables IX to XX directly reflect the property change caused through an exchange of silica by 1 wt% of the considered glass component. For example if 1 wt% Li_2O is introduced into a TV panel glass in exchange for silica, the Littleton softening point at $\log(\eta/\mathrm{Pa{\cdot}s}) = 6.6$ would drop 43°C (Table IX), and the Littleton softening point for a soda lime glass would drop 53°C (Table XIV). At the same time, the high temperature electrical resistivity at 1200°C of a TV panel glass would drop $\log(\rho/\Omega{\cdot}\mathrm{cm}) = 0.118$ (Table XV), and the electrical resistivity at 1200°C of a soda lime glass would drop $\log(\rho/\Omega{\cdot}\mathrm{cm}) = 0.123$ (Table XIX). The model intercepts in the Tables IX to XX are supposed to represent the property of the residual comprising mainly SiO_2, and all insignificant components.

Alternatively, the viscosity model coefficients in the Tables IX to XIV can be interpreted as the temperature change (+/-) needed by an increase of 1 wt% oxide in exchange for 1 wt% SiO_2 to maintain the same viscosity. For instance if at the Littleton softening point 1 wt% Na_2O is introduced into a textile fiber E glass in exchange for 1 wt% SiO_2, the temperature may be decreased 15°C for maintaining the same viscosity. For comparison, the component influences on the high temperature electrical resistivity were converted from resistivity influences at constant temperature (Tables XV to XX) to approximately equivalent temperature influences at constant resistivity in Appendix A at the end of this chapter (Tables XXI to XXVI). It should be noted that the Tables XXI to XXVI should be used only as a guide. Exact model calculations for the high temperature electrical resistivity have to be based on the coefficients in the Tables XV to XX.

In addition to the effects of the glass components on the viscosity and high temperature electrical resistivity, the glass "length" may be examined. A "long" glass is one that undergoes relatively small property changes if the temperature is increased, where a "short" glass shows large property changes through increased temperature. In this study influences of components on the glass "length" can be derived from the coefficient differences at low and high temperatures. For example in Table XIX Na_2O, Li_2O, and CaO decrease the "length" the most, and Al_2O_3 and K_2O have less influence.

Glass former (SiO_2, B_2O_3, F)

SiO_2

The influence of silica on the viscosity and high temperature electrical resistivity is always the difference to the influences of the remaining glass components. In general, silica increases the viscosity and high temperature electrical resistivity, and it increases the "length" of glasses most significantly. Specific values can be derived from the models in the Tables IX to XX..

B_2O_3

If silica is exchanged for boron oxide it decreases the viscosity more strongly than it decreases the high temperature electrical resistivity, and it "shortens" a glass. This behavior could be expected since boron oxide does not add additional mobile ions to the glass that can transport electric current. Boron oxide weakens the molecular glass-forming network; the required activation energy for breaking chemical bonds is decreased. For example the fiber wool glasses have similar compositions to the soda lime container and float glasses, except that some silica is exchanged by boron oxide. The fiber wool glasses are "short" and have low viscosities, but the electrical resistivities are only slightly reduced compared to the soda lime glasses.

F (fluoride ions)

Fluoride ions strongly decrease the viscosity and "length" of TV panel glasses (Table IX), but no effect on the high temperature electrical resistivity could be detected. This interesting behavior cannot be explained through data scattering, because the estimated model standard errors in the Tables IX, XV, and XXI are comparable on the temperature scale. It is assumed that the fluoride ions partially "break" the silica glass-forming molecular network through the formation of terminal (non-bridging) Si-F bonds (i.e. the viscosity decreases), however, at the same time the fluoride ions impede the movement of electrically conducting alkali ions through electrostatic interaction. Fluoride ions seem to show the largest discrepancy between the influence on the viscosity and the high temperature electrical resistivity among all glass components studied in this work. Further investigations are desirable since no research is known to the authors where the influence of fluoride ions on the high temperature electrical resistivity was systematically measured for commercial glasses.

Intermediate glass oxides (Al_2O_3, TiO_2, ZrO_2, Fe_2O_3, Cr_2O_3)

Al_2O_3

Alumina moderately increases the viscosity, it has small and variable influences on the high temperature electrical resistivity, and it does not change the glass length significantly for the studied compositions. The negative values in the Tables XVII and XIX presumably occur because addition of an aluminum ion in glass-forming mode absorbs a non-bridging oxygen ion making alkali ions more mobile [65].

TiO_2

Titanium oxide does not influence the high temperature electrical resistivity for the studied compositions. The influence on the viscosity listed in the Tables IX and XIV appears very interesting. It seems that titanium oxide is changing its influence the most among all glass components in the way that it makes the glasses "short" the most. At high temperatures titanium oxide decreases the viscosity, and at low temperatures it has no or the opposite effect. Further investigations are desirable because due to the low TiO_2 concentrations in all glasses an estimation of its influence may be a result of the experimental uncertainty. The TiO_2 coefficients in the Table IX are very close to being insignificant, but the TiO_2 coefficients in the Table XIV should not be neglected.

ZrO_2

Zirconium oxide increases the viscosity of TV panel glasses, particularly at low temperatures (Table IX).

Fe_2O_3

For iron oxide, model coefficients are difficult to estimate due to its variable oxidation states, +2 (FeO) and +3 (Fe_2O_3). Oxidizing furnace atmospheres increase the Fe_2O_3 / FeO ratio, and reducing atmospheres decrease it. The reader is referred to studies where relatively high iron oxide concentrations were analyzed [6]. In this work a significant negative influence of iron oxide on the viscosity was observed for soda lime glasses (Table XIV) at high temperatures.

Cr_2O_3

Chromium oxide seems to increase the glass viscosity at low temperatures (Table XIV). This effect appears to be significant beyond experimental error (Table). However, the investigation of glasses with increased chromium oxide concentration would lead to more precise values.

Glass modifying oxides (Na_2O, K_2O, Li_2O, CaO, MgO, BaO, SrO, PbO, ZnO, water)

Na_2O

Sodium oxide "shortens" glasses, and strongly decreases the viscosity and high temperature electrical resistivity. For most common technical glass compositions it is the component that influences the electrical resistivity the most.

K_2O

References [66, 67] and the average values in the Tables VI (system SiO_2-Na_2O) and VII (system SiO_2-K_2O) make it evident that K_2O significantly decreases the high temperature electrical resistivity as long as Na_2O is *not* present. In the presence of Na_2O, however, an increase in electrical resistivity due to the mixed-alkali effect [66, 67, 68] may counteract the decrease. Although, the mixed-alkali effect vanishes at high temperatures, the coefficients in Table XIX showing little or no influence of K_2O on the electrical resistivity of soda lime glasses at 1000°C, and slight decrease at higher temperatures are not a surprise. In this study the combined influences of K_2O alone and the resulting Na_2O-K_2O mixed-alkali effect may be simplified and reduced to a linear influence of K_2O because the K_2O concentrations are relatively low. All K_2O coefficients in the Tables IX to XIX are significant and should be used within the specified limits. The K_2O coefficients can not be applied to glasses with higher than the investigated K_2O concentrations in this study, or to glasses where the product $Na_2O \times K_2O$ exceeds the limits according to the source data in the Tables III to VIII in Chapter 2.

K_2O has a similar or sometimes slightly lower influence on lowering the glass viscosity as Na_2O on the molar basis, where on the wt% basis in the Tables IX to XIV the influence of K_2O is significantly lower than of Na_2O. There appears to be a trend in the Tables XII to XIV according to which the influences of K_2O and Na_2O approach each other at high viscosities, a similar trend observed by Lakatos et al [4].

Li_2O

Lithium oxide has by far the strongest influence on the viscosity and high temperature electrical resistivity among all glass components studied on wt% basis. This effect can be explained partially through the low density of lithium oxide. However, even on the molar basis, the influence of lithium oxide is one of the strongest compared to other oxides [48]. The coefficients for Li_2O in the Tables IX to XX may be the sum of the Li_2O influence, plus any contributions from mixed alkali effects [66].

Lithium oxide decreases the viscosity and high temperature electrical resistivity very significantly. In addition it strongly decreases the glass "length".

CaO

The effect of calcium oxide varies from positive to negative with increasing temperature. The viscosity is increased at low temperatures, and decreased at high temperatures. Similarly, its effect on the high temperature electrical resistivity decreases from positive values to zero in the Tables XV and XIX, and to negative values with increasing temperature in Table XVI.

MgO

In most cases magnesium oxide has similar effects on the glass viscosity and high temperature electrical resistivity as calcium oxide, however, the magnitude of the influence is less pronounced. The interesting behavior of MgO concerning the low temperature viscosity of soda lime glasses in Table XIV should be noted. MgO appears to decrease the viscosity in the glass transformation range. This result is supported by other studies [4, 48].

BaO, SrO

The effects of barium oxide and strontium oxide can be studied in the Tables IX, XV (TV panel glasses), and XVII (low-expansion borosilicate glasses). They tend to decrease the glass viscosity more than does calcium oxide, and the decrease of the glass "length" may be less pronounced than for CaO. The influence on the high temperature electrical resistivity can vary with the glass composition.

It is interesting to note that the influences of the alkali oxides Li_2O, Na_2O, and K_2O on the high temperature electrical resistivity of TV panel glasses decrease with the atomic weight, whereas the opposite is the case for the alkali earth oxides MgO, CaO, SrO, and BaO (Table XV). This effect may be related to the interference of various influences: the ionic radii, the ionic charges, the size of coordination spheres around the ions, and component interactions.

PbO

Lead oxide decreases the viscosity of TV panel glasses (Table IX), the high temperature electrical resistivity (Table XV), and the glass "length".

ZnO

In this study no influence of zinc oxide on the TV panel glass viscosity could be detected. This finding, however, may be related to data scattering and low ZnO concentrations. Based on the data reported in reference [48], it can be concluded that ZnO decreases the viscosity at high temperatures. In Table XV it is demonstrated that ZnO decreases the high temperature electrical resistivity.

Water

The water concentrations of all transparent glasses in this study were analyzed by infrared spectroscopy (see Appendix I at the end of this book, Tables I to VI). The concentrations varied from 80 to 640 ppm by weight (wt ppm), with an average of 280 wt ppm. No influence of water on the viscosity and the high temperature electrical resistivity could be detected, probably due to its low concentration and data scattering. It is known from the literature, however, that water decreases the viscosity significantly at higher concentrations [69, 70, 77].

Further influences of the glass composition on the viscosity are discussed in detail by Priven [13, 14, 71], including a viscosity modeling approach based on chemical equilibria of structural groups in glass.

Viscosity – Electrical Resistivity Relation

For a given composition, the glass viscosity is mainly controlled by molecular bonds of glass forming oxides, and the electrical resistivity is mainly controlled by ionic interactions of glass modifying oxides and steric effects due to the molecular bonds of glass forming oxides. Therefore it can be expected that the viscosity and the electrical resistivity are partially associated, although different effects might occur. The model coefficients in Tables IX to XX demonstrate that glass components may influence the viscosity and electrical resistivity sometimes similarly (Na_2O) and sometimes differently (Al_2O_3, F). Both the molecular bonds and ionic interactions are thermally activated. In general it can be stated that for a given glass composition the electrical resistivity decreases with decreasing viscosity. At temperatures above the Littleton softening point, a linear relation with positive slope can be constructed between the logarithm of the glass viscosity and the logarithm of the electrical resistivity [72, 73]. Figure 7 demonstrates this relation for the six base glasses from the current study, derived from the experimental data in Chapters 7 and 8:

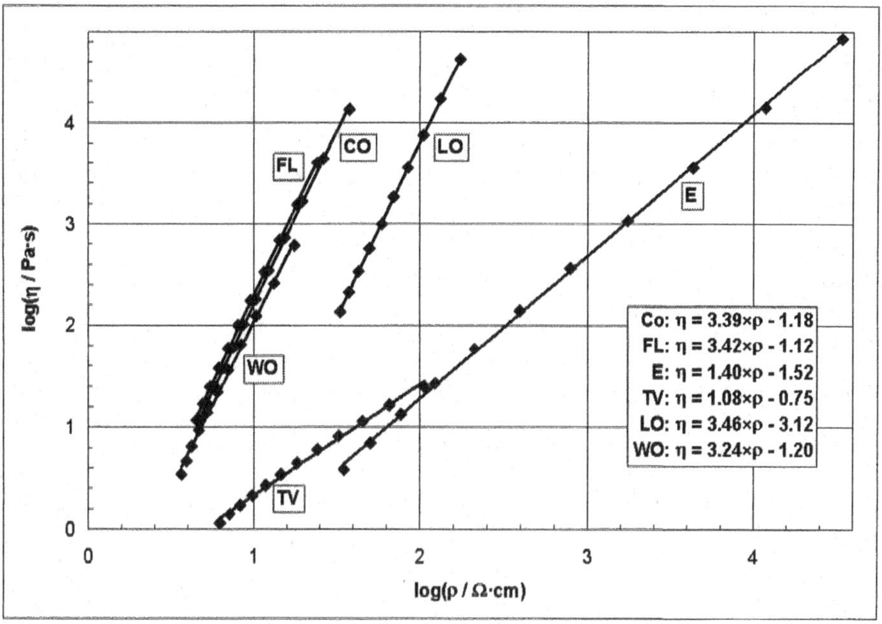

Figure 7: Relation between the high temperature electrical resistivity and viscosity for the investigated base glasses at T = 900-1450°C (TV - TV panel glass, E - textile fiber E glass, LO - low-expansion borosilicate glass, CO - container glass, FL - float glass, WO - fiber wool glass)

The slope appears to be influenced by the glass network-modifying oxides (e.g. alkali, alkaline earth, PbO), and the intercept by SiO_2 and B_2O_3. In essence: SiO_2 and B_2O_3 "shift" the line (e.g. fiber wool glasses) and modifiers "bend" it or change the slope (e.g. TV panel glasses). This behavior may be explained through the fact that SiO_2 and B_2O_3 mainly influence the viscosity, and modifiers influence viscosity and electrical resistivity simultaneously.

Assuming that the electrical resistivity in a binary soda silica glass is influenced mainly by the Na^+ diffusion described through the Nernst-Einstein equation, and the viscosity is influenced mainly by the SiO_4^{2-} diffusion described through the Stokes-Einstein equation it follows:

$$\eta / \rho = (D_{Na} / D_{Si}) \times \frac{z_{Na}^{\ 2} \cdot c_{Na} \cdot F^2}{6 \cdot \pi \cdot N_A \cdot r_{Si}} \qquad \text{Eq. (20)}$$

with z being the number of the electric charges, c the concentration, r_{Si} the silica SiO_4^{2-} tetrahedron radius, D the diffusion coefficients, N_A the Avogadro number, and F the Faraday constant. For understanding the principles of the viscosity – electrical resistivity relation in glass melts Equation (20) may be generalized in the way that Na_2O represents glass modifying oxides, and SiO_2 represents glass formers. An advanced understanding requires the extension of the Stokes-Einstein equation through statistical mechanics [74], and a consideration of the thermodynamics and kinetics of chemical interactions.

Recommendations For Practical Model Application

To the authors' knowledge the accuracy and precision of the models developed in this work is matched only by few publications [4, 5, 6, 7, 44, 48]. However, some of those models have either a very limited application range [4, 5, 7], or are valid for glass systems not investigated here [6]. The model publication by Lyon [44] in general allows accurate predictions for Na_2O concentrations higher than 11 wt%, but the required statistical data analysis is not mentioned, e.g., factor correlation, factor significance, outlier tests, and data leverage. The accuracy of the models by Priven [13] is less compared to all regression models listed above, but the compositional application range is nearly unlimited. The models by Priven can be used through the SciGlass program [59].

For all estimations of the high temperature electrical resistivity, the models in the Tables XV to XX and reference [6] are recommended. For predicting viscosities of soda-lime glasses the models in Table XIV should be used. Regarding the high temperature viscosity of TV panel and borosilicate glasses (LO, WO, E) the models in reference [48] are considered most reliable, where at lower temperatures models in the Tables IX and X and [48] can be applied. For all special composition waste glass properties the reader is referred to Hrma et al. [6]. Electrode polarization effects always need to be considered if the applied frequency is less than 1-2 kHz [61, 62].

For glass design through property modeling, evaporation losses during the glass batch melting, and possible influences of the oxidation states of transition metal oxides must be taken into account. To comply with the appropriate concentration limits in the Tables IX to XX, care should be used to make sure that all component combinations do not exceed those of the original data in Tables III to VIII in Chapter 2, and in Tables III and IV of this Chapter ("interaction limits"). For example the maximum Na_2O concentration listed in Table XIV is 15.3 wt%, and the maximum K_2O concentration is 8.7 wt%. However, the source data show that Na_2O and K_2O are never at their maximum concentrations simultaneously. In case of strong component interactions the model predictions may be incorrect for glasses with extreme component combinations, i.e., outside the interaction limits.

Most model calculations in this study can be performed using property calculators [54].

Importance of Statistical Modeling

Experimental data given, models based on multiple regression are more precise than most models based on other approaches summarized in the introduction, because of their empirical character, i.e., no "a priori" knowledge of the details of bonding is used. This statement is not only true for viscosity and high temperature electrical resistivity predictions, but for other properties as well. For example the SO_3 solubility in soda-lime glass melts at 1370°C may be predicted thermodynamically with a standard deviation of the residuals of 0.18 mol%, using advanced procedures [19, p 188]. Common statistical modeling will result in a standard deviation of the residuals of 0.13 mol% with $SO_3 = 0.0587 * x_{Ca} + 0.142 * x_{Na} - 0.0194$.

Statistical analysis is required for accuracy evaluation. In various publications modeling results are compared to measurements by some investigators, but many other experimental data in the literature are not considered. It would be useful to take all available data into account when accuracy evaluations of new models are made.

Due to its empirical character statistical modeling can not be used if no experimental data are given. For extrapolations to so far unknown areas alternative techniques should be applied.

CONCLUSIONS

Models based on multiple regression using linear functions were provided for estimating the viscosity and high temperature electrical resistivity of commercial glasses from their chemical composition with high accuracy. The influences of specific glass components could be quantified.

In future, experimental investigations multi-component glasses should be reduced to the significant components established in this work, especially TV panel glasses. For borosilicate compositions (LO, WO, E) that may phase separate or crystallize, an experimental approach is recommended with several concentration levels to allow for the analysis of component interactions and areas of phase separation or crystallization.

Comparative modeling studies based on multiple regression using non-linear functions should be performed for establishing reliable model accuracy where no direct comparison of experimental data is possible.

ACKNOWLEDGEMENTS

Special thank is extended to Oleg V. Mazurin for providing electrical resistivity data in the systems SiO_2-Na_2O, SiO_2-K_2O, and SiO_2-Na_2O-CaO from the SciGlass database, and for making his so far unpublished data [62] available.

The viscosity and electrical resistivity data for the commercial base glasses and the statistically-designed variations (CO, E, FL, LO, TV, WO) were provided by the NSF Industry-University Center for Glass Research, New York State College of Ceramics at Alfred University, Alfred, New York.

APPENDIX A

The equivalents in the Tables XXI to XXVI were estimated based on the average slope of the T-log(ρ) curve at 1000°C, 1200°C, and 1400°C. They should be used only as a guide. Exact model calculations for the high temperature electrical resistivity have to be performed on basis of the coefficients in the Tables XV to XX above.

The listed resistivity-temperature equivalents in the Tables XXI to XXVI can be interpreted similarly to the viscosity model coefficients in the Tables IX to XIV: The equivalents reflect the temperature change (+/-) needed by an increase of 1 wt% oxide in exchange for 1 wt% SiO_2 to maintain the same resistivity. For example, if a specific resistivity value established at 1400°C for TV panel glasses has to remain constant but the temperature needs to be decreased by 100°C, it can be achieved through a replacement of approximately 1 wt% SiO_2 by 1 wt% Li_2O.

TABLE XXI: ESTIMATED RESISTIVITY-TEMPERATURE EQUIVALENTS FOR THE TV PANEL GLASSES

TV panel glasses	**Estimated resistivity-temperature equivalents in °C**		
Components	**1000°C**	**1200°C**	**1400°C**
CaO	7	0	0
SrO	0	0	-6
BaO	0	-4	-8
Li_2O	-47	-59	-101
Na_2O	-46	-52	-72
K_2O	-19	-23	-34
PbO	-6	-10	-19
ZnO	-14	-17	-27
Model standard error in °C	21	23	37

TABLE XXII: ESTIMATED RESISTIVITY-TEMPERATURE EQUIVALENTS FOR THE TEXTILE FIBER E GLASSES

Textile fiber E glasses	**Estimated resistivity-temperature equivalents in °C**		
Components	**1000°C**	**1200°C**	**1400°C**
B_2O_3	-8	-10	-14
MgO	0	-8	-17
CaO	-4	-10	-19
Na_2O	-64	-45	-50
Model standard error in °C	31	27	36

TABLE XXIII: ESTIMATED RESISTIVITY-TEMPERATURE EQUIVALENTS FOR THE LOW-EXPANSION BOROSILICATE GLASSES

Low-expansion borosilicate glasses	Estimated resistivity-temperature equivalents in °C		
Components	**1000°C**	**1200°C**	**1400°C**
Al_2O_3	-9	-7	0
Na_2O	-52	-71	-115
K_2O	12	0	0
BaO	22	24	28
Model standard error in °C	33	41	65

TABLE XXIV: ESTIMATED RESISTIVITY-TEMPERATURE EQUIVALENTS FOR THE FIBER WOOL GLASSES

Fiber wool glasses	Estimated resistivity-temperature equivalents in °C		
Components	**1000°C**	**1200°C**	**1400°C**
Na_2O	-31	-37	-61
K_2O	-14	-18	-29
Model standard error in °C	27	36	79

TABLE XXV: ESTIMATED RESISTIVITY-TEMPERATURE EQUIVALENTS FOR THE SODA LIME CONTAINER AND FLOAT GLASSES

Soda lime container and float glasses	Estimated resistivity-temperature equivalents in °C		
Components	**1000°C**	**1200°C**	**1400°C**
Al_2O_3	-6	-6	-7
CaO	9	4	0
Li_2O	-64	-84	-148
Na_2O	-37	-46	-69
K_2O	0	-5	-14
Model standard error in °C	11	18	29

TABLE XXVI: ESTIMATED RESISTIVITY-TEMPERATURE EQUIVALENTS FOR HIGH LEVEL WASTE GLASSES [6] AT AC FREQUENCY 10 KHZ

High level waste glasses	**Estimated resistivity-temperature equivalents in °C**
Components	**1200°C**
Al_2O_3	3
Li_2O	-70
Na_2O	-37
Fe_2O_3	-6
Bi_2O_3	-4
Model standard error in °C	28

APPENDIX B

Figures 8 to 13 show VFT fits (see Equation (12)) to the original viscosity data of the six base glasses. Deviations from the fitted curve at high temperatures for the textile fiber E and fiber wool base glasses (Figures 9 and 11) are random errors.

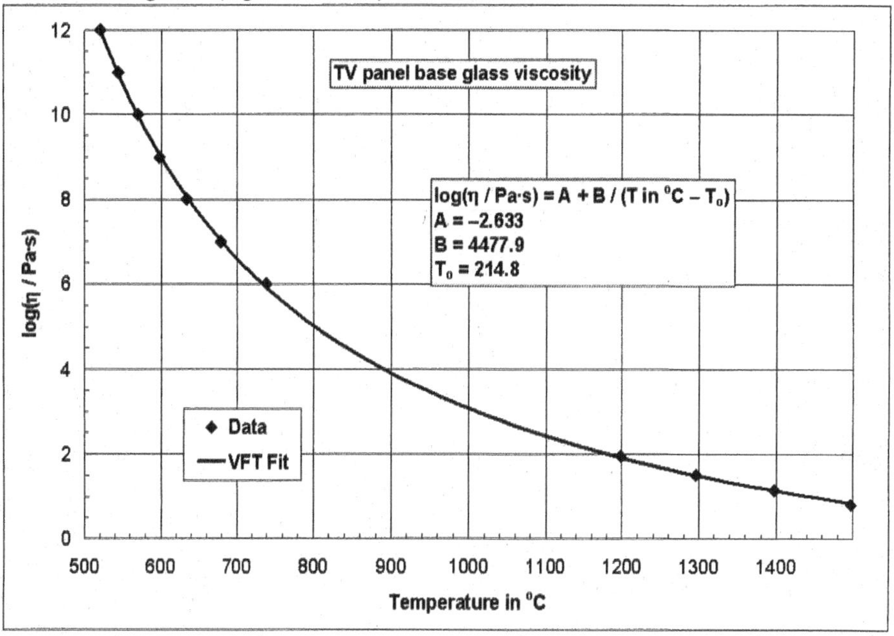

Figure 8: Vogel-Fulcher-Tammann fit to the original data of the TV panel base glass

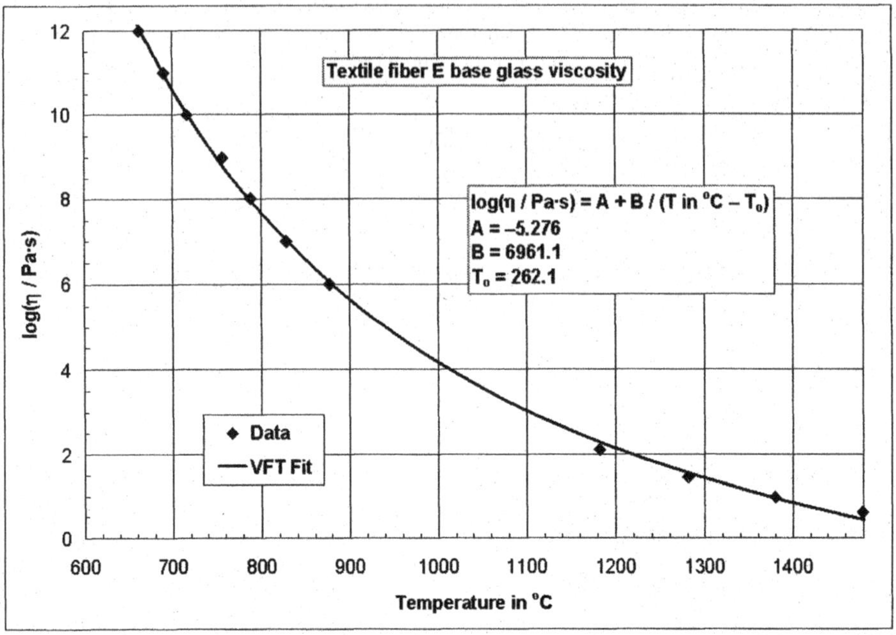

Figure 9: Vogel-Fulcher-Tammann fit to the original data of the textile fiber E base glass

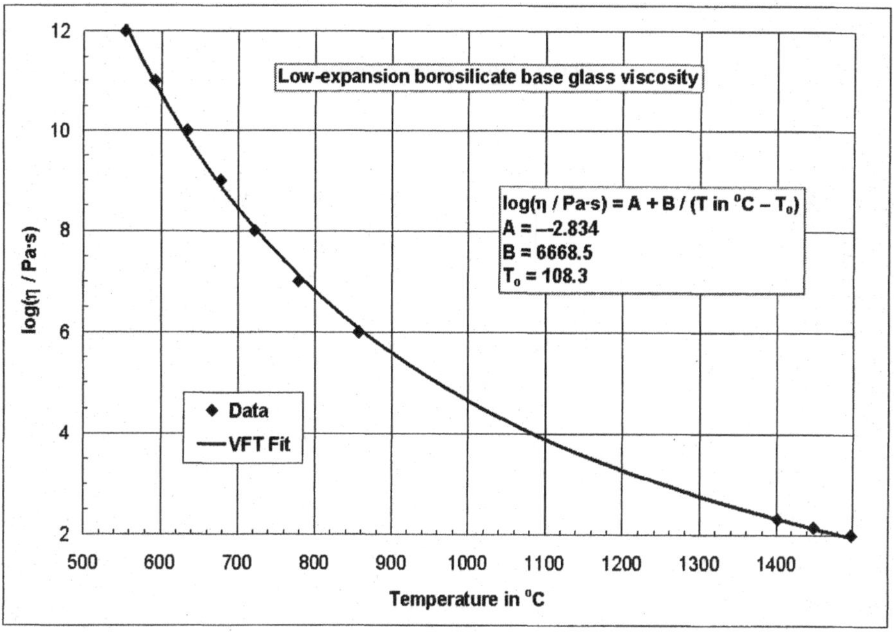

Figure 10: Vogel-Fulcher-Tammann fit to the original data of the low-expansion borosilicate base glass

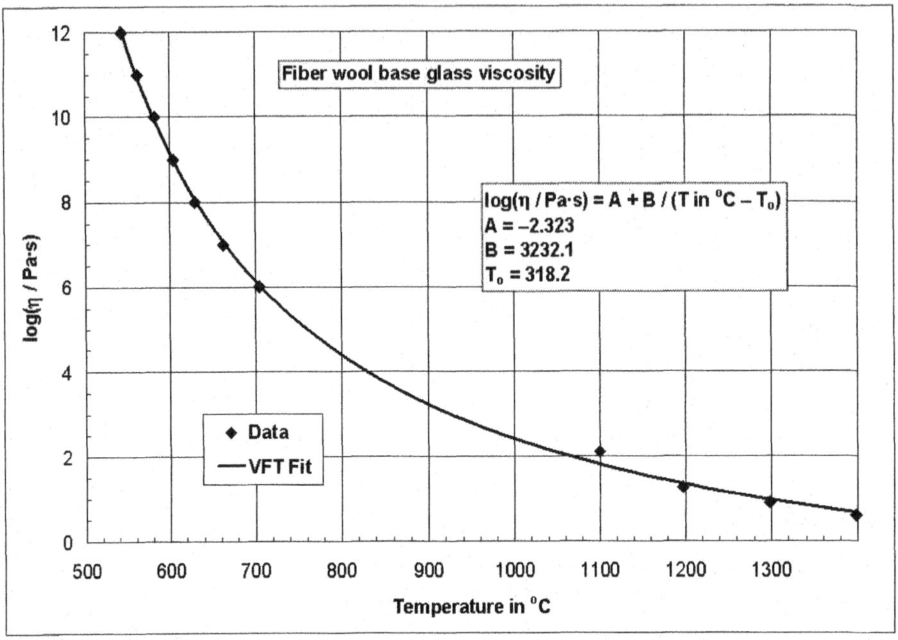

Figure 11: Vogel-Fulcher-Tammann fit to the original data of the fiber wool base glass

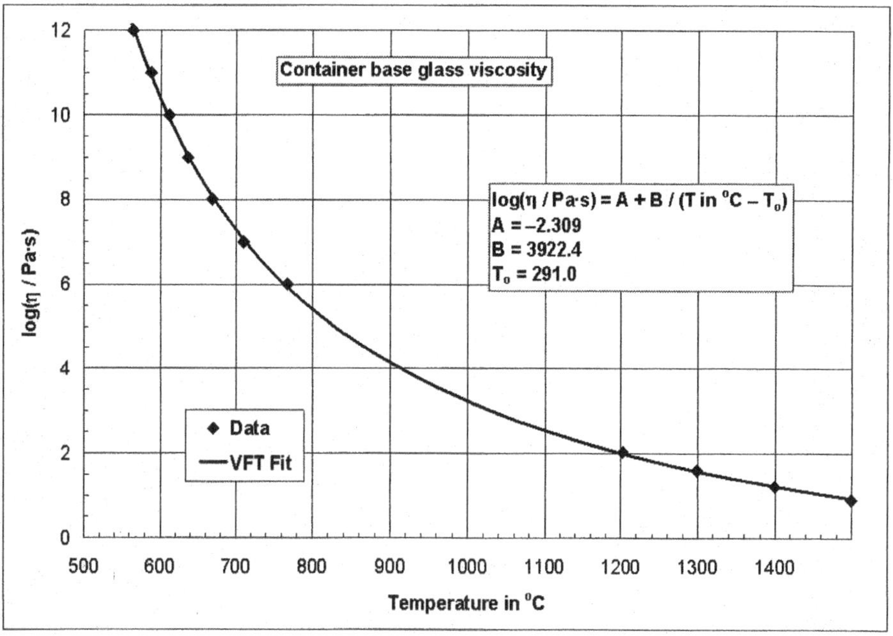

Figure 12: Vogel-Fulcher-Tammann fit to the original data of the container base glass

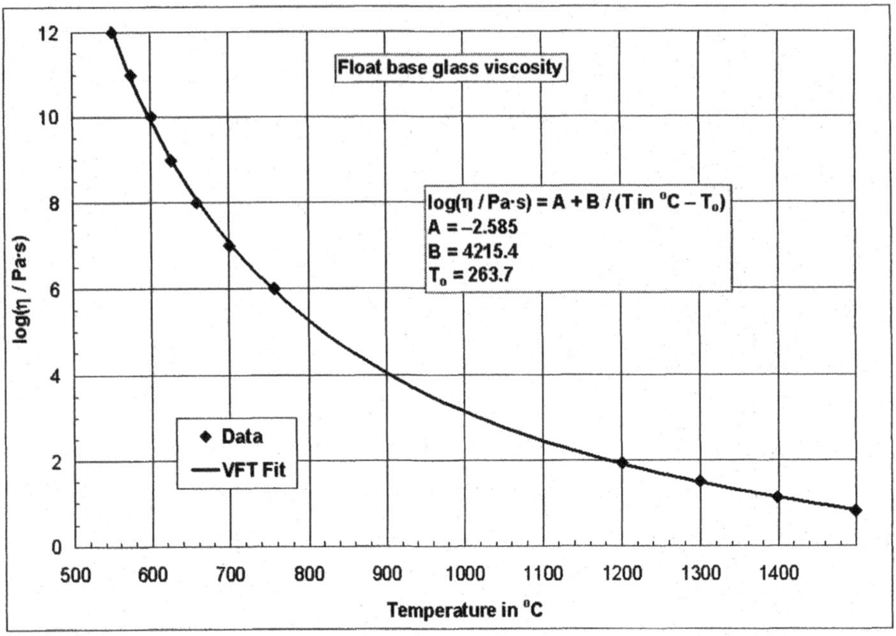

Figure 13: Vogel-Fulcher-Tammann fit to the original data of the float base glass

APPENDIX C, MODELING OF OTHER GLASS MELT PROPERTIES

GAS SOLUBILITY

Inert gases solubility

A recent publication by Tournour et al. [75] gives an overview of inert gas solubilities in glass available in the literature, including Helium, Neon, Argon, and Nitrogen. The inert gas solubilities of glass melts tend to increase slightly with increasing temperature (Nitrogen is not discussed), increasing glass former content, and decreasing molecular weight of the inert gas (i.e. Helium is dissolves the most).

Water solubility

No systematic influences can be derived from the data in this study due to the few reported values and scattering. Therefore, additional values from the literature of water solubilities in the glass system SiO_2-Al_2O_3-RO-R_2O at 1200°C were used for comparative modeling, including 30 datapoints by Rapp and McGinnis et al. [76, 77, 78], 35 datapoints by Franz and Scholze [79], and one float glass value by McGinnis et al. [80, 81]. Early publications by Tomlinson [82], Kurkjian et al. [83], and Garbe [84] were not considered due to inconsistencies, e.g. a minimum in the water solubility curve with increasing Na_2O concentration. The data by Franz and Scholze reported for 1250°C to 1480°C were extrapolated to 1200°C. The modeling result is shown in Table XXVII. Except for the TV panel base glass, all reported solubilities are comparable. No systematic differences between datasets could be found due to scattering. However, it should be noted that for selected soda-lime glasses studied in detail McGinnis et al. [78] claim that their findings tend to be 50 ppm by weight higher than Franz and Scholze [79]. Alkali oxides and boron oxide increase the water solubility of glasses, and alkali earth oxides and alumina decrease it. No influence of the temperature on the water solubility could be detected in this work, but Franz and Scholze [79], Rapp [76], and McGinnis et al. [78] found that the water solubility increases with increasing temperature. The measurement precision should be increased in future studies.

TABLE XXVII: WATER SOLUBILITY MODEL FOR SILICATE GLASSES AT 1200°C UNDER ATMOSPHERIC PRESSURE

Factors	Coefficients based on mol% for H_2O solubility in ppm by weight at 1200°C
B_2O_3	77.92
Al_2O_3	-18.94
CaO	-8.35
SrO	-20.26
BaO	-13.84
Li_2O	21.10
Na_2O	21.27
K_2O	35.29
Intercept	775.94
Prediction error in ppm weight*, depending on composition	18-160
Model standard error in ppm weight	115.8
Std. dev. of residuals in ppm weight	108.9
R^2	0.79
Average solubility in ppm weight	1190.8
Model standard error in %	9.7
Outlier	TV base glass
Approximate concentration limits in mol% for silicate glass compositions after glass melting: SiO_2: 57-82, B_2O_3: 0-6, Al_2O_3: 0-14, MgO: 0-10, CaO: 0-20, SrO: 0-10, BaO: 0-10, Li_2O 0-38:, Na_2O: 0-38, K_2O: 0-38, TiO_2: 0-0.4, F: 0-0.7, Fe total as Fe_2O_3: 0-0.7	

*Exact prediction errors can be determined using property calculators [54].
The excluded insignificant factors are not mentioned.

Surface tension of glass melts

The surface tensions of the six base glasses in this study were measured from 950°C to 1400°C using sessile and pendant drop methods [85, 86]. Regression studies were performed at 1400°C by Kucuk et al. [8, 45] for evaluating the measurement accuracy and for estimating the surface tension of various compositions. The model is summarized in Table XXVIII, and a discussion of the results can be found in the references [8, 45]. The source data from SciGlass [59] for the model are not specified by Kucuk in detail [8, 45], but an extensive evaluation of the values of various investigators is given in reference [45]. It can be assumed that the model estimates derived from Table XXVIII are more accurate than most individual measurements because the model is based on 300 data (out of a total of 800 in 90 papers), carefully selected from authors that tend to agree in their results. Compared to previous surface tension models [87, 88, 89, 90] the model by Kucuk et al. is based on more data than all except a model by Goleus et al. [91]. In future studies it would be desirable to exactly specify the source data. In addition, it may be interesting to extend the statistical analysis of the surface tension of glass melts to wider temperature ranges.

TABLE XXVIII: SURFACE TENSION MODEL BY KUCUK ET AL. [8, 45]

Factors	Coefficients based on mol% for surface tension in mN/m at 1400°C
Li_2O	1.48
K_2O	-2.22
Rb_2O	-3.43
MgO	1.96
CaO	3.34
BaO	1.28
SrO	3.32
FeO	2.68
MnO	2.92
PbO	-1.38
B_2O_3	-2.86
Al_2O_3	3.47
MoO_3	-24.5
Intercept	271.2
Model standard error in °C	18.8
Number of data in model	303
R^2	0.965
Mean prediction in mN/m	299.4
Model standard error in %	6.3
Excluded factors with insignificant effect	Na_2O, Fe_2O_3, ZnO, Cs_2O, Al_2O_3×SiO_2, Na_2O×Al_2O_3, MgO×Al_2O_3, PbO×SiO_2, B_2O_3×SiO_2, K_2O×B_2O_3, CaO×B_2O_3, BaO×B_2O_3
Outlier	not given
The concentration limits are not reported by the authors. The sign × denotes multiplication, e.g., Na_2O×Al_2O_3 is an interaction factor between the components Na_2O and Al_2O_3.	

Table XXIX shows a comparison of experimental data from Chapter 6 and corresponding modeling results. It appears that experiments concerning the surface tension of the low-expansion borosilicate base glass should be repeated. It is especially surprising that it shows a relatively high surface tension in the experiment, even though it is well known that boron oxide decreases the surface tension of glass melts [8, 45, 86, 92, 93]. It needs to be noted however at this point that surface tension data in the literature are subject to scattering up to ~100 mN/m [59] due to the difficulty of the measurements. The experimental results in this study appear satisfactory for most glasses.

In general, the surface tension of glass melts tends to increase slightly with increasing temperature, as demonstrated through the data in this study and through findings of other investigators [86, 94].

TABLE XXIX: COMPARISON OF EXPERIMENTAL SURFACE TENSION DATA AT 1400°C AND MODEL ESTIMATES BY KUCUK ET AL. [8, 45] AND GOLEUS ET AL. [91]

	Experimental data	**Model estimates**		**Residuals**	
		Kucuk	Goleus	Kucuk	Goleus
TV	283	290	288	-7	-5
E	328	363	348	-34	-20
LO	376*	244	244	132*	133*
CO	328	312	307	16	21
FL	/	313	319	/	/
WO	288	312	318	-24	-29

All surface tension data in mN/m. The outlier is marked *

TV - TV panel glasses, E - textile fiber E glasses, LO - low-expansion borosilicate glasses, CO - container glasses, FL - float glasses, WO - fiber wool glasses

APPENDIX D

References as given in SciGlass [59] for high temperature electrical resistivity data in the systems SiO_2-Na_2O, SiO_2-K_2O, and SiO_2-CaO-Na_2O, that the models in Tables VI to VIII are based on:

System SiO_2-Na_2O:

Ashizuka M. and Ohtani M., J.Jpn Inst.Metals, 1969, vol. 33, No. 4, p. 498.

Bockris J.O'M., Kitchener J.A., Ignatowicz S. and Tomlinson J.W., Trans.Faraday Soc., 1952, vol. 48, No. 1, p. 75.

Bonetti G. and Lazzari S., Vetro e Silicati, 1969, vol. 13, No. 5, p. 5.

Bonetti G., Riv.Staz.Sper.Vetro, 1976, vol. 6, No. 6, p. 241.

Boricheva V.N., Issledovanie Elektroprovodnosti Nekotorykh Prostykh Silikatnykh Stekol v Shirokom Intervale Temperatur. Thesis. Leningrad, 1956.

Botvinkin O.K. and Okhotin M.V., In: Noveishie Raboty po Fizicheskoi Khimii Stekla, Moskva, 1936, p. 72.

Buchanan R.C. and Kingery W.D., Compt.Rend.VII Congr.Intern. du Verre, Bruxelles, 1965, vol. 2, p. 368

Evstropiev K.S., In: Fiziko-Khimicheskie Svoistva Troinoi Sistemy Na_2O-PbO-SiO_2, Moskva, 1949, p. 83.

Gukasyan S.B. and Kostanyan K.A., Dokl.Akad.Nauk Arm.SSR, 1978, vol. 67, No. 2, p. 101.

Kawahara M., Ozima Y., Morinaga K. and Yanagase T., J.Jpn Inst.Metals, 1978, vol. 42, No. 6, p. 618

Kim K.D. and Lee Seung-Heun, J.Ceram.Soc.Jpn, 1997, vol. 105, No. 10, p. 827.

Kim K.D., Glass Technology, 1995, vol. 36, No. 1, p. 27.

Kim K.D., J.Am.Ceram.Soc., 1996, vol. 79, No. 9, p. 2422.

Kim K.D., Proc.XVIIth Intern.Congr.on Glass, Beijing, 1995, vol. 3, p. 747.

Kirakosyan S.Sh., In: Stekloobraznoe Sostoyanie, Erevan, 1974, p. 78.

Kostanyan K.A. and Erznkyan E.A., Arm.Khim.Zh., 1967, vol. 20, No. 9, p. 686.

Kostanyan K.A. and Erznkyan E.A., Dokl.Akad.Nauk Arm.SSR, 1966, vol. 43, No. 5, p. 279.

Kostanyan K.A. and Kirakosyan S.Sh., Arm.Khim.Zh., 1974, vol. 27, No. 7, p. 547.

Kostanyan K.A. and Saakyan K.S., Izv.Akad.Nauk Arm.SSR,Khim.Nauki, 1961, vol. 14, No. 5, p. 409.

Kostanyan K.A. and Shakhmuradyan G.T., Arm.Khim.Zh., 1975, vol. 28, No. 9, p. 692.

Kostanyan K.A., Erznkyan E.A. and Avetisyan E.M., Arm.Khim.Zh., 1967, vol. 20, No. 8, p. 592.

Kostanyan K.A., Erznkyan E.A. and Loryan Yu.G., In: Stekloobraznoe Sostoyanie, Erevan, 1970, p. 206.

Kostanyan K.A., Fizika i Khimiya Stekla, 1982, vol. 8, No. 6, p. 650.

Kroger C. and Weisgerber P., Z.Phys.Chem., 1958, vol. 18, No. 1/2, p. 90.

Lotto B. and Lazzari S., Vetro e Silicati, 1965, vol. 9, No. 50, p. 5.

Mori K., J.Iron Steel Inst.Jpn, 1956, vol. 42, No. 8, p. 633.

Morinaga K., Suginohara Y. and Yanagase T., J.Jpn Inst.Metals, 1975, vol. 39, No. 12, p. 1312.

Piechurowski A., 4[e] Conference sur la fusion electrique du verre, Tchecoslovaque, 1972, vol. 2, p. 16.

Shakhmuradyan G.T., Kostanyan K.A. and Dzhavuktsyan S.G., Arm.Khim.Zh., 1976, vol.

29, No. 3, p. 218.
Shinozaki N., Okusu H., Mizoguchi K. and Suginohara Y., J.Jpn Inst.Metals, 1977, vol. 41, No. 6, p. 607.
Stanek J., Sasek L. and Meissnerova H., Sprechsaal, 1966, No. 5, p. 151.
Tickle R.E., Phys.Chem.Glasses, 1967, vol. 8, No. 3, p. 101.
Urnes S., Glass Industry, 1959, vol. 40, No. 5, p. 237.
Wakabayashi H. and Terai R., Bull.Governm.Ind.Res.Inst.Osaka, 1984, vol. 35, No. 1, p. 58.
Wakabayashi H. and Terai R., J.Ceram.Soc.Jpn, 1983, vol. 91, No. 7, p. 334.
Wakabayashi H. and Terai R., J.Ceram.Soc.Jpn, 1985, vol. 93, No. 1, p. 13.
Excluded from the calculation due to significant inconsistency with other investigators: Endell K. and Hellbrugge J., Glastech.Ber., 1942, vol. 20, no. 10, p. 277

System SiO_2-K_2O:

Ashizuka M., J.Ceram.Soc.Jpn, 1989, vol. 97, No. 4, p. 489.
Bockris J.O`M., Kitchener J.A., Ignatowicz S. and Tomlinson J.W., Trans.Faraday Soc., 1952, vol. 48, No. 1, p. 75.
Boricheva V.N., Issledovanie Elektroprovodnosti Nekotorykh Prostykh Silikatnykh Stekol v Shirokom Intervale Temperatur. Thesis. Leningrad, 1956.
Kim K.D. and Lee Seung-Heun, J.Ceram.Soc.Jpn, 1997, vol. 105, No. 10, p. 827.
Kim K.D., Glass Technology, 1995, vol. 36, No. 1, p. 27.
Kim K.D., J.Am.Ceram.Soc., 1996, vol. 79, No. 9, p. 2422.
Kim K.D., Proc.XVIIth Intern.Congr.on Glass, Beijing, 1995, vol. 3, p. 747.
Kostanyan K.A. and Erznkyan E.A., Arm.Khim.Zh., 1967, vol. 20, No. 9, p. 686.
Kostanyan K.A. and Erznkyan E.A., Dokl.Akad.Nauk Arm.SSR, 1966, vol. 43, No. 5, p. 279.
Kostanyan K.A. and Erznkyan E.A., Izv.Akad.Nauk Arm.SSR,Khim.Nauki, 1964, vol. 17, No. 6, p. 613.
Kostanyan K.A. and Saringyulyan R.S., Elektrokhimiya i Rasplavy, Moskva, 1974, p. 193.
Kostanyan K.A. and Shakhmuradyan G.T., Arm.Khim.Zh., 1975, vol. 28, No. 9, p. 692.
Kostanyan K.A., Erznkyan E.A. and Loryan Yu.G., In: Stekloobraznoe Sostoyanie, Erevan, 1970, p. 206.
Mazurina E.K. and Evstropiev K.S., In: Stekloobraznoe Sostoyanie, Erevan, 1970, p. 195.
Saringyulyan R.S. and Kostanyan K.A., Arm.Khim.Zh., 1970, vol. 23, No. 10, p. 928.
Saringyulyan R.S. and Kostanyan K.A., In: Stekloobraznoe Sostoyanie, Erevan, 1974, p. 31.
Saringyulyan R.S. and Kostanyan K.A., In: Stekloobraznoe Sostoyanie, Moskva, 1971, p. 289.
Saringyulyan R.S., Vyazkost i Elektroprovodnost Stekol Sistemy KâO-PbO-SiOâ v Shirokom Intervale Temperatur. Thesis. Erevan, 1972.
Sasek L., Meissnerova H. and Persin J., Sb.Vys.Sk.Chem.Technol.Praze,Chem.Technol.Silik., 1973, vol. L4, p. 87.
Tickle R.E., Phys.Chem.Glasses, 1967, vol. 8, No. 3, p. 101.
Urnes S., Glass Industry, 1959, vol. 40, No. 5, p. 237.
Wakabayashi H. and Terai R., Bull.Governm.Ind.Res.Inst.Osaka, 1984, vol. 35, No. 1, p. 58.
Wakabayashi H. and Terai R., J.Ceram.Soc.Jpn, 1983, vol. 91, No. 7, p. 334.
Excluded from the calculation due to significant inconsistency with other investigators: Endell K. and Hellbrugge J., Glastech.Ber., 1942, vol. 20, no. 10, p. 277

System SiO_2-Na_2O-CaO:

Bonetti G. and Lazzari S., Vetro e Silicati, 1969, vol. 13, No. 5, p. 5.

Bonetti G., Riv.Staz.Sper.Vetro, 1976, vol. 6, No. 6, p. 241.

Boricheva V.N., Issledovanie Elektroprovodnosti Nekotorykh Prostykh Silikatnykh Stekol v Shirokom Intervale Temperatur. Thesis. Leningrad, 1956.

Botvinkin O.K. and Okhotin M.V., In: Noveishie Raboty po Fizicheskoi Khimii Stekla, Moskva, 1936, p. 72.

Kostanyan K.A. and Geokchyan O.K., Steklo Keram., 1964, No. 4, p. 5.

Kostanyan K.A., Issledovanie Elektroprovodnosti Natrii-Kaltsii-Magnii-Alyumosilikatnykh Stekol v Rasplavlennom Sostoyanii. Thesis. Leningrad, 1952.

Kostanyan K.A., Saakyan K.S. and Geokchyan O.K., Izv.Akad.Nauk Arm.SSR,Khim.Nauki, 1964, vol. 17, No. 4, p. 357.

Kroger K. and Heckmann H., Glastech.Ber., 1966, vol. 39, No. 11, p. 479.

Mazurin O.V., Tr.Leningr.Tekhnol.Inst., 1954, No. 29, p. 72.

Mazurina E.K. and Evstropiev K.S., Izv.Vyssh.Uchebn.Zaved., Khimiya i Khim.Tekhnol., 1967, No. 6, p. 673.

Mazurina E.K., Steklo, 1967, No. 1, p. 129.

Stanek J., Sasek L. and Meissnerova H., Sprechsaal, 1966, No. 5, p. 151.

Wakabayashi H. and Terai R., Bull.Governm.Ind.Res.Inst.Osaka, 1984, vol. 35, No. 1, p. 58.

Wakabayashi H. and Terai R., J.Ceram.Soc.Jpn, 1983, vol. 91, No. 7, p. 334.

NOTES

[1] A. Winkelmann, O. Schott: "Über die Elastizität und über die Druckfestigkeit verschiedener neuer Gläser in ihrer Abhängigkeit von der chemischen Zusammensetzung (Dependence of the elasticity and strength of various new glasses from the chemical composition)"; Ann. Physik Chemie, vol. 51, 1894, p 697; and "Über thermische Widerstandscoefficienten verschiedener Gläser in ihrer Abhängigkeit von der chemischen Zusammensetzung (Dependence of the thermal resistance of various glasses from the chemical composition)", vol. 51, 1894, p 730; and "Über die specifischen Wärmen verschieden zusammengesetzter Gläser (Heat capacity of various glasses)", vol. 49, 1893, p 401 (in German)

[2] M. B. Volf: "Mathematical Approach to Glass"; Glass Science and Technology, vol. 9, Elsevier, 1988, ISBN 0-444-98951-X

[3] G. Gehlhoff, M. Thomas; Z. techn. Physik 6 (1925), p 544; Z. techn. Physik 7 (1926), p 105 and p 260; "Lehrbuch der technischen Physik (Applied Physics)", J. A. Barth-Verlag, Leipzig, 1924, p 376 (in German)

[4] T. Lakatos, L.-G. Johansson, B. Simmingsköld, "Viscosity temperature relations in the glass system SiO_2-Al_2O_3-Na_2O-K_2O-CaO-MgO in the composition range of technical glasses"; Glass Technology, vol. 13, no. 3, June 1972, p 88-95

[5] D. Öksoy, D. L. Pye, E. N. Boulos: "Statistical analysis of viscosity-composition data in glassmaking"; Glastech. Ber. Glass Sci. Technol., vol. 67, no. 7, 1994, p 189-195

[6] P. R. Hrma, G. F. Piepel et al.: "Property/Composition Relationships for Hanford High-Level Waste Glasses Melting at 1150°C"; PNL Report 10359 to the US Department of Energy, vol. 1 and 2, Contract DE-AC06-76RLO 1830, December 1994
http://www.osti.gov/dublincore/gpo/servlets/purl/10121755-P8oQTl/webviewable/ (18 MB)
http://www.osti.gov/dublincore/gpo/servlets/purl/10121752-cDjMo0/webviewable/ (24 MB)

J. D. Vienna, P. R. Hrma et al.: "Effect of Composition and Temperature on the Properties of High Level Waste (HLW) Glass Melting above 1200°C (Draft)"; PNNL Report 10987 to the US Department of Energy, Contract DE-AC06-76RLO 1830, February 1996
http://www.osti.gov/dublincore/gpo/servlets/purl/212394-mv0A6T/webviewable/ (3 MB)

P. Hrma, R. J. Robertus: "Waste glass design based on property composition functions"; Ceram. Eng. Sci. Proc., vol. 14, no. 11/12, 1993, p 187-203

P. Hrma, G. F. Piepel, P. E. Redgate, D. E. Smith, M. J. Schweiger, J. D. Vienna, D. S. Kim: "Prediction of processing properties for nuclear waste glasses"; Ceramic Transactions, vol. 61, p 505-513

[7] A. Fluegel, A. K. Varshneya, T. P. Seward, D. A. Earl: "Viscosity of commercial glasses in the softening range"; Ceramic Transactions, vol. 141; Eds.: J. R. Varner, T. P. Seward, H. Schaeffer), The American Ceramic Society, 2004, p 379-386

[8] A. Kucuk, A. G. Clare, L. Jones: "An estimation of the surface tension of silicate glass melts at 1400°C using statistical analysis"; Glass Technol., vol. 40, no. 5, Oct 1999, p 149-153

[9] R. Blume, C. Drummond III: "Modeling and optimization of solar-control glasses"; J. Am. Ceram. Soc., vol 85, no. 5, May 2002, p 1070-1076

[10] A. A. Appen: "Chemistry of Glass", 1970, Khimiya, Leningrad (In Russian)

[11] Gan Fuxi, Scientia Sinica, vol. 12 (1963), p 1365; vol. 17 (1974), 533-551

[12] L. I. Demkina: "Physicochemical Principle of the Manufacture of Optical Glass" (in Russian), Izd. Chimia, Leningrad, 1976

[13] A. I. Priven: "General method for calculating the properties of oxide glasses and glass-forming melts from their composition and temperature"; Glass Technol., vol. 45, no. 6, 2004, in print; and: A. I. Priven, Doctoral Thesis, St. Petersburg, 2002 (in Russian)

[14] A. I. Priven, O. V. Mazurin: "Comparison of methods used for the calculation of density, refractive index and thermal expansion of oxide glasses"; Glass Technol., vol. 44, no. 4, August 2003, p 156-166
http://www.ingenta.com/isis/searching/Expand/ingenta?pub=infobike://sgt/gt/2003/00000044/00000004/art00003

[15] A. Makishima, J. D. Mackenzie: "Direct calculation of Young's modulus of glass", J. Non-Cryst. Solids, vol. 12, 1973, p 35-45; and : "Calculation of bulk modulus, shear modulus, and Poisson's ratio of glass", vol. 17, 1975, p 147-157; and: "Calculation of thermal expansion coefficient of glasses", vol. 22, 1976, p 305-313

[16] Y. Bottinga, D. F. Weill, P. Richet: "Thermodynamic modeling of silicate melts" in "Thermodynamics of minerals and melts", Ed. by R. C. Newton, A. Navrotsky, and B. J. Wood, New York - Heidelberg - Berlin, Springer-Verlag, 1981, p. 207-245

[17] S. V. Nemilov: "Thermodynamic and kinetic aspects of the vitreous state", CRC Press, Boca Raton - Ann - Arbor - London - Tokyo 1995

[18] A. D. Pelton, M. Blander: "Thermodynamic analysis of ordered liquid solutions by a modified quasichemical approach - application to silicate slags"; Metallurgical Transactions B (Process Metallurgy), vol. 17B, no. 4, Dec 1986, p 805-815

[19] A. D. Pelton, P. Wu: "Thermodynamic modeling in glass-forming melts"; J. Non-Cryst. Solids, vol. 253, no. 1-3, 1999, p 178-191
http://dx.doi.org/10.1016/S0022-3093(99)00352-X (0.3 MB)

[20] P. Chartrand, A. D. Pelton: "Modeling the charge compensation effect in silica-rich Na_2O-K_2O-Al_2O_3-SiO_2 melts"; Calphad: Computer Coupling of Phase Diagrams and Thermochemistry, vol. 23, no. 2, June 1999, p 219-230
http://dx.doi.org/10.1016/S0364-5916(99)00026-7 (0.7 MB)

[21] V. C. Kress: "On the mathematics of associated solutions"; Am. J. Sci., vol. 303, Oct 2003, p 708-722

[22] R. Conradt: "Thermodynamic Approach to viscosity in the glass transition"; Glastech. Ber. Glass Sci. Technol., vol. 67, no. 11, 1994, p 304-311

[23] R. Conradt: "Modeling of the thermochemical properties of multicomponent oxide melts"; Zeitschrift für Metallkunde/Materials Research and Advanced Techniques, vol. 92, no. 10, October 2001, p 1158-1162

[24] J. Björkvall, Du Sichen, V. Stolyarova, S. Seetharaman: "A model description of the thermochemical properties of multicomponent slags and its application to slag viscosities"; Glass Phys. Chem., vol. 27, no. 2, March-April 2001, p 132-147
http://dx.doi.org/10.1023/A:1011332410674

[25] N. M. Vedishcheva, B. A. Shakhmatkin, M. M. Shultz: "A simulation of the thermodynamic properties of oxide melts and glasses"; Ceramic Transactions, vol. 29, Eds.: A. K. Varshneya, D. F. Bickford, P. P. Bihuniak; The American Ceramic Society, 1993, p 283-288

[26] N. M. Vedishcheva, B. A. Shakhmatkin: "Thermodynamic studies of oxide glass-forming liquids by the electromotive force method"; J. Non-Cryst. Solids, vol. 171, no. 1, July 1994, p 1-30

[27] B. A. Shakhmatkin, N. M. Vedishcheva, M. M. Shultz, A. C. Wright: "Thermodynamic properties of oxide glasses and glass-forming liquids and their chemical structure"; J. Non-Cryst. Solids, vol. 177, no. pt 1, 1994, p 249-256

[28] N. M. Vedishcheva, B. A. Shakhmatkin, M. M. Shultz, A. C. Wright: "Thermodynamic modelling of glass properties: A practical proposition?"; J. Non-Cryst. Solids, vol. 196, March 1996, p 239-243

[29] B. A. Shakhmatkin, N. M. Vedishcheva, A. C. Wright: "Can thermodynamics relate the properties of melts and glasses to their structure?"; J. Non-Cryst. Solids, vol. 293-295, no. 1, November 2001, p 220-226
http://dx.doi.org/10.1016/S0022-3093(01)00674-3 (0.2 MB)

[30] B. A. Shakhmatkin, N. M. Vedishcheva, A.C. Wright: "Thermodynamic modeling of the structure of glasses and melts: single-component, binary and ternary systems"; J. Non-Cryst. Solids, vol. 293-295, November 2001, p 312-317
http://dx.doi.org/10.1016/S0022-3093(01)00683-4 (0.3 MB)

[31] J. Schneider, V. R. Mastelaro, E. D. Zanotto, B. A. Shakhmatkin, N. M. Vedishcheva, A. C. Wright, H. Panepucci: "Qn distribution in stoichiometric silicate glasses: Thermodynamic calculations and 29Si high resolution NMR measurements"; J. Non-Cryst Solids, vol. 325, no. 1-3, Sep 2003, p 164-178
http://dx.doi.org/10.1016/S0022-3093(03)00332-6 (0.3 MB)

[32] T. M. Besmann, K. E. Spear, E. C. Beahm: "Thermochemical models for nuclear waste glass subsystems - MgO-CaO and $MgO-Al_2O_3$"; Materials Research Society Symposium - Proceedings, vol. 556, 1999, p 383-389

[33] K. E. Spear, T. M. Besmann, E. C. Beahm: "Thermochemical modeling of glass: Application to high-level nuclear waste glass"; MRS Bulletin, vol. 24, no. 4, April 1999, p 37-44

[34] T. M. Besmann, K. E. Spear: "Thermochemical modeling of oxide glasses"; J. Am. Ceram. Soc., vol. 85, no. 12, Dec 2002, p 2887-2894

[35] T. M. Besmann, K. E. Spear, J. D. Vienna: "Extension of the modified associate species thermochemical model for high-level nuclear waste: Inclusion of chromia"; Materials Research Society Symposium - Proceedings, vol. 757, 2003, p 195-200

[36] M. D. Allendorf, K. E. Spear: "Thermodynamic analysis of silica refractory corrosion in

glass-melting furnaces"; J. Electrochem. Soc., vol. 148, no. 2, Feb 2001, p B59-B67
http://dx.doi.org/10.1149/1.1337603

[37] R. H. Nilson, S. K. Griffiths, N. Yang, P. M. Walsh, M. D. Allendorf, B. Bugeat, O. Marin, K. E. Spear, G. Pecoraro: "Analytical models for high-temperature corrosion of silica refractories in glass-melting furnaces"; Glass Sci. Technol., vol. 76, no. 3, May/June 2003, p 136-151

[38] B. C. Tischendorf, T. M. Alam, R. T. Cygan et al.: "The structure and properties of binary zinc phosphate glasses studied by molecular dynamics simulations"; J. Non-Cryst. Solids, vol. 316, no. 2-3, Feb 2003, p 261-272
http://dx.doi.org/10.1016/S0022-3093(02)01795-7 (0.3 MB)

[39] G. Kresse: "Ab initio molecular dynamics: recent progresses and limitations"; J. Non-Cryst. Solids, vol. 312-314, Oct 2002, p 52-59
http://dx.doi.org/10.1016/S0022-3093(02)01649-6 (0.4 MB)

[40] A. N. Cormack, Jincheng Du: "Molecular dynamics simulations of soda-lime-silicate glasses"; J. Non-Cryst. Solids, vol. 293-295, Nov 2001, p 283-289
http://dx.doi.org/10.1016/S0022-3093(01)00831-6 (0.2)

[41] D. G. Montgomery: "Design and Analysis of Experiments", John Wiley & Sons, 2001

[42] S. Dowdy and S. Wearden: "Statistics for Research", John Wiley & Sons, 1983

[43] V. K. Leko, N. K. Gusakova, E. V. Meshcheryakova, T. I. Prokhorova: "The effect of impurity alkali oxides, hydroxyl groups, Al_2O_3, and Ga_2O_3 on the viscosity of vitreous silica"; Glass Phys. Chem., vol. 3, no. 3, May-June 1977, p 204-219

[44] K. C. Lyon, "Prediction of the Viscosities of Soda-Lime Silica Glasses"; J. Res. Nat. Bur. Standards A, Physics and Chemistry, vol. 78A, no. 4, July-Aug 1974, p 497-504

[45] A. Kucuk: "Structure and the physicochemical properties of glasses and glass melts"; Doctoral Thesis, Alfred University, New York, 1999

[46] Q. Rao, G. F. Piepel, P. Hrma, J. V. Crum: "Liquidus temperatures of HLW glasses with zirconium-containing primary crystalline phases"; J. Non-Cryst. Solids, vol. 220, no. 1, Oct 1997, p 17-29
http://dx.doi.org/10.1016/S0022-3093(97)00227-5 (1 MB)

[47] R. Backman, K. H. Karlsson, M. Cable, N. P. Pennington: "Model for liquidus temperature of multi-component silicate glasses"; Phys. Chem. Glasses, vol. 38, no. 3, 1997, p 103-109

[48] A. Fluegel, A. K. Varshneya, D. A. Earl, T. P. Seward, D. Oksoy: "Improved composition-property relations in silicate glasses, part I: viscosity"; Ceramic Transactions, vol. 170, The American Ceramic Society, 2004 (in print)
http://glassproperties.com

[49] C. Dreyfus, G. Dreyfus: "A machine learning approach to the estimation of the liquidus temperature of glass-forming oxide blends"; J. Non-Cryst. Solids, vol. 318, 2003, p 63-78
http://dx.doi.org/10.1016/S0022-3093(02)01859-8 (1 MB)

[50] C. Bishop: "Neural Networks for Pattern Recognition"; Oxford University, 1995

[51] A. Fluegel, A. K. Varshneya, D. A. Earl: "Viscosity Prediction of Industrial Glasses"; Proceedings, Glass & Optical Materials Division Fall 2004 Meeting, November 7-12, 2004, Cape Canaveral, Florida, USA

[52] User guide of the MCA software, Haller Information Technology System, Harold S Haller & Company, Cleveland, Ohio

[53] R. H. Myers, D. C. Montgomery: "Response surface methodology"; 2nd edition, John Wiley & Sons, Inc. 2002, p 40

[54] Microsoft Excel based programs on accompanying CD, published with this monograph

[55] P. J. Rousseeuw, A. M. Leroy: "Robust Regression & Outlier Detection"; Wiley, 1987

[56] P. J. Huber: "Robust Statistics"; Wiley, 1981

[57] O. V. Mazurin: "Phase separation in glass", North-Holland, 1984

[58] O. V. Mazurin et al.: "Handbook of Glass Data"; Elsevier, Amsterdam, 1983 - 1993

[59] SciGlass 4.0 Database and Information System, Academic Press, 1999, ISBN 0126063117

[60] International Glass Database System INTERGLAD Ver.6; New Glass Forum, Tokyo, Japan; http://www.ngf.or.jp

[61] J. Stanek: "Electric Melting of Glass"; Elsevier Scientific Publishing, Amsterdam 1977, p 31

[62] O. V. Mazurin, O. A. Prokhorenko: "Electrical conductivity of glass melts"; Chapter 9 in: "Handbook on the Properties of Glass-Forming Melts" ed. by D. L. Pye, I. Joseph, A. Montenaro; CRC Press, Boca Raton, Florida, 2005 (in print)

[63] Owens-Illinois Glass Company, General Research Laboratory: "Effect of Barium Oxide and Zinc Oxide on the Properties of Soda-Dolomite Lime-Silica Glass" J. Am. Ceram. Soc., vol. 25, no. 3 (1942), p 61-69; "Effect of Iron Oxide on the Properties of Soda-Dolomite Lime-Silica Glass" J. Am. Ceram. Soc., vol. 25, no. 14 (1942), p 401-408; "Effect of Substituting MgO for CaO on Properties of typical Soda-Lime Glasses" J. Am. Ceram. Soc., vol. 27, no. 8 (1944), p 221-225; "Effect of Fluorine and Phosphorus Pentoxide on Properties of Soda-Dolomite Lime-Silica Glass" J. Am. Ceram. Soc., vol. 27, no. 12 (1944), p 369-372; "Effect of Boric Oxide on Properties of Soda-Dolomite Lime-Silica Glass" J. Am. Ceram. Soc., vol. 31, no. 1 (1948), p 8-14; "Effect of K_2O and Li_2O on Properties of Soda-Dolomite Lime-Silica Glasses" J. Am. Ceram. Soc., vol. 33, June (1950), p 181-186

[64] A. Fluegel, D. A. Earl, O. V. Mazurin, A. K. Varshneya, T. P. Seward, T. Vascott: "Improved composition-property relations in silicate glasses, part II: High temperature electrical conductivity"; Proceedings, Glass & Optical Materials Division Fall 2004 Meeting, November 7-12, 2004, Cape Canaveral, Florida USA

[65] J. O. Isard: "Electrical conduction in the aluminosilicate glasses"; J. Soc. Glass Technol., vol. 43, 1959, p 113T-123T

[66] R. E. Tickle: "The electrical conductance of molten alkali silicates"; Phys. Chem. Glasses, 1967, vol. 8, no. 3, p 101-124

[67] K. D. Kim: "Resistivity measurement of molten glass and mixed alkali effect in sodium and potassium silicate glass melts"; Glass Technology, 1995, vol. 36, no. 1, p 27-31

[68] J.-H. Hwang, J.-H. Park, K. D. Kim, S.-S. Choi: "Influence of mixed alkali oxides on some melt properties of TV screen glass"; Glass Sci. Technol. Glastech. Ber., vol. 73, no. 1, Jan 2000, p 8-17

[69] C. J. R. Gonzalez-Oliver, P. S. Johnson, P. F. James: "Influence of water content on the rates of crystal nucleation and growth in lithia-silica and soda-lime-silica glasses"; J. Mater. Sci., vol. 14, no. 5, 1979, p 1159

[70] E. M. Birtch, J. E. Shelby: "Effect of water in the melting atmosphere on the transformation temperature of commercial glasses"; Ceramic Transactions, vol. 141, The American Ceramic Society, 2004, p 347-354

[71] A. I. Priven: "Calculation of the Viscosity of Glass-forming Melts I: The Li_2O-Na_2O-K_2O-SiO_2 System"; Glass Physics and Chemistry, vol. 23, no. 5, 1997, p 344-353
http://www.maik.rssi.ru/cgi-bin/search.pl?type=abstract&name=physglas&number=5&year=97&page=344

A. I. Priven: "Calculation of the Viscosity of Glass-forming Melts: II. The MgO-CaO-SrO-BaO-Al_2O_3-SiO_2 System"; Glass Physics and Chemistry, vol. 23, no. 6, 1997, p 416-428
http://www.maik.rssi.ru/cgi-bin/search.pl?type=abstract&name=physglas&number=6&year=97&page=416

A. I. Priven: "Calculation of the Viscosity of Glass-forming Melts: III. The Alk_2O-RO-Al_2O_3-SiO_2 System"; Glass Physics and Chemistry, vol. 24, no. 1, 1998, p 19-30
http://www.maik.rssi.ru/cgi-bin/search.pl?type=abstract&name=physglas&number=1&year=98&page=19

A. I. Priven: "Calculation of the Viscosity of Glass-forming Melts: IV. A Unified Method for Calculating the Viscosity of Silicate and Aluminate Melts"; Glass Physics and Chemistry, vol. 24, no. 1, 1998, p 31-40
http://www.maik.rssi.ru/cgi-bin/search.pl?type=abstract&name=physglas&number=1&year=98&page=31

A. I. Priven: "Calculation of the Viscosity of Glass-Forming Melts: V. Binary Borate Systems"; Glass Physics and Chemistry, vol. 26, no. 6, 2000, p 541-558
http://www.maik.rssi.ru/cgi-bin/search.pl?type=abstract&name=physglas&number=6&year=0&page=541
http://dx.doi.org/10.1023/A:1007196013464

A. I. Priven: "Calculation of the Viscosity of Glass-Forming Melts: VI. The R_2O-B_2O_3-SiO_2 and RO-B_2O_3-SiO_2 Ternary Borosilicate Systems"; Glass Physics and Chemistry, vol. 27, no. 4, 2001, p 360-370
http://www.maik.rssi.ru/cgi-bin/search.pl?type=abstract&name=physglas&number=4&year=1&page=360
http://dx.doi.org/10.1023/A:1011372328059

A. I. Priven: "Calculation of the Temperature Coefficient of Viscosity Logarithm $\delta\log\eta/\delta\log T$ for Oxide Melts in the Glass Transition Range from Their Composition"; Glass Physics and Chemistry, vol. 26, no. 5, 2000, p 455-469
http://www.maik.rssi.ru/cgi-bin/search.pl?type=abstract&name=physglas&number=5&year=0&page=455

A. I. Priven: "Comparative Characterization of Methods for Calculating the Viscosity of Silicate Glass-forming Melts"; Glass Physics and Chemistry, vol. 23, no. 5, 1997, p 333-343
http://www.maik.rssi.ru/cgi-bin/search.pl?type=abstract&name=physglas&number=5&year=97&page=333

A. I. Priven: "A New Equation for Describing the Temperature Dependence of the Viscosity of Glass-forming Melts"; Glass Physics and Chemistry, vol. 25, no. 6, 1999, p 491-497
http://www.maik.rssi.ru/cgi-bin/search.pl?type=abstract&name=physglas&number=6&year=99&page=491

A. I. Priven: "Calculation of Temperature Dependences of the Viscosity and Volume Relaxation Time for Oxide Glass-Forming Melts from Chemical Composition and Dilatometric Glass"; Glass Physics and Chemistry, vol. 27, no. 6, 2001, p 527-542
http://www.maik.rssi.ru/cgi-bin/search.pl?type=abstract&name=physglas&number=6&year=1&page=527
http://dx.doi.org/10.1023/A:1013298225368

[72] C. L. Babcock: "Viscosity and electrical conductivity of molten glasses"; J. Am. Ceram. Soc. vol. 17, 1934, p 329-342

[73] T. Pfeiffer: "Viscosities and electrical conductivities of oxidic glass-forming melts"; Solid State Ionics, vol. 105, 1998, p 277-287

[74] K. Rah, B. C. Eu: "Theory of the viscosity of supercooled liquids and the glass transition: Fragile liquids"; Physical Review E, vol. 68, November 2003, article# 51204, p 1-8
http://link.aps.org/abstract/PRE/v68/e051204
http://dx.doi.org/10.1103/PhysRevE.68.051204

[75] C. C. Tournour, M. G. Mesko, J. E. Shelby: "Inert gas solubility in glasses and melts of commercial compositions"; Ceramic Transactions, vol. 141, The American Ceramic Society, 2004, p 119-126

[76] D. B. Rapp: "Water solubility and diffusion in glassforming melts"; Masters Thesis, Alfred University, New York, 2001

[77] D. B. Rapp, J. E. Shelby: "Water diffusion and solubility in soda-lime-silica melts"; Phys. Chem. Glasses, vol. 44, no. 6, 2003, p 393-400

[78] P. B. McGinnis, M. G. Mesko, D. B. Rapp, J. E. Shelby: "Water diffusion and solubility in glasses and melts of float, container, and other commercial compositions"; Ceramic Transactions, vol. 141, The American Ceramic Society, 2004, p 127-134

[79] H. Franz, H. Scholze: "Die Löslichkeit von H_2O-Dampf in Glasschmelzen verschiedener Basizität (Solubility of H_2O in glass melts of various basicity)"; Glastechn. Ber., vol. 36, 1963, p 347-356 (in German)

[80] P. B. McGinnis: "The diffusion kinetics of water in glasses and glass-forming melts"; Doctoral Thesis, Alfred University, New York, 1994

[81] P. B. McGinnis, J. E. Shelby: "Diffusion of water in float glass melts"; J. Non-Cryst-Solids, 177 (1994), 381-388

[82] J. W. Tomlinson: "A note on the solubility of water in a molten sodium silicate"; J. Soc. Glass Technol., vol. 40, 1956, p 25T-31T

[83] C. R. Kurkjian, L. E. Russell: "Solubility of water in molten silicates"; J. Soc. Glass Technol., vol. 42, 1958, p 130T-144T

[84] S. Garbe: "Zur Wasserabgabe von Natriumsilikatgläsern (Water release of silicate glasses)"; Glastech. Ber., vol. 34, 1961, p 413-417 (in German)

[85] A. G. Clare: "Using sessile and pendant drop methods to measure density and surface tension"; Glass Researcher, vol. 10-11, no. 2-1, 2001, p 30-32+52

[86] A. G. Clare, D. Wing, L. E. Jones, A. Kucuk: "Density and surface tension of borate containing silicate glass melts"; Glass Technology, vol. 44, no. 2, April 2003, p 59-62

[87] A. Dietzel: "Practical importance and calculation of surface tension of glasses, glazes, and enamels"; Sprechsaal 75 (1942), p 82-91

[88] K. C. Lyon: "Calculation of the surface tensions of glasses"; J. A. Ceram. Soc. 27 (1944), p 186-189

[89] R. E. Boni, G. Derge: "Surface tension of silicates"; J. Metals, vol. 206, 1956, p 53-59

[90] C. Rubenstein: "Factors for the calculation of the surface tension of glasses at 1200°C"; Glass Technol. 5 (1964), p 36-40

[91] V. I. Goleus, A. Ya. Belyi, E. M. Sardak, Ya. I. Belyi: "Calculation of the surface tension of molten borosilicate glasses"; Trans. Stek. Keram. 53 (1996), p 226-228

[92] N. P. Bansal, R. H. Doremus: "Handbook of Glass Properties"; Academic Press, Inc., Orlando, Florida, 1986, p 680

[93] S. Fujino, C. Hwang, K. Morinaga: "Density, surface tension, and viscosity of PbO-B_2O_3-SiO_2 glass melts"; J. Am. Ceram. Soc., vol. 87, no. 1, January 2004, p 10-16

[94] J. W. Nowok, J. P. Hurley, J. A. Bierbe: "The cause of surface tension increase with temperature in multicomponent aluminosilicates derived from coal-ash slags"; J. Mater. Sci., vol. 30, no. 2, 1995, p 361-364

Chapter 10

SUMMARY

Thomas P. Seward III
New York State College of Ceramics at Alfred University
Alfred, New York 14802

In this book, we have provided new glass melt property data on several sets of glass compositions potentially useful to the commercial glass industry. A perhaps unique aspect of the study is that measurements of a variety of meaningful properties have been made on the same, relatively large set of compositions. There are of course gaps in the property data, the most obvious being that density, surface tension, gas solubility and radiative conductivity were measured on only the six commercial base glasses. Those property measurements could be extended to the 144 experimental composition variations, if suitable interest and funding were to become available.

Gas solubility

As pointed out in Chapter 1, the extent and reliability of the solubility data available for N_2, CO_2 and SO_2 as a function of glass composition and temperature, whether from this study or elsewhere in the literature, may not yet be sufficient for process modelers' needs. To be of greater value, the sensitivities of the reported gas solubility and diffusivity studies need to be enhanced. An obvious method for doing so would be to saturate the glasses at much higher pressures. With moderate investment and suitable safety precautions the saturation pressures could be increased by orders of magnitude, as is often done in geophysical studies. However, it would also be necessary to determine the pressure dependencies of the solubility coefficients, particularly for the reactive species CO_2 and SO_2.

Also, the time required to reach chemical equilibrium during gas absorption or desorption deserves further study because reaction kinetics as well as diffusion kinetics may need to be incorporated into realistic models and simulations.

Heat capacity

Heat capacity of glass melts was not measured in this study because the reliability of DSC (differential scanning calorimetry) techniques for glass melts at temperatures above 1000°C was deemed inadequate. However, as discussed in Chapter 1, heat capacity of melts can exceed the Dulong-Petite limit by as much as 30% at temperatures above the glass transition. This deviation is due to variations in the structural configuration states available to the melt and is not predictable by current theory. Development of precise calorimetric techniques for glass melts and their application to determination of compositional dependencies of commercial glass melt heat capacities, particularly with respect to boron and aluminum oxides, could prove valuable and should be encouraged. In particular, an improved understanding of the temperature and heating rate dependencies of heat capacity throughout the glass transition region could be important for improved modeling of glass forming operations.

Viscosity

There is a gap in the reported viscosity data for temperatures from ~800-1000°C because we did not measure viscosity in the range between 10^2 and 10^4 Pa·s for most of the compositions in this study. This viscosity range is at the high end for the rotating spindle method used for the measurements, and at the low end, where the glasses begin to flow rapidly, for the parallel plate method. The gap could be closed by further measurements. In particular, the parallel plate viscosity measurements could be extended to the range of 10^4 to 10^{10} Pa·s. Multiple measurements per sample using different loads would be necessary. The rotating spindle technique allows reliable measurements up to 10^4 Pa·s, but requires use of different spindles and additional experimental time. Fortunately, as discussed in Chapters 1 and 9, for the purposes of modeling the melting process alone, this gap should not be a significant drawback.

Statistical analysis

In Chapter 9, the authors statistically analyzed most of the data and provided predictive statistical models for viscosity and electrical resistivity. Although limited in basis to about 150 different glass compositions, this procedure allows many useful comparisons to be made, which will facilitate ongoing model creation and criticism. Obviously, if extensive data sets for the other measured properties were to be developed for this set of compositions, then similar analysis and modeling could be performed for them as well.

There is of course opportunity to add data from other sources. For example, the soda-lime-silica viscosity model was extended in Chapter 9 to include the data of Lakatos [1]. Since that writing, those investigators have pulled other viscosity and resistivity data from the literature into their database and are continuing to do so. While the outcome of those efforts lies outside the intended scope of this book, the initial results have been very encouraging and some have already been published elsewhere [2].

Cautionary note

It may be very tempting for melting and forming process designers and modelers to suggest composition changes, based on our data or those of others, for the purpose of decreasing cost or improving product quality and yield. This is to be encouraged, but should be approached with caution for several reasons.

Primarily, of course, compositional changes can affect the physical and chemical properties of the final glass product, and often the glass composition has been carefully designed to provide rather tightly specified properties.

Further, viscosity-temperature relationships reign supreme in glass manufacturing. Almost every process step is performed most effectively within a certain range of melt viscosity, as noted in Chapter 1, Figure 2. Any change in composition made to meet a melting process need (or to solve a melting process problem) might adversely affect a downstream forming process, and perhaps the product quality or applicability.

Liquidus temperatures

Another key factor of concern, but one that has not been addressed in the studies contained in this book, is the melt liquidus temperature, and in particular the viscosity of the melt at that temperature. The liquidus temperature of a melt is defined thermodynamically as the temperature above which the melt is stable as a liquid. Whether this liquid is very fluid or very viscous is not of thermodynamic concern. At temperatures below the liquidus temperature, the melt begins to devitrify (i.e., develop crystals); the extent of devitrification depends upon the chemical composition of the glass, the temperature and the time at that temperature.

If, in order to operate a certain manufacturing process at the required viscosity, the glass temperature is decreased below the liquidus temperature, then there is the risk that given enough time, the molten glass will crystallize. See References [3-5], Kingery et al., Chapter 8; Varshneya, Chapter 3 and Shelby, Chapter 2, for discussions of crystallization, kinetics of glass formation and TTT (time-temperature-transformation) diagrams. The point here is that, for any melting and forming process, the glass composition must be designed so that the liquidus temperature is lower than any processing temperature at which some or all of the molten glass will be held for long periods of time. Further, any composition changes recommended by process designers or modelers must be made with potential effects of crystallization in mind.

Therefore, a future extension of the present work should be to measure the liquidus temperature of all the glass compositions in the study and to compare the results with predictions based on thermodynamic theory and thermodynamic data. Examples of the latter approach can be found in References [6, 7].

Conclusion

In conclusion, we hope that the work presented in this book will be valuable for glass scientists and engineers in their endeavors beyond modeling of glass melting and forming processes, perhaps in the creation of new and useful oxide glass materials and products.

REFERENCES

1. Lakatos, T., L.-G. Johansson, and B. Simminsköld, *Viscosity-temperature relations in the glass system SiO_2-Al_2O_3-Na_2O-K_2O-CaO-MgO in the composition range of technical glasses.* Glass Technol., 1972. **13**: p. 88-95.
2. Fluegel, A., et al., *Improved composition-property relations in silicate glasses, part I: viscosity.* Ceramic Transactions, 2004 - in press. **170**.
3. Kingery, W.D., H.K. Bowen, and D.R. Uhlmann, *Introduction to Ceramics.* Second ed. Wiley Series on the Science and Technology of Materials. 1976, New York, NY: John Wiley & Sons. 1032.
4. Varshneya, A.K., *Fundamentals of Inorganic Glasses.* 1994, San Diego, CA: Academic Press, Inc. 570.
5. Shelby, J.E., *Introduction to Glass Science and Technology.* 1997, Cambridge, England: The Royal Society of Chemistry.
6. Shakhmatkin, B.A., N.M. Vedischeva, and A.C. Wright, *Thermodynamic modeling of the structure of glasses and melts: single-component, binary and ternary systems.* J. Non-Crystalline Solids, 2001. **293-295**: p. 312-317.
7. Conradt, R., *Modeling of the Thermochemical Properties of Multicomponent Oxide Melts.* Z. Metallkd., 2001. **92**.

Appendix I

GLASS PROPERTIES

James E. Shelby
New York State College of Ceramics at Alfred University
Alfred, New York 14802

INTRODUCTION

A number of properties have been determined for the glasses comprising the DOE Glass Melt Properties Data Base project. The six base compositions used in this study are typical of float, container, wool, E-glass, TV panel, and hard borosilicate glasses. Compositional variations based on these typical compositions were determined and used to produce a large set of glasses. Glasses were produced at Corning, Inc. and at the New York State College of Ceramics at Alfred University. Analyzed compositions and a discussion of glass preparation are presented elsewhere in this book (Chapter 2).

Although they are not melt properties, additional measurements were made to expand the utility of this data base and to aid in characterizing the glasses. These additional properties include the density, refractive index, thermal expansion coefficient (TEC), glass transformation temperature (T_g), dilatometric softening temperature (T_d), and direct current electrical conductivity. Two values of the glass transformation temperature are reported; one was obtained from the thermal expansion curve (indicated as "dil") and one using differential scanning calorimetry (DSC).

Since it has been shown that the properties of glasses are strongly influenced by their water content, in the form of hydroxyl, the water content of each glass was determined using infrared spectroscopy. The extinction coefficients for hydroxyl for each base glass were used for all of the samples in a given compositional set. Most of these extinction coefficients were determined in this study and are discussed elsewhere in this book (Chapter 4). In general, the water content of all of the compositional variation glasses is less than that of the commercially produced base glasses. This difference is due to the difference in melting environment. Commercial glasses are produced in a gas fired tank, while the experimental glasses were produced in electrically heated laboratory furnaces, where the atmosphere normally contains less water vapor.

EXPERIMENTAL PROCEDURES

Samples

The samples of glasses melted at Alfred were provided in the form of broken chunks. Approximately 20 grams of each composition were remelted in a Pt/5Au crucible in an electrically heated furnace to form a single piece. The piece of glass was removed from the crucible and the glass transformation temperature (T_g) was measured using a differential scanning calorimeter. The sample was then annealed by heating to this value of T_g, holding for 30 minutes, and cooling at 1 K/min for 100 K. The sample was then cooled at an uncontrolled rate to room temperature. This annealing cycle produces a well annealed glass for further measurements.

Samples melted at Corning, Inc. were provided in the form of bars ≈25 by 25 by 100 mm which were cut from much larger cast slabs of glass. The samples were annealed at Corning before shipping to Alfred, so no further heat treatment was needed before cutting samples for the property measurements. The details of the annealing procedure are not known.

Each large piece of glass was cut into a number of smaller pieces for various property measurements. Bars for thermal expansion measurements were ≈3 mm wide, ≈3 mm thick, and ≈25 mm long. The thermal expansion bars were also used for the density measurements. Plates ≈1 mm thick and 1 to 2 cm^2 in area were cut from the larger pieces, ground and polished to an optical finish, and used for the refractive index, water content, and electrical conductivity measurements. Small chips of glass weighing 10 to 50 mg were used for the DSC measurements.

Thermal Expansion Curves

Thermal expansion coefficients, glass transformation temperatures, and dilatometric softening temperatures were obtained using a horizontal vitreous silica dilatometer with a heating rate of 4 K/min using the method described elsewhere by the author [1]. The dilatometer was calibrated for expansion coefficient using a NIST borosilicate glass and for temperature using a quartz bar. The expansion curve obtained for the quartz bar is shown in Figure 1. Note that the α–β transformation occurs at 570°C in this curve, as compared to the normally accepted transformation temperature of 573°C.

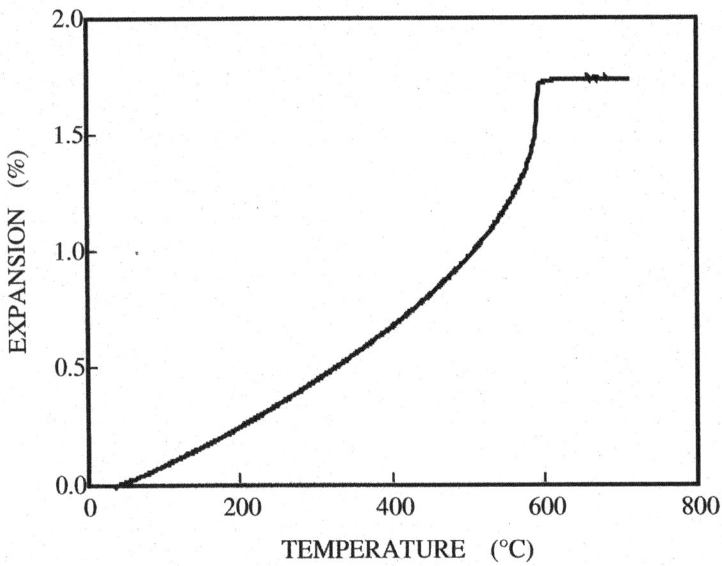

Figure 1. Calibration curve based on the α–β transition of quartz.

The dilatometer has been automated so that a computer controls the heating ramp and records the time from the beginning of the experiment, temperature, and change in length of the sample throughout the experiment. The program automatically divides each change in length by the original sample length, which was measured using a micrometer and entered into the program at the start of the experiment, and lists the expansion as a % of the original length for each set of points. A typical file contains 500 to 700 sets of time, temperature, expansion, % expansion values for glasses with dilatometric softening temperatures in the range of 500 to 700°C. The thermal expansion curve for the wool base glass shown in Figure 2 is typical of the results of this study.

Thermal expansion coefficients (TEC) are reported as average values for the temperature range 100 to 400°C. The TEC was calculated using the total change in length over this temperature range and the original length of the sample. Since the dilatometer actually measures the difference between the TEC of the sample and that of the vitreous silica used for the sample system, a correction for the thermal expansion of the silica apparatus is added to the measured TEC of the sample to yield the reported value. The glass transformation temperature was determined using the traditional "intercept of slopes" method [1] and is reproducible to within ±2 K. The dilatometric softening temperature, which is altered slightly by variations in the load applied to the sample is less reproducible, with a repeatability of ±4 K. The values listed for T_g and T_d are those actually measured and have not been corrected for the small difference between the measured and reported α–β transformation temperature for the quartz calibration bar.

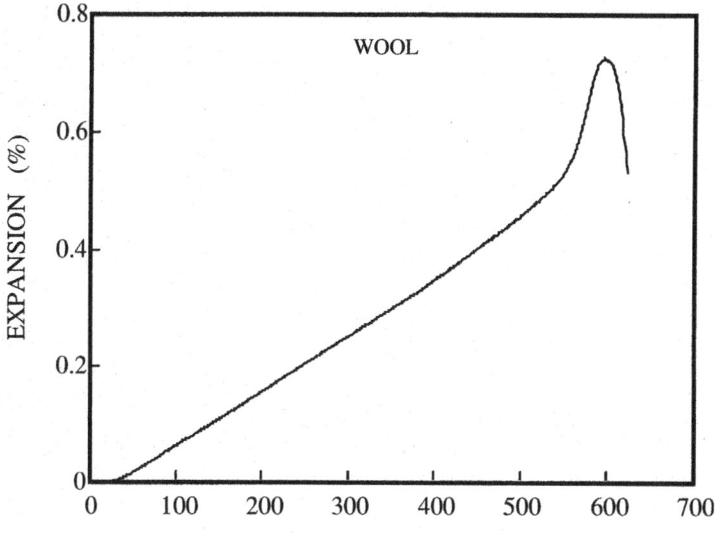

Figure 2. Thermal expansion curve for the base wool glass.

Differential Scanning Calorimeter

The differential scanning calorimeter (DSC) was used to measure the glass transformation temperature of the annealed glasses. A heating rate of 20 K/min was used for all measurements. Samples consisted of small chips of glass weighing 10 to 50 mg. Aluminum pans were used whenever possible, i.e. for upper heating temperatures below 600°C. Platinum pans were used for measurements which required heating to between 600 and 720°C. Measurements could not be made for glasses with glass transformation temperatures exceeding ≈680°C due to limitations on the upper use temperature of the instrument. Only a few such glasses were included in this study.

The DSC was calibrated using a sample cut from the same piece of quartz used to produce the bar used to calibrate the dilatometer. The average $\alpha-\beta$ transformation temperature for the quartz standard for several repetitions of the measurement was 575 ± 2 °C. The values listed in the tables are the actual values and are not corrected for the small difference found for the standard.

Density

Densities were measured using the Archimedes Method and a commercial "density apparatus" supplied by Mettler for use with their analytical balances. Kerosene was used as the immersion fluid for all measurements. The temperature of the kerosene was monitored during each set of measurements; density of the kerosene was then corrected for variations in temperature using published data. Listed densities are based on the average of at least 3 measurements for each glass. Measurements of the density of a sample of vitreous silica yielded a value of 2.001 ± 0.002 g/cm^3, which compares quite favorably with the published value of 2.001 g/cm^3 for this particular type of vitreous silica.

Refractive Index

Refractive indices (n_d) were measured using a digital Abbe Refractometer. Measurements of a standard glass sample with a reported index of 1.5173 yielded an average value of 1.5175 ± 0.0005. Indices could not be measured for a few of the float compositions due to their extremely high content of coloring ions, which rendered the samples opaque at this wavelength.

Electrical Conductivity

Electrical conductivities were measured using a direct current, guard ring configuration. Measurements were made in ≈10 K increments over a range of 150 to 200 K for each sample wherever possible. Resistivity values were converted to conductivities and plotted as log conductivity versus reciprocal temperature (see Figure 3 for an example). Conductivities were determined at 200 and 300°C from each curve and listed in the Tables XIII to XVIII. While most of the conductivities reported were interpolated within the range of the data, values for very conductive and for highly resistive glasses are extrapolated to these temperatures.

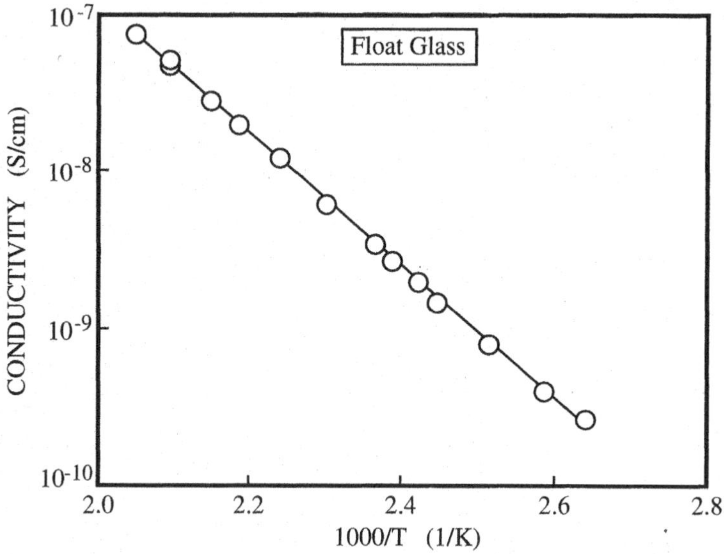

Figure 3. Effect of temperature on the electrical conductivity of a float glass sample.

In a few cases, samples were so resistive that measurements could not be obtained using this method. This problem was primarily encountered for some of the E-glass compositions containing very little, if any, alkali oxide. No conductivities are reported for those glasses.

Calibration of the electrical conductivity apparatus is limited due to the lack of any suitable standards within the temperature range (<300°C) useful for measurements for most of the glasses of this study. The NIST glass standard for d.c. conductivity is a lead silicate glass with a very low electrical conductivity at temperature below 400°C. Measurements for this glass indicate that the measured resistance is within ±10% of the published values at any given temperature. While this may appear to be a significant error, the conductivity of glasses changes so rapidly with temperature that this error range is representative of an error in iso-conductivity temperature of no more than ± 1 °C.

Water Concentrations

Water concentrations were determined from the infrared spectra of the samples used for the refractive index and electrical conductivity measurements. Where possible, the infrared absorption data were converted to concentrations (C) using the extinction coefficients (ε) determined earlier in our laboratory for the base glasses (presented in Chapter 4 of this book) and the Beer-Lambert Law, i.e.

$$C = 18(A/L) \div \varepsilon\rho \tag{1}$$

where A is the absorption measure at the maximum of the first water band minus the background absorption at 4000 cm^{-1}, L is the sample thickness in mm, and ρ is the density of the glass in g/cm^3. The concentration obtained from this expression is given in wt%, but has been converted to wtppm here by multiplying by 10^4. Since no extinction coefficient is available for the borosilicate glass, the data are reported as values of A/L for those compositions.

The "single band" method was used for determination of water concentrations from infrared spectra in this study. Results of multiple measurements indicate that this method yields water concentrations which are reproducible to within ± 1%. The accuracy of these values, however, is dependent on the accuracy of the extinction coefficient used in the calculation. Extinction coefficients are difficult to measure. Reported values for float glasses, for example, vary from about 39 to 43 L/mol-cm, i.e. by ≈10%. Extinction coefficients also vary by some unknown amount with glass composition. The assumption used here that all of the glasses based on a single glass type will have the same extinction coefficient is known to be incorrect. Since we do not believe, however, that the extinction coefficients for similar glasses vary by a significant amount, the values reported here should be correct to within ±10% relative to those for the base glasses.

RESULTS AND DISCUSSION

The complex compositions of these glasses does not allow a detailed analysis of trends in behavior without statistical analysis, which will be provided by other investigators. A number of observations were made regarding the data, which will be discussed below.

Chemical Composition of Glasses

The chemical compositions of the glasses, as analyzed at Corning, Inc., are listed elsewhere in this book (Chapter 2). Significant variations between actual and target compositions were evident for many of the glasses. The effect of these differences on the statistical analysis of the results of this study is unknown.

In most cases, the variations from glass to glass appear to be reasonable. Several glasses appear to have compositions which are not consistent with those of the other glasses in that family. Since the statistical analysis may assume that the compositions are those of the target values, it may be necessary to take some care in including these glasses in the analysis. Glasses which appear to deviate by a significant amount from their targets include: E-glass #19, borosilicate glasses # 13 and 16, wool glasses #6 and 10, float glass #15, and container glasses # 13 and 23.

Thermal Expansion Coefficients, T_g, and T_d

The thermal expansion coefficients for these glasses typically vary by less than a factor of 2 within a given family. Values for the wool glasses, for example, vary from ≈8.5 to 12.1 ppm/K.

While the thermal expansion curves obtained for most of the glasses were typical of annealed, homogeneous glasses, several curves for the borosilicate glasses differ considerably from the others. The curves for the borosilicate glasses #2, 4, 12, 16, 19, and 20 have the characteristic shape expected for phase separated borosilicate glasses. As a result, the glass transformation temperatures of these glasses are not indicative of those of the bulk composition, but rather that of the lower viscosity phase. Care should be taken in considering these values in any statistical analysis.

Density and Refractive Index

The refractive index of glasses usually increases with increasing density. This general trend was observed for the glasses considered here. With the exception of the TV panel glass, which exhibited a different curve, the density/ refractive index data for all of the other glasses lie, within a reasonable error, on a single curve (Figure 4). A separate curve (Figure 5) is needed for the TV panel glasses, which are the only glasses to contain large concentrations of SrO, BaO, and PbO. The presence of the heavy metal cations apparently alters the density, which depends on the mass of the nucleus, more than the refractive index, which depends on the electron density of the ion.

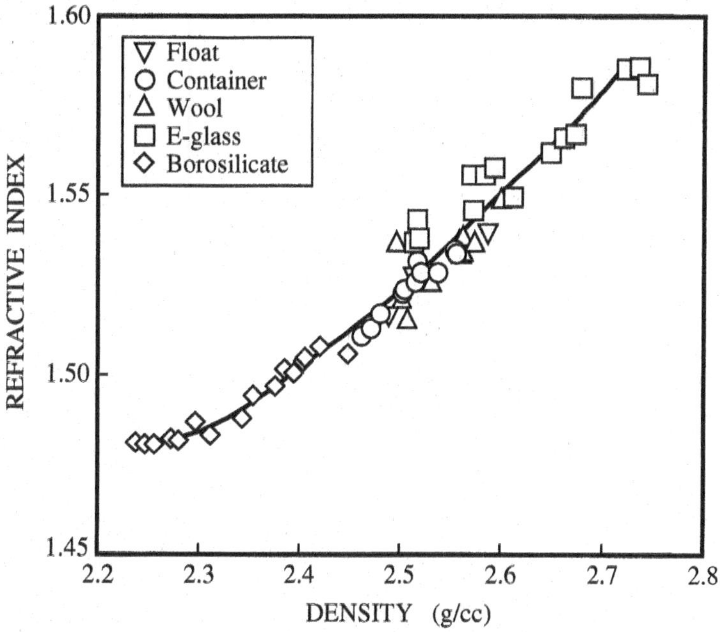

Figure 4. Relation between refractive index and density for DOE glasses except TV panel glasses.

Electrical Conductivity

Alkali ions are the current carriers in all of these glasses. Since the float, container, and wool glasses all contain large concentration of soda, glasses in these families are relatively conductive when compared to the other families of glasses. The E-glasses, which contain very little alkali oxide (and, in several cases, no deliberate additions of alkali oxides), are much more resistive than the other glasses. In most cases, the conductivity of the E-glasses was not detectable using our measurement technique. The borosilicate glasses, which contain less alkali oxide than the float, container, and wool glasses, exhibit an intermediate electrical conductivity.

At first glance, the relatively low electrical conductivities of the TV panel glasses appear to be somewhat surprising in view of their high alkali oxide contents. It should be noted, however, that, while the other glasses primarily contain only soda, the TV panel glasses contain similar quantities of both soda and potash. The relatively low electrical conductivities of these glasses are attributed to the mixed alkali effect, which is expected to be much greater for these glasses than for those of the other families of glasses.

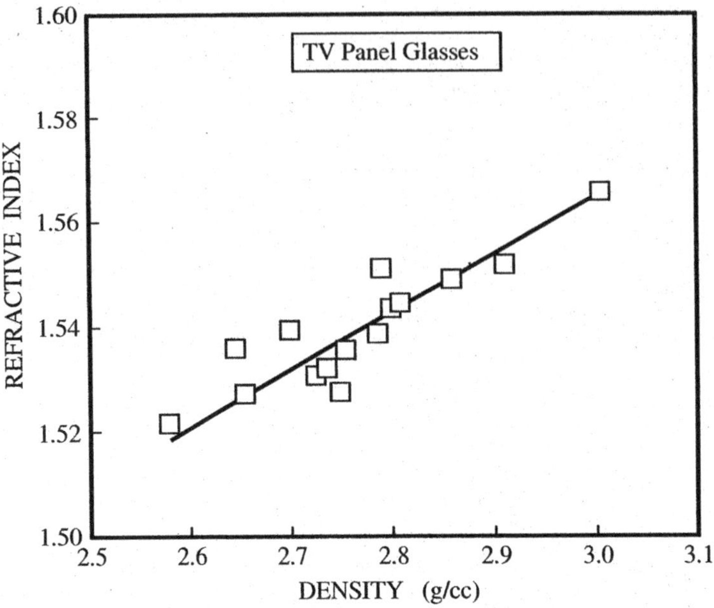

Figure 5. Relation between refractive index and density for TV panel glasses.

Water Concentrations

In some cases (wool, E-glass, borosilicate), the water concentration in the base glass exceeds that in any of the variations based on that glass. In others (container, float, TV panel), the water concentration in the base glass is similar to that of the variations. Since the base glasses are all produced in gas fired tanks, while the variations were melted in laboratory-sized electrically heated furnaces, which should have a drier atmosphere, it was expected that all of the base glasses would be wetter than their variation glasses. This difference in behavior cannot be attributed to the production site for the glasses, since the wool, float, and container glasses were melted at Alfred, whereas the E-glass, borosilicate, and TV panel glasses were made at Corning, Inc.

Variation in water content is exceptionally large for the TV panel glasses. This wide variation in water content may be due to a particularly large compositional dependence of the extinction coefficient for these glasses as compared to the other glasses. The TV panel glasses have very large differences in concentration for SrO and BaO. Since the concentration of the alkaline earth oxides significantly alters the infrared spectrum of the glass, it is expected that the use of a single value for the extinction coefficient for hydroxyl for all of these glasses is not adequate. The lack of information regarding the relation between alkaline earth identity and content and the extinction coefficient for hydroxyl prevents further consideration of this question at this time.

REFERENCES

1. J. E. Shelby, INTRODUCTION TO GLASS SCIENCE AND TECHNOLOGY, CH. 7, Royal Society of Chemistry, Cambridge, 1997.

TABLE I: PROPERTIES OF E-GLASSES

Number	Density (g/cm^3)	Refractive Index	TEC (ppm/K)	T_d (°C)	T_g dil (°C)	T_g DSC (°C)	H_2O Conc (wtppm)	Elect. Cond. 200°C-S/cm	Elect. Cond. 300°C-S/cm
BASE	2.62	1.5582	6.00	740	686	684	592	nd	4.6e-11
1	2.500	1.5302	5.00	843	803	n.d.	80	nd	•1e-11
2	2.558	1.5454	5.67	821	767	n.d.	119	nd	nd
3	2.734	1.5859	8.01	760	716	n.d.	130	nd	nd
4	2.572	1.5459	5.79	772	722	n.d.	141	nd	8.4e-10
5	2.671	1.5671	7.63	791	732	n.d.	113	nd	nd
6	2.743	1.5812	7.56	774	728	n.d.	134	nd	nd
7	2.518	1.5381	6.41	707	660	653	300	5.4e-10	3.9e-8
9	2.682	1.5752	8.40	685	648	645	222	nd	nd
10	2.662	1.5736	6.93	719	683	n.d.	263	nd	nd
11	2.593	1.5575	6.71	717	676	679	268	nd	nd
12	2.583	1.5558	5.95	752	709	n.d.	364	nd	nd
13	2.660	1.5662	7.92	760	725	n.d.	130	nd	nd
14	2.647	1.5620	6.85	800	737	n.d.	147	nd	nd
15	2.586	1.5466	6.31	776	725	n.d.	127	•1e-11	1.0e-9
16	2.561	1.5441	5.20	811	765	n.d.	112	nd	nd
17	2.610	1.5491	6.49	760	720	n.d.	171	nd	nd
18	2.721	1.5852	7.03	794	743	n.d.	130	nd	nd
19	2.517	1.5429	6.35	722	684	683	312	5.0e-10	2.5e-8
20	2.570	1.5554	5.97	728	676	677	325	nd	nd
21	2.677	1.5801	6.64	734	692	676	308	nd	nd
22	2.513	1.5366	5.16	761	704	n.d.	253	nd	nd
23	2.665	1.5746	7.51	715	675	n.d.	237	nd	nd
24	2.730	1.5874	8.41	692	648	644	127	nd	nd

*Based on assumption of constant extinction coefficient equal to that for the base glass, i.e. 51.5 L/mol-cm.

TABLE II: PROPERTIES OF BOROSILICATE GLASSES

Number	Density (g/cm^3)	Refractive Index	TEC (ppm/K)	T_d (°C)	T_g dil (°C)	T_g DSC (°C)	H_2O Conc (wtppm)	Elect. Cond. 200°C-S/cm	Elect. Cond. 300°C-S/cm
BASE	2.235	1.4727	3.50	613	540	532	0.642	1.2e-9	7.8e-8
1	2.250	1.4762	3.69	646	576	562	0.378	9.5e-9	5.8e-7
2	2.342	1.4880	4.12	696	596	580	0.266	4.3e-9	2.0e-7
3	2.405	1.5044	6.94	645	600	589	0.167	2.0e-9	1.5e-7
4	2.280	1.4815	5.39	640	575	558	0.333	5.5e-9	2.9e-7
5	2.385	1.5015	7.23	649	597	592	0.200	9.2e-8	4.5e-6
6	2.394	1.5003	6.02	639	600	596	0.203	2.0e-9	1.5e-7
7	2.319	1.4991	5.17	611	561	556	0.398	2.0e-10	1.7e-8
8	2.385	1.4897	4.58	625	586	582	0.288	6.9e-9	2.7e-7
9	2.420	1.5079	6.96	631	590	586	0.205	9.0e-10	7.9e-8
10	2.295	1.4868	5.69	607	556	552	0.295	3.4e-8	1.1e-6
11	2.338	1.4892	4.95	631	577	565	0.392	2.1e-9	1.3e-7
12	2.245	1.4804	4.49	663	584	566	0.341	7.2e-9	3.1e-7
13	2.447	1.5056	7.22	654	602	593	0.159	2.6e-9	1.8e-7
14	2.353	1.4941	5.68	661	610	588	0.227	2.0e-8	8.4e-7
15	2.376	1.4965	5.64	660	598	593	0.244	2.5e-10	2.4e-8
16	2.310	1.4830	4.61	658	586	577	0.309	4.1e-8	1.4e-6
17	2.304	1.4870	4.89	655	598	590	0.330	9.3e-9	5.1e-7
18	2.360	1.4951	5.87	649	600	592	0.257	8.5e-8	8.7e-6
19	2.272	1.4821	4.85	612	545	539	0.372	1.4e-10	2.0e-8
20	2.256	1.4807	3.97	674	577	536	0.339	1.7e-9	9.2e-8
21	2.404	1.5034	6.18	631	590	579	0.221	3.9e-9	2.1e-7
22	2.237	1.4812	4.34	608	557	543	0.398	7.3e-9	3.5e-7
23	2.398	1.5004	6.84	619	574	575	0.239	5.1e-8	5.0e-6
24	2.345	1.5085	7.55	615	579	579	0.234	3.1e-8	1.9e-6

TABLE III: PROPERTIES OF TV PANEL GLASSES

Number	Density (g/cm^3)	Refractive Index	TEC (ppm/K)	T_d (°C)	T_g dil (°C)	T_g DSC (°C)	H_2O Conc (wtppm)	Elect. Cond. 200°C-S/cm	Elect. Cond. 300°C-S/cm
BASE	2.82	1.5372	11.2	571	524	527	585	5.3e-11	9.8e-9
1	2.724	1.5308	10.5	554	506	508	257	3.0e-10	4.7e-8
2	2.529	1.5097	8.90	535	471	464	285	1.3e-10	2.7e-8
3	2.517	1.5005	9.22	539	482	465	266	1.1e-9	1.2e-7
4	2.498	1.5094	9.27	608	547	562	592	2.5e-10	2.9e-8
5	2.699	1.5396	9.58	656	602	592	582	4.9e-11	8.0e-9
6	2.978	1.5602	11.4	594	546	526	390	1.1e-9	1.3e-7
7	3.004	1.5658	9.77	567	519	517	406	nd	2.3e-10
8	3.038	1.5695	11.6	570	527	528	286	1.0e-10	2.4e-8
9	2.909	1.5520	14.6	521	471	472	430	1.2e-11	5.7e-9
10	2.735	1.5322	13.3	520	471	470	439	3.5e-11	1.1e-8
11	2.654	1.5274	11.9	544	492	498	572	1.0e-10	2.6e-8
12	2.807	1.5448	12.8	549	492	495	513	4.2e-11	1.0e-8
13	2.789	1.5512	11.5	596	539	537	509	nd	1.8e-9
14	2.989	1.5629	13.2	548	501	502	264	1.1e-11	2.5e-9
15	2.857	1.5493	11.4	570	501	502	501	1.1e-11	3.6e-9
16	2.645	1.5359	10.5	570	509	509	624	2.8e-10	4.3e-8
17	2.786	1.5388	13.0	544	493	490	542	4.4e-11	1.1e-8
18	2.945	1.5540	11.8	573	522	533	261	1.0e-11	1.2e-9
19	2.748	1.5276	10.7	531	472	466	364	2.1e-11	5.9e-9
20	2.577	1.5219	11.2	567	504	496	621	4.0e-10	8.7e-8
21	2.753	1.5356	12.5	541	489	485	363	7.4e-11	2.2e-8
22	2.798	1.5436	11.1	565	510	518	504	1.4e-11	3.3e-9
23	2.598	1.5126	9.60	548	492	490	343	4.1e-11	6.9e-9
24	2.366	1.4827	7.37	562	503	497	266	3.0e-10	3.1e-8

*Assumption of constant e of 18 L/mol-cm - questionable due to variation in K_2O content in these glasses.

TABLE IV: PROPERTIES OF WOOL GLASSES

Number	Density (g/cm^3)	Refractive Index	TEC (ppm/K)	T_d (°C)	T_g dil (°C)	T_g DSC (°C)	H_2O Conc (wtppm)*	Elect. Cond. 200°C-S/cm	Elect. Cond. 300°C-S/cm
BASE	2.55	1.5310	10.2	596	557	545	412	2.0e-8	9.4e-7
1	2.435	1.5076	8.40	618	564	557	284	3.0e-8	---
2	2.476	1.5105	9.47	590	545	541	283	5.0e-9	---
3	2.589	1.5396	11.8	575	535	525	282	2.6e-8	---
3A	2.574	1.5370	11.3	579	546	548	160	1.8e-8	9.6e-7
4	2.483	1.5148	9.28	596	549	549	290	1.3e-8	---
5	2.560	1.5338	12.1	600	551	557	153	1.7e-8	8.7e-7
6	2.599	1.5487	11.1	602	560	554	117	1.6e-9	1.5e-7
7	2.531	1.5259	11.3	602	558	557	138	2.7e-8	1.3e-6
8	2.563	1.5386	9.81	614	566	557	156	1.0e-9	8.7e-8
9	2.569	1.5367	10.3	611	571	553	129	8.0e-9	4.7e-7
10	2.558	1.5376	9.83	611	569	572	257	7.0e-9	3.7e-7
11	2.533	1.5240	11.0	577	541	535	307	1.5e-8	---
12	2.481	1.5241	8.54	615	575	574	443	3.7e-8	---
13	2.507	1.5155	11.7	562	516	520	104	9.3e-8	---
14	2.524	1.5312	9.12	612	577	562	340	6.3e-9	3.3e-7
15	2.512	1.5200	10.1	580	532	528	369	2.2e-8	1.7e-6
16	2.463	1.5116	9.35	587	546	551	470	1.7e-8	---
17	2.502	1.5212	10.6	601	556	548	173	1.7e-7	---
18	2.562	1.5343	9.87	621	572	576	233	1.1e-8	5.5e-7
19	2.515	1.5218	9.22	597	556	552	258	6.5e-8	---
20	2.500	1.5205	8.78	610	565	563	471	6.5e-9	---
21	2.496	1.5366	8.47	616	579	556	251	1.6e-8	8.4e-7
22	2.487	1.5185	7.99	593	566	567	428	1.5e-8	9.1e-7
23	2.583	1.5397	11.1	591	560	562	307	2.1e-8	---
24	2.598	1.5434	11.9	555	522	529	166	3.2e-9	---

*Based on assumption of constant extinction coefficient equal to that for the base glass, i.e. 46.8 L/mol-cm.

TABLE V: PROPERTIES OF CONTAINER GLASSES

Number	Density (g/cm^3)	Refractive Index	TEC (ppm/K)	T_d (°C)	T_g dil (°C)	T_g DSC (°C)	H_2O Conc (wtppm)*	Elect. Cond. 200°C-S/cm	Elect. Cond. 300°C-S/cm
BASE	2.52	1.5290	9.05	612	572	575	232	2.5e-8	1.1e-6
1	2.405	1.5019	7.93	618	574	580	212	2.2e-8	---
2	2.503	1.5239	9.47	621	595	595	214	1.9e-9	1.6e-7
3	2.534	1.5336	11.1	573	539	545	151	7.1e-9	---
4	2.487	1.5266	9.59	595	557	561	239	6.2e-8	---
5	2.494	1.5205	11.1	544	504	509	213	4.5e-9	---
6	2.545	1.5351	10.1	580	536	545	280	---	3.5e-10
7	2.477	1.5150	10.3	592	551	558	127	2.8e-8	---
8	2.444	1.5111	9.23	576	537	539	172	2.8e-9	---
9	2.554	1.5350	11.5	582	535	547	183	1.1e-8	7.4e-7
10	2.477	1.5201	8.77	588	548	553	231	1.2e-8	---
11	2.559	1.5334	11.5	591	556	566	310	3.5e-9	3.5e-7
12	2.520	1.5287	8.88	644	615	612	204	1.0e-8	4.8e-7
13	2.497	1.5170	10.4	561	527	528	226	1.0e-8	8.7e-7
14	2.430	1.5109	8.79	580	529	531	180	8.2e-9	5.8e-7
15	2.536	1.5287	11.7	606	557	560	186	2.7e-8	1.8e-6
16	2.470	1.5130	8.99	616	578	567	296	8.3e-9	5.1e-7
17	2.556	1.5336	10.6	609	579	567	219	2.6e-8	1.3e-6
18	2.538	1.5347	10.2	584	553	556	301	1.3e-9	1.4e-7
19	2.461	1.5106	8.90	602	562	568	174	1.5e-7	---
20	2.517	1.5318	8.69	657	622	623	256	1.1e-8	4.8e-7
21	2.514	1.5261	10.1	597	546	552	290	1.5e-9	1.6e-7
22	2.469	1.5138	8.94	622	570	575	387	1.2e-8	7.0e-7
23	2.502	1.5228	9.82	564	516	526	263	1.6e-8	9.9e-7
24	2.590	1.5439	12.3	557	523	535	182	1.0e-9	1.6e-7

*Based on assumption of constant extinction coefficient equal to that for the base glass, i.e. 42.4 L/mol-cm.

TABLE VI: PROPERTIES OF FLOAT GLASSES

Number	Density (g/cm^3)	Refractive Index**	TEC (ppm/K)	T_d (°C)	T_g dil (°C)	T_g DSC (°C)	H_2O Conc (wtppm)*	Elect. Cond. 200°C-S/cm	Elect. Cond. 300°C-S/cm
BASE	2.53	1.5198	9.45	615	562	565	305	3.9e-8	1.4e-6
1	2.462	1.515	9.05	613	565	572	290	2.9e-8	---
2	2.524	1.528	10.3	605	540	548	238	4.3e-8	---
3	2.492	1.517	11.6	591	541	538	267	1.7e-8	---
4	2.529	1.527	---	---	---	551	233	1.1e-8	---
5	2.520	1.524	10.6	585	545	548	266	1.1e-8	---
6	2.586	1.539	11.9	592	538	549	224	2.0e-8	1.0e-6
7	2.526	1.528	9.85	609	555	570	274	7.9e-9	4.7e-7
8	2.536	1.531	10.4	593	557	560	233	2.5e-8	---
9	2.513	1.527	9.64	621	563	568	279	1.3e-8	7.0e-7
10	2.548	1.534	10.3	591	561	560	301	2.0e-8	---
11	2.533	1.531	11.0	585	543	544	275	1.4e-8	---
12	2.522	1.529	10.2	604	560	568	347	5.3e-10	3.9e-7
13	2.530	opaque	9.22	621	572	572	205	2.6e-8	1.5e-6
14	2.490	1.520	10.4	579	545	551	406	4.5e-8	---
15	2.471	opaque	9.19	601	551	558	150	2.9e-8	1.4e-6
16	2.515	1.527	9.24	591	565	559	354	6.7e-9	4.2e-7
17	2.505	1.523	9.38	614	577	578	324	1.1e-8	6.7e-7
18	2.511	1.524	10.2	589	551	555	392	4.0e-8	---
19	2.484	1.519	9.35	598	567	566	304	1.2e-8	6.8e-7
20	2.554	opaque	10.6	592	543	556	195	1.5e-8	7.1e-7
21	2.489	1.516	9.72	599	557	560	252	9.5e-9	1.7e-7
22	2.514	opaque	9.50	619	567	580	282	1.6e-8	7.9e-7
23	2.474	1.519	9.18	609	563	566	310	1.7e-8	---
24	2.443	1.508	8.85	597	561	565	325	1.7e-8	---

*Based on assumption of constant extinction coefficient equal to that for the base glass, i.e. 41.3 L/mol-cm.

**Dark colors of several samples decrease accuracy of measurement. Some samples were opaque could not be measured using the Abbe refractometer.

Appendix II

GLASS COMPOSITIONS IN MOL %

TABLE I. CONTAINER GLASS COMPOSITIONS AS ANALYZED (MOL %)

ID	SiO_2	Al_2O_3	MgO	CaO	Li_2O	Na_2O	K_2O	Fe_2O_3	Cr_2O_3	TiO_2	SO_3	Total
base	74.418	0.755	0.298	11.266	0.000	12.900	0.185	0.014	0.000	0.006	0.158	100
1	81.274	0.673	0.090	7.451	0.000	10.326	0.160	0.011	0.000	0.008	0.008	100
2	74.662	0.601	0.151	12.744	0.000	9.846	1.341	0.011	0.104	0.357	0.182	100
3	68.801	0.681	0.149	13.179	1.864	14.795	0.011	0.173	0.122	0.007	0.217	100
4	72.971	0.639	4.372	7.254	0.000	14.090	0.011	0.161	0.114	0.374	0.015	100
5	70.224	0.653	4.543	7.177	1.811	14.110	1.243	0.015	0.000	0.007	0.215	100
6	68.290	0.659	4.778	12.338	1.770	10.281	1.319	0.171	0.000	0.380	0.015	100
7	73.891	1.885	0.092	7.438	0.000	14.573	1.334	0.177	0.000	0.393	0.215	100
8	76.314	1.941	0.091	7.408	1.881	10.695	1.349	0.176	0.121	0.008	0.015	100
9	68.357	1.884	0.164	13.035	1.834	14.301	0.006	0.011	0.000	0.392	0.015	100
10	73.392	1.859	4.644	7.399	1.815	10.273	0.018	0.019	0.118	0.381	0.082	100
11	65.148	1.846	4.715	12.447	0.000	14.329	1.358	0.019	0.115	0.008	0.015	100
12	70.588	1.797	4.234	12.996	0.000	10.024	0.013	0.147	0.000	0.008	0.195	100
13	75.164	0.678	0.015	7.740	1.136	13.360	1.371	0.008	0.112	0.410	0.008	100
14	79.327	0.641	0.089	7.256	1.845	10.163	0.038	0.161	0.000	0.375	0.105	100
15	69.683	0.616	0.148	12.827	0.000	13.950	1.327	1.419	0.000	0.007	0.022	100
16	75.428	0.603	4.191	7.476	0.000	10.667	1.319	0.015	0.119	0.008	0.173	100
17	67.249	0.640	5.032	12.045	0.000	14.465	0.012	0.011	0.000	0.368	0.176	100
18	69.500	0.663	4.885	12.629	1.810	10.346	0.017	0.021	0.108	0.007	0.015	100
19	75.331	1.926	0.091	7.650	0.000	14.645	0.013	0.015	0.129	0.008	0.192	100
20	74.061	1.860	0.151	13.254	0.000	9.985	0.019	0.145	0.112	0.382	0.030	100
21	71.752	1.933	0.120	12.823	2.009	9.764	1.358	0.014	0.000	0.008	0.220	100
22	74.114	1.738	4.216	7.482	0.000	10.681	1.338	0.011	0.000	0.396	0.023	100
23	70.545	1.873	4.119	7.905	2.131	13.240	0.008	0.164	0.000	0.007	0.007	100
24	62.899	1.843	4.693	12.206	1.929	14.280	1.288	0.165	0.119	0.376	0.203	100

TABLE II. E GLASS COMPOSITIONS AS ANALYZED (MOL %)

ID	SiO_2	B_2O_3	Al_2O_3	MgO	CaO	Na_2O	K_2O	Fe_2O_3	FeO	TiO_2	F	Total
base	56.993	6.118	8.782	6.500	19.635	0.615	0.000	0.133	0.084	0.439	0.701	100
1	73.605	0.000	7.647	0.856	17.892	0.000	0.000	0.000	0.000	0.000	0.000	100
2	68.730	0.000	9.247	2.386	17.709	0.000	0.347	0.232	0.000	0.787	0.562	100
3	56.091	0.000	6.646	7.551	26.650	2.001	0.000	0.303	0.000	0.757	0.000	100
4	65.731	0.000	9.296	0.878	19.156	2.090	0.000	0.321	0.000	0.799	1.730	100
5	59.363	0.000	9.292	0.891	28.047	2.038	0.368	0.000	0.000	0.000	0.000	100
6	53.058	0.000	8.688	8.792	27.188	0.000	0.339	0.317	0.000	0.000	1.618	100
7	61.605	7.743	6.822	0.817	18.605	2.044	0.356	0.307	0.000	0.000	1.700	100
8	55.529	7.677	6.915	0.828	27.614	0.000	0.334	0.331	0.000	0.772	0.000	100
9	51.018	7.630	7.244	6.002	24.806	2.036	0.000	0.000	0.000	0.000	1.264	100
10	52.987	8.184	9.953	0.757	26.257	0.000	0.000	0.000	0.000	0.781	1.082	100
11	52.836	8.190	9.341	7.373	19.021	2.080	0.357	0.000	0.000	0.803	0.000	100
12	56.383	7.987	9.202	7.249	18.864	0.000	0.000	0.315	0.000	0.000	0.000	100
13	60.233	0.000	6.749	0.838	27.391	1.993	0.352	0.000	0.000	0.762	1.682	100
14	62.866	0.000	6.665	1.018	27.401	0.000	0.000	0.311	0.000	0.000	1.740	100
15	65.500	0.000	7.451	6.489	17.767	2.140	0.337	0.316	0.000	0.000	0.000	100
16	68.839	0.000	9.138	3.108	16.727	0.000	0.341	0.197	0.000	0.788	0.862	100
17	61.893	0.000	8.937	6.951	18.555	2.076	0.000	0.000	0.000	0.000	1.588	100
18	55.384	0.000	8.841	7.935	27.083	0.000	0.000	0.000	0.000	0.757	0.000	100
19	67.764	7.994	5.767	0.688	15.173	1.983	0.000	0.000	0.000	0.631	0.000	100
20	56.539	7.774	6.868	7.225	18.951	0.000	0.000	0.315	0.000	0.808	1.520	100
21	54.107	7.562	6.703	6.386	24.923	0.000	0.320	0.000	0.000	0.000	0.000	100
22	60.222	8.110	9.509	0.912	19.323	0.000	0.357	0.000	0.000	0.000	1.568	100
23	52.200	8.450	9.994	0.804	25.958	2.274	0.000	0.319	0.000	0.000	0.000	100
24	42.620	8.247	9.730	6.738	27.199	2.131	0.303	0.322	0.000	0.784	1.926	100

TABLE III. FLOAT GLASS COMPOSITIONS TARGET/ANALYZED* (MOL %)

ID	SiO_2	Al_2O_3	MgO	CaO	MnO_2	Na_2O	K_2O
base	71.858	0.075	5.641	9.228	0.000	13.126	0.019
1	74.641	1.178	4.471	7.497	0.173	11.629	0.319
2*	72.677	0.131	3.917	7.648	0.000	13.888	0.353
3	72.026	0.059	4.452	7.466	0.172	14.476	1.270
4	69.264	0.059	4.464	9.624	0.000	14.513	1.273
5	68.790	0.058	5.896	9.535	0.000	14.379	1.261
6*	66.949	1.236	5.464	10.058	0.000	14.638	1.389
7*	69.914	1.170	5.467	9.812	0.000	11.677	1.338
8	67.619	1.177	5.956	9.631	0.173	14.524	0.319
9*	70.866	1.185	5.555	7.941	0.175	11.565	1.354
10	68.113	1.188	4.506	9.716	0.174	14.652	0.321
11	68.981	0.059	5.996	7.542	0.174	14.622	1.283
12	70.595	1.187	4.506	9.715	0.174	11.720	1.285
13*	70.192	0.129	5.596	10.391	0.000	11.854	0.375
14	70.315	1.175	5.946	7.478	0.172	14.499	0.318
15*	72.004	1.114	3.738	7.124	0.000	13.291	1.324
16	72.076	0.059	4.482	9.664	0.173	11.658	1.278
17	71.756	0.058	5.890	9.524	0.171	11.490	0.315
18	69.858	1.172	5.932	7.461	0.000	14.467	0.317
19	72.935	1.196	4.537	7.608	0.000	11.801	1.294
20*	69.640	0.137	4.182	10.628	0.160	14.087	0.360
21*	72.546	0.128	5.584	7.697	0.185	11.624	1.353
22*	71.683	1.253	4.185	9.808	0.000	11.791	0.377
23	74.007	0.059	5.940	7.470	0.000	11.587	0.318
24	76.252	0.058	4.415	7.403	0.000	11.483	0.315

TABLE III (CONT'D). FLOAT GLASS COMPOSITIONS TARGET/ANALYZED* (MOL %)

ID	Fe_2O_3	Cr_2O_3	Co_3O_4	TiO_2	SO_3	Se	Total
base	0.040	0.000	0.000	0.013	0.000	0.000	100.000
1	0.038	0.016	0.000	0.038	0.000	0.002	99.998
2*	0.553	0.000	0.000	0.803	0.023	0.008	99.992
3	0.037	0.000	0.004	0.037	0.000	0.000	100.000
4	0.038	0.016	0.000	0.751	0.000	0.000	100.000
5	0.037	0.000	0.004	0.037	0.000	0.002	99.998
6*	0.045	0.154	0.000	0.045	0.022	0.000	100.000
7*	0.537	0.000	0.000	0.046	0.030	0.008	99.992
8	0.564	0.000	0.000	0.038	0.000	0.000	100.000
9*	0.542	0.000	0.000	0.794	0.023	0.000	100.000
10	0.569	0.000	0.004	0.758	0.000	0.000	100.000
11	0.568	0.016	0.004	0.756	0.000	0.000	100.000
12	0.038	0.016	0.004	0.758	0.000	0.002	99.998
13*	0.554	0.004	0.050	0.816	0.030	0.008	99.992
14	0.038	0.016	0.004	0.038	0.000	0.002	99.998
15*	0.491	0.154	0.000	0.727	0.023	0.008	99.992
16	0.566	0.000	0.004	0.038	0.000	0.002	99.998
17	0.037	0.016	0.000	0.743	0.000	0.000	100.000
18	0.037	0.000	0.004	0.748	0.000	0.002	99.998
19	0.572	0.016	0.004	0.038	0.000	0.000	100.000
20*	0.562	0.160	0.000	0.046	0.030	0.008	99.992
21*	0.048	0.000	0.000	0.798	0.022	0.015	99.985
22*	0.045	0.000	0.050	0.784	0.023	0.000	100.000
23	0.562	0.016	0.004	0.037	0.000	0.000	100.000
24	0.037	0.000	0.000	0.037	0.000	0.000	100.000

TABLE IV. LOW-EXPANSION BOROSILICATE GLASS COMPOSITIONS AS ANALYZED (MOL %)

ID	SiO_2	B_2O_3	Al_2O_3	CaO	Na_2O	K_2O	BaO	Total
base	83.338	11.192	1.328	0.025	4.081	0.036	0.000	100
1	86.164	8.775	1.337	0.000	3.724	0.000	0.000	100
2	82.498	9.307	1.345	2.142	3.907	0.000	0.800	100
3	77.517	8.968	1.384	2.160	8.049	1.922	0.000	100
4	80.267	9.467	4.270	0.000	3.978	2.018	0.000	100
5	74.116	9.401	4.493	0.000	8.945	2.187	0.858	100
6	74.288	9.341	4.415	2.263	8.842	0.000	0.852	100
7	77.973	13.605	1.359	0.000	4.215	1.998	0.850	100
8	76.167	13.362	1.353	0.000	8.270	0.000	0.847	100
9	73.344	13.317	1.364	2.138	7.928	1.909	0.000	100
10	73.795	13.893	4.429	0.000	7.884	0.000	0.000	100
11	72.066	13.842	4.606	2.326	4.250	2.060	0.851	100
12	75.232	13.655	4.427	2.256	4.429	0.000	0.000	100
13	80.312	8.051	1.309	0.000	7.735	1.800	0.793	100
14	82.016	8.980	1.250	0.000	7.753	0.000	0.000	100
15	80.139	9.468	1.311	2.127	4.153	1.999	0.802	100
16	82.946	8.992	4.290	0.000	3.073	0.000	0.700	100
17	78.242	9.095	4.511	2.217	3.982	1.953	0.000	100
18	75.821	9.008	4.450	2.297	8.425	0.000	0.000	100
19	78.986	13.634	1.372	0.000	4.000	2.009	0.000	100
20	79.191	13.401	1.403	1.941	4.064	0.000	0.000	100
21	73.852	13.760	1.340	2.182	8.057	0.000	0.810	100
22	75.971	13.933	4.508	0.000	4.716	0.000	0.872	100
23	70.189	13.844	4.517	0.000	8.631	1.984	0.835	100
24	66.837	14.351	4.670	2.429	8.719	2.101	0.893	100

TABLE V. TV PANEL GLASS COMPOSITIONS AS ANALYZED (MOL %)

ID	SiO_2	Al_2O_3	MgO	CaO	SrO	BaO	Li_2O	Na_2O	K_2O	TiO_2
base	71.931	1.419	0.000	0.063	6.230	4.228	0.024	8.658	5.630	0.378
1	75.282	0.945	0.017	0.050	0.627	5.786	0.000	10.268	4.494	0.087
2	77.488	2.260	0.000	0.000	0.645	0.876	1.118	6.499	6.325	0.067
3	76.271	2.119	2.329	0.000	0.658	0.805	0.000	8.580	4.150	0.459
4	76.311	2.279	2.418	3.580	0.642	0.813	0.000	5.832	6.037	0.065
5	71.196	2.360	2.385	4.027	6.272	0.885	0.000	6.354	4.238	0.421
6	62.222	2.493	2.676	4.071	6.785	6.033	0.000	7.570	4.319	0.070
7	62.165	0.923	2.404	3.938	6.520	5.792	1.158	6.757	4.468	0.087
8	62.471	2.577	0.000	4.226	6.834	6.192	1.290	10.884	4.396	0.071
9	64.006	0.960	2.411	0.076	6.817	6.013	1.187	10.218	6.693	0.089
10	66.480	2.320	0.000	4.002	0.624	5.706	1.150	9.984	6.462	0.439
11	70.029	2.391	2.248	0.083	6.339	0.895	1.115	9.566	6.259	0.425
12	64.785	0.902	2.283	3.840	0.618	5.707	0.000	9.940	6.482	0.444
13	68.963	0.885	0.119	4.194	6.730	0.927	1.167	6.762	6.583	0.445
14	63.445	2.619	0.000	0.000	7.034	6.309	0.000	11.335	4.695	0.433
15	68.534	2.474	2.525	0.064	0.643	5.983	1.199	7.076	6.840	0.090
16	72.166	0.869	2.328	3.755	0.633	0.860	1.098	9.792	4.186	0.419
17	65.534	0.887	0.067	4.164	6.572	0.911	0.000	9.910	6.406	0.084
18	64.398	2.456	0.000	0.000	6.759	6.106	0.000	7.671	6.796	0.436
19	73.317	0.892	2.408	0.097	0.622	5.700	1.135	6.722	4.380	0.433
20	72.117	2.168	0.000	3.805	0.612	0.827	1.072	9.553	4.109	0.072
21	67.317	0.888	2.363	0.060	6.666	0.914	0.000	9.858	6.415	0.084
22	71.487	0.944	0.070	4.264	0.649	5.905	0.000	7.065	6.736	0.460
23	75.909	0.888	0.000	0.000	6.360	0.829	1.118	5.948	4.078	0.450
24	87.103	0.764	0.000	0.000	0.639	0.800	0.000	6.180	4.383	0.073

TABLE V (CONT'D). TV PANEL GLASS COMPOSITIONS AS ANALYZED (MOL %)

ID	CeO_2	ZrO_2	PbO	ZnO	As_2O_3	Sb_2O_3	F	Fe_2O_3	FeO	Total
base	0.114	0.792	0.000	0.440	0.000	0.072	0.000	0.016	0.004	99.996
1	0.276	1.684	0.003	0.000	0.088	0.170	0.221	0.000	0.000	100.000
2	0.000	1.550	0.850	0.000	0.000	0.126	2.180	0.017	0.000	100.000
3	0.000	0.000	0.853	1.651	0.000	0.044	2.065	0.016	0.000	100.000
4	0.304	0.000	0.000	1.562	0.101	0.040	0.000	0.016	0.000	100.000
5	0.004	1.604	0.009	0.000	0.092	0.154	0.000	0.000	0.000	100.000
6	0.232	0.000	0.922	0.000	0.000	0.130	2.469	0.009	0.000	100.000
7	0.032	1.657	0.006	1.250	0.014	0.059	2.769	0.000	0.000	100.000
8	0.000	0.000	0.931	0.000	0.072	0.047	0.000	0.009	0.000	100.000
9	0.029	0.012	0.006	1.307	0.011	0.168	0.000	0.000	0.000	100.000
10	0.020	0.000	0.000	0.000	0.092	0.060	2.661	0.000	0.000	100.000
11	0.240	0.000	0.003	0.000	0.000	0.160	0.245	0.000	0.000	100.000
12	0.269	1.648	0.000	0.008	0.000	0.061	3.013	0.000	0.000	100.000
13	0.258	1.692	0.956	0.000	0.010	0.059	0.252	0.000	0.000	100.000
14	0.258	1.764	0.940	1.109	0.000	0.049	0.000	0.009	0.000	100.000
15	0.291	1.727	0.909	1.294	0.101	0.061	0.189	0.000	0.000	100.000
16	0.008	1.587	0.867	1.185	0.093	0.155	0.000	0.000	0.000	100.000
17	0.259	0.011	0.940	1.219	0.102	0.162	2.771	0.000	0.000	100.000
18	0.000	1.673	0.000	1.079	0.077	0.132	2.407	0.009	0.000	100.000
19	0.276	0.000	0.891	0.008	0.096	0.165	2.857	0.000	0.000	100.000
20	0.223	1.498	0.003	1.126	0.000	0.149	2.665	0.000	0.000	100.000
21	0.012	1.676	0.965	0.008	0.105	0.058	2.612	0.000	0.000	100.000
22	0.033	0.000	0.925	1.286	0.004	0.172	0.000	0.000	0.000	100.000
23	0.287	0.000	0.000	1.626	0.114	0.044	2.333	0.016	0.000	100.000
24	0.000	0.000	0.000	0.000	0.000	0.042	0.000	0.016	0.000	100.000

TABLE VI. FIBER WOOL GLASS COMPOSITIONS TARGET/ANALYZED* (MOL %)

ID	SiO_2	B_2O_3	Al_2O_3	MgO	CaO	Na_2O	K_2O	Fe_2O_3	F	SO_3	Total
base	63.514	2.891	2.923	5.221	8.968	15.700	0.536	0.118	0.000	0.129	100
1	77.969	2.588	0.000	1.490	5.355	12.598	0.000	0.000	0.000	0.000	100
2	69.446	2.525	0.000	7.270	5.225	12.292	1.244	0.000	1.851	0.146	100
3	60.834	2.508	0.000	7.221	11.417	15.964	0.000	0.219	1.838	0.000	100
4	67.734	2.633	3.596	1.516	5.448	16.761	0.000	0.230	1.930	0.153	100
5*	60.886	2.709	3.638	1.539	12.415	17.462	1.351	0.000	0.000	0.000	100
6*	55.993	2.966	3.836	7.473	13.753	14.109	1.425	0.262	0.032	0.152	100
7*	65.851	8.009	0.000	1.438	5.393	17.586	1.344	0.245	0.000	0.134	100
8*	61.905	7.842	0.000	1.495	12.388	13.094	1.312	0.250	1.715	0.000	100
9	56.981	7.657	0.000	7.348	11.619	16.247	0.000	0.000	0.000	0.148	100
10*	54.722	8.312	3.785	1.712	14.962	13.739	0.000	0.000	2.552	0.217	100
11	55.865	7.833	3.565	7.516	5.402	16.619	1.286	0.000	1.913	0.000	100
12	62.435	7.898	3.595	7.580	5.447	12.815	0.000	0.230	0.000	0.000	100
13*	68.348	2.789	0.000	1.524	5.929	18.316	1.390	0.000	1.535	0.170	100
14	71.169	2.588	0.000	1.490	11.780	12.597	0.000	0.226	0.000	0.150	100
15	66.877	2.569	0.000	7.396	5.316	16.352	1.266	0.224	0.000	0.000	100
16	70.366	2.648	3.616	1.525	5.479	12.890	1.305	0.231	1.941	0.000	100
17*	63.156	2.637	3.812	6.894	5.763	17.578	0.000	0.000	0.000	0.161	100
18*	60.407	2.611	3.432	6.529	12.287	12.843	0.000	0.015	1.876	0.000	100
19	67.136	7.737	0.000	1.485	5.336	16.416	0.000	0.000	1.890	0.000	100
20	65.260	7.611	0.000	7.304	5.249	12.349	0.000	0.221	1.859	0.147	100
21*	68.179	8.533	0.000	1.592	5.958	14.340	1.398	0.000	0.000	0.000	100
22	66.473	8.093	3.684	1.553	5.581	13.131	1.329	0.000	0.000	0.156	100
23	57.299	8.034	3.657	1.542	12.190	17.045	0.000	0.233	0.000	0.000	100
24	49.006	7.832	3.565	7.516	11.884	16.618	1.286	0.228	1.913	0.151	100

Appendix III

UNITS, CONVERSIONS, AND MOLECULAR WEIGHTS

TABLE I. BASIC AND DERIVED SI UNITS

Quantity	Unit	Symbol
Length	meter	m
Mass	kilogram	kg
Time	second	s
Electric current	Ampere	A
Thermodynamic temperature	kelvin	K
Amount of substance	mole	mol
Area	square meter	m^2
Volume	cubic meter	m^3
Density	kilogram per cubic meter	$kg{\cdot}m^{-3}$
Force	newton	$N = kg{\cdot}m{\cdot}s^{-2} = J{\cdot}m^{-1}$
Pressure (stress)	Pascal=newton per square meter	$Pa = N{\cdot}m^{-2}$
Energy	Joule	$J = kg{\cdot}m^2{\cdot}s^{-2} = N{\cdot}m$
Power	watt	$W = kg{\cdot}m^2{\cdot}s^{-3} = J{\cdot}s^{-1}$
Electric charge	coulomb	$C = A{\cdot}s$
Electric potential difference	volt	$V = kg{\cdot}m^2{\cdot}s^{-3}{\cdot}A^{-1} = J{\cdot}A^{-1}{\cdot}s^{-1}$
Electric resistance	ohm	$\Omega = V{\cdot}A^{-1}$
Electrical conductance	sieman	$S = \Omega^{-1}$

TABLE II. SOME NON-SI UNITS AND CONVERSIONS

Quantity	Name	SI equivalent
Length	angstrom (Å) inch	10^{-10}m 0.0254 m
Volume	liter (L)	$10^{-3}m^3$
Mass	pound (lb)	0.454 kg
Force	dyne (dyn) kilogram force (kgf = kp) pound force (lbf) poundal	10^{-5}N 9.807 N 4.448 N 0.138 N
Pressure	atmosphere (atm) torr (mm Hg) bar pounds per square inch (psi)	1.013×10^5 Pa 133.3 Pa 10^6dyn/cm^2 = 10^5 Pa 6895 Pa
Energy	erg calorie (cal) electronvolt (eV) kilowatt-hour (kWh)	10^{-7} J 4.187 J 0.1602×10^{-18} J 3.600×10^6 J
Power	horsepower (hp), metric	735.5 W
Frequency	hertz (Hz)	s^{-1}
Viscosity	poise (P)	10^{-1} Pa·s

TABLE III. CONSTANTS

Constant	Usual notation	Value
Speed of light	c	2.998×10^8 m·s^{-1}
Acceleration due to gravity	g	9.806 m·s^{-2}
Avogadro's number	N	6.022×10^{23} mol^{-1}
Boltzmann constant	k_B	1.381×10^{-23} J·K^{-1}
Gas constant	R (= k_BN)	8.314 J·K^{-1}·mol^{-1} 1.987 cal·K^{-1}·mol^{-1}
Charge of electron	e	-1.602×10^{-19} C
Faraday constant	F (= eN)	96,487 C·mol^{-1}
Planck constant	h	6.626×10^{-34} J·s
Stefan-Boltzmann constant	σ	5.67×10^{-8} W·m^{-2}·K^{-4}

TABLE IV. MOLECULAR WEIGHTS OF OXIDES USED IN GLASS COMPOSITIONS

Oxide	Mol. Wt. (g)
Al_2O_3	101.96128
As_2O_3	197.8414
BaO	153.3294
B_2O_3	69.6182
CaO	56.0794
Cr_2O_3	151.997
CeO_2	172.1188
Co_3O_4	240.8
F	18.998403
FeO	71.8464
Fe_2O_3	159.6922
K_2O	94.196
Li_2O	29.8814
MgO	40.3044
MnO_2	86.94
Na_2O	61.97894
PbO	223.1994
Sb_2O_3	291.4982
Se	78.96
SiO_2	60.0843
SO_3	80.057
SrO	103.6194
TiO_2	79.8788
ZnO	81.3894
ZrO_2	123.2228

Author Index

Clare, A.G., 119
Earl, D.A., 187
Fluegel, A., 131, 151, 187
Hrma, P., 131, 133
Jones, J.M., 173
Jones, L.E., 119
Karuppanan, R., 173
Kucuk, A., 119
Lam, O.P., 131, 133
Minister, K.B.C., 131, 133
Öskoy, D., 187
Prokhorenko, O.A., 75, 95
See, C.A., 131, 133
Seward, T.P., III, 1, 27, 257
Shelby, J.E., 47, 131, 163, 261
Varshneya, A.K., 131, 151, 173, 187
Vascott, T., 1, 27, 173
Wing, D.R., 119

Keyword Index

absorption spectra, 75
Arrhenius coefficients
  tables, 147
Arrhenius equation, 133, 138

ball penetration
  viscometry, 153
borosilicate glass data
  tables, 27, 35, 44, 61, 73, 111, 125, 141, 159, 168, 184, 209, 216, 235, 270, 280
Brookfield Digital Viscometer DV-III+ Rheometer, 134
Brookfield Digital Viscometer Model LVTD, 134,

canonical model, 188
carbon dioxide solubility, 62
CGR research activity, 5
classification of property dependencies, 8
commercial glass comparison tables, 7, 17, 27, 49, 56, 62, 135, 178, 197
commercial glass compositions, 27
constants, 286
container glass data tables, 29, 43, 58, 71, 110, 125, 145, 160, 171, 181, 211, 212, 218, 235, 273, 276
conversions, 285
Cook value, 191
correlation analysis, 190

density of glass melts, 15, 119
differential scanning caliometer, 264
diffusion coefficients (SO3), 82
Dulong-Petit limit, 16

E-glass data tables, 31, 43, 59, 72, 110, 112, 125, 142, 158, 166, 182, 208, 215, 234, 269, 277
electrical conductivity, 264
electrical resistivity, 20, 187

fiber elongation viscometry, 154
fiber wool glass data tables, 39, 45, 59, 72, 109, 126, 143, 162, 167, 186, 210, 217, 235, 272, 283
float glass data tables, 33, 43, 57, 71, 125, 146, 161, 170, 183, 211, 218, 235, 274, 278

gas diffusivity, 11, 75
gas saturation system, 50
gas solubility, 9, 47, 75, 243, 257
glass furnace modeling, 2
glass manufacturing, 1
glass melting procedures, 41
glass melting process modeling, 4
glass modifying oxides, 227

heat capacity, 16, 257
high-temperature spectrophotometry, 95, 100

inert gas solubility, 57, 243
infrared absorption, 47, 98
intermediate glass oxides, 227

liquidus temperatures, 258

mathematical modeling, 76
molecular weights, 285

optical densities, 106
outgassing procedure, 51

parallel plate viscometry, 151
phonon conductivity, 13
photon conductivity, 13
Plackett-Burman designs, 192
properties and measurement techniques, 8

quality assurance, 23

radiative conductivity of melts, 12, 95
refractive index, 264
regression analysis, 188
robust regression, 192
Rosseland formulas, 114
rotating spindle technique, 133

Scheffé type model, 188
sessile drop method, 119
slack variable technique, 188
SO_3 diffusion, 75, 82
SO_3 solubility, 75, 85
soda-lime-silicate melts, 75, 124
statistically designed glass compositions, 27
statistical modeling, 22, 187, 232
Stephan-Boltzmann constant, 113
surface tension of glass melts, 15, 121, 187, 244

thermal diffusivity, 14
thermal expansion, 15
thermal expansion curves, 262
transformation range viscosity, 163
TV panel glass data tables, 36, 44, 60, 72, 111, 126, 144, 157, 169, 185, 207, 214, 234, 271, 281

viscosity, 18, 131, 133, 187, 258
viscosity-electrical resistivity relation, 230
Vogel-Fulcher-Tammann equation, 133, 195, 237

water concentrations, 265
water diffusion, 47, 65
water solubility, 55, 63, 187, 243

Printed and bound by CPI Group (UK) Ltd, Croydon, CR0 4YY

15/09/2026

14962410-0002